# 64 Structure and Bonding

# Metal Complexes with Tetrapyrrole Ligands I

Editor: J. W. Buchler

With Contributions by
D. Dolphin   R. Guilard   K. M. Kadish
T. Kitagawa   C. Lecomte   Y. J. Lee
B. Morgan   Y. Ozaki   W. R. Scheidt

With 73 Figures and 58 Tables

Springer-Verlag
Berlin Heidelberg GmbH

ISBN 978-3-662-15140-2     ISBN 978-3-540-47446-3 (eBook)
DOI 10.1007/978-3-540-47446-3

Library of Congress Catalog Card Number 67-11280

© Springer-Verlag Berlin Heidelberg 1987
Originally published by Springer-Verlag Berlin Heidelberg New York in 1987.
Softcover reprint of the hardcover 1st edition 1987

The use of general descriptive names, trade marks, etc. in this publication, even if the former are not especially identified, is not to be taken as a sign that such names, as understood by the Trade Marks and Merchandise Marks Act, may accordingly be used freely by anyone.

Typesetting: Mitterweger Werksatz GmbH, 6831 Plankstadt, Germany

2151/3140-543210

# Foreword

Metal complexes of tetrapyrrole ligands are a large class of coordination compounds that are at least as important and versatile as other classes of coordination compounds, e.g. the metal carbonyls, despite the apparent rigidity of the tetradentate macrocycle. This class comprises not only the biologically important heme-, chlorophyll-, and vitamin $B_{12}$ derivatives, but also the phthalocyanines, which have found wide application as pigments and are presently being studied as organic materials for a variety of electronic devices. The state of knowledge in this field is well documented up to the end of the seventies in several notable editions: the single volume *"Porphyrins and Metalloporphyrins"*, K. M. Smith (Ed.), Elsevier, Amsterdam, 1975, the seven volumes *"The Porphyrins"*, D. Dolphin (Ed.), Academic Press, New York, 1978/79, the two volumes *"Vitamin $B_{12}$"*, D. Dolphin (Ed.), Wiley, New York 1982, and the volume *"Coordination Compounds of Porphyrins and Phthalocyanines"*, B. D. Berezin, Wiley, Chichester 1981. The abundance of research on heme derivatives justified two further volumes *"Iron Porphyrins"*, H. B. Gray, A. B. Lever (Eds.), Addison-Wesley, Reading/Mass., 1983.

Progress in this field, however, is so rapid that Professor *James A. Ibers,* member of the editorial board of this series, felt it necessary to publish a special volume on metalloporphyrins and suggested me as a guest editor. Living very close to Springer-Verlag, I agreed to start the venture; good cooperation with Dr. Rainer Stumpe and his team in Heidelberg was greatly facilitated. Fortunately, more authors were willing to contribute than could be included in this volume, therefore, there will be a second one. Many thanks are due to the authors who have postponed a good deal of urgent research, presented their current expertise and who gave the impression to the reading guest editor that he was participating in the opening night of an exciting play.

The first two articles are very relevant to the title of the series. The reviews on *Stereochemistry* (W. Robert Scheidt and Young Ja Lee) and *Vibrational Spectra* (Teizo Kitagawa) demonstrate how subtle the interactions between the metal and the tetrapyrrole ligand are, and that an enormous amount of data is available which provide a deep insight into the intra- and intermolecular bonding of metal tetrapyrrole systems which will stimulate further studies of structure-function relationships in the biochemistry of tetrapyrrole systems.

The second two articles represent novel developments in the synthetic chemistry of tetrapyrrole complexes. Following the first papers of *T. G.*

*Traylor, J. P. Collman,* and *J. E. Baldwin* describing porphyrins with peripheral modifications meant to implant properties to a heme molecule that otherwise are only found in the macromolecular heme proteins, ingenious organic chemists have created a large variety of substituted porphyrins that can no longer be kept in mind by coordination chemists. Therefore, the article on *Biomimetic Porphyrins* (Brian Morgan and David Dolphin) is a very timely and useful aid for everybody working or interested in the field. The article on *Metal Carbon Sigma Bonds and Metal Metal Bonds* (Roger Guilard, Karl Kadish, and Claude Lecomte) documents how building blocks known from organometallic chemistry may be combined with metalloporphyrin systems producing some very fascinating novel classes of molecules.

Contributions covering the properties of metal tetrapyrroles as organic electrical materials, the noble metal porphyrins, photophysical aspects, progress in the chemistry of cytochrome P-450 models, and extended x-ray absorption fine structure are planned for the second volume.

Darmstadt, April 1, 1987                                     Johann W. Buchler

# Table of Contents

# Recent Advances
# in the Stereochemistry of Metallotetrapyrroles

**W. Robert Scheidt and Young Ja Lee**

Department of Chemistry, University of Notre Dame, Notre Dame, Indiana 46556, USA

Information pertaining to the stereochemistry of metalloporphyrins and other tetrapyrroles continues to expand. The present article reviews important developments of this structural chemistry. Detailed updates on the relationship of the structure and physical properties of iron derivatives are given. Metalloporphyrins with unusually high or low oxidation states are reviewed. Surveys of recent work on $\pi$-cation radical complexes, bound $O_2$ species, tetrapyrroles with N-substituents, the stereochemistry of ring-reduced tetrapyrroles and a variety of novel species are given. Newly developed data concerning experimental electron density studies are summarized. Detailed reviews of porphyrin-porphyrin ($\pi$-$\pi$) interactions in the solid state are given. A variety of important conformational aspects of metalloporphyrins and their consequent results on physical properties are presented. Finally, a number of isomorphous series and crystal packing effects in tetrapyrroles have been detailed.

Structure and Bonding 64
© Springer-Verlag Berlin Heidelberg 1987

# A. Introduction

Our knowledge of the stereochemistry of porphyrins and related tetrapyrrole macro-cycles has expanded rapidly since the first reported x-ray structure determination in 1959[1]. The structures of metallotetrapyrrole complexes are of interest because of the common occurrence of this type of macrocycle in biological systems. As is well known, foremost among these are the heme proteins (iron derivatives), the various photosynthe-tic pigments (magnesium complexes), the vitamin $B_{12}$ coenzyme (cobalt corrinoids), and coenzyme F430 (nickel corphinoids) of the methanogenic bacteria.

In the past 15 years, a number of reviews have appeared. Two general reviews appeared in the mid 70s[2, 3]. Both of these reviews attempted to comprehensively survey the topic of porphyrin stereochemistry up to the time of publication. These two reviews are appropriately consulted for complete information of all work completed to that time. In addition, there have been a number of more specialized reviews pertaining to tetrapyr-role macrocyclic structure. An excellent article by Glusker[4] has detailed the structural work on vitamin $B_{12}$ derivatives. An early classic review examined the stereochemistry of hemes (iron porphyrinates) and their relationship to the function of the hemoproteins[5]. A review of trends in metalloporphyrin stereochemistry as a function of electronic state and position in the periodic table was written by the author in 1977[6]. There are also two subsequent reviews in which the senior author has participated: a 1983 article (with Martin Gouterman)[7] that attempted to reach an understanding of control of spin state in metalloporphyrins and a 1981 article (with Christopher A. Reed)[8] that catalogues spin-state/stereochemical relationships of the iron porphyrinates and the implications of these structures for the hemoproteins. Articles by Hoffman and Ibers[9, 10] have discussed the use of oxidized porphyrins and phthalocyanine derivatives as molecular metals. It is not the intention of the present review to attempt to supplant any of these earlier reviews but rather to extend them when appropriate, new information is available. Further, we will review some additional topics that have not been considered previously.

Hence, the scope of this review is twofold. Firstly, an attempt to integrate and extend new information into the concepts and patterns given in the previous pertinent reviews will be made in Sect. B. Secondly, a number of new topics that have not been considered in previous reviews will be broached in the article. The first component will thus cover the literature for only the past 5–10 years; the second portion of the review may require

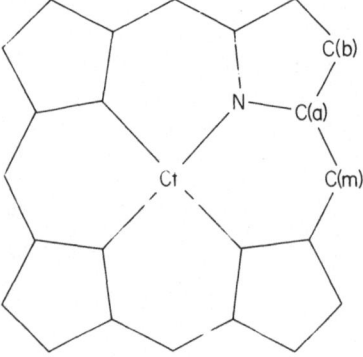

**Fig. 1.** Formal diagram of a porphinato core showing the labeling of the chemically distinct types of atoms. $Ct$ represents the center of the hole defined by the four nitrogen atoms

may require that the entire body of available stereochemical data be included. Only compounds with the porphine nucleus will be considered; all hydroporphyrin species are thus included but not the corrins or phthalocyanines.

The abbreviations for porphyrins and other tetrapyrroles, axial ligands and other species that will be used in this article are listed in Sect. E. A generalized description of the porphine nucleus is given in Fig. 1. The generalized notation for the atoms of the porphine nucleus shown was originally suggested by Hoard[11]. Structural parameters of interest include bond distances and angles, the size of the central hole (Ct···N), the position of the metal atom with respect to the mean plane of the core and a conformational description of the core itself.

# B. Recent Highlights

## I. Iron Complexes

In the 1981 review of Scheidt and Reed[8], the utility of categorizing the stereochemistry of iron porphyrinates by defining spin state, coordination number, and oxidation state was enunciated. In their tabulations, comprehensive to that time, they showed that the specification of these three properties defined the structural parameters of the complex within narrow limits. Here we will expand (slightly) on this type of categorization and describe new work that further illustrates these principles. We will also describe new work that shows some additional "pigeonholing" of structural types. We first consider iron(III) examples followed by iron(II) species and finally two compounds with unusually low oxidation states.

Scheidt and Reed emphasized that the specifications above lead to the prediction, within narrow limits, of the value of the Fe–$N_p$ bond length, the displacement of the iron atom out of the porphyrin plane, and the axial bond lengths. Values of the parameters given by these spin-state/stereochemical relationships were collected in Tables I and II of their review. To these specifications, we can now add one additional parameter needed for the precise prediction of structure: the charge on the complex. There is now an adequate number of data to demonstrate effects for complexes that have an overall negative charge (six-coordinate species) or overall positive charge (five-coordinate species). Thus for six-coordinate iron(III) porphyrinate species the determination of structures for low-spin [Fe(TPP)(CN)$_2$]$^-$ [12], low-spin [Fe(TPP)(4-MeIm$^-$)$_2$]$^-$ [13], and high-spin [Fe(TPP)F$_2$]$^-$ [14] provides suffcent data to note charge effects. The negative charge leads to an apparent decrease in the interaction between metal and the porphinato ligand. Thus there are small but real increases in the size of the cental hole and in the Fe–$N_p$ bond distances. For the low-spin species this increase in Fe–$N_p$ is ~ 0.01 Å and for the high-spin species the increase is ~ 0.02 Å compared to derivatives with analogous spin states but with an overall neutral or + 1 charge for the complex. Thus for low-spin anionic complexes Fe–$N_p$ will be ~ 2.000 A and in the high-spin complexes, the value will be 2.065 Å. It is to be noted that the charge effect in the high-spin iron(III) derivative leads to larger values of Fe–$N_p$ than found for the sole known high-spin six-coordinate iron(II) complex[15]. In all of these derivatives, the iron atom is centered in the porphinato plane. It is also to be noted that for the [Fe(TPP)(4-MeIm$^-$)$_2$]$^-$ complex, the use of the imidazolate ligand leads to a 0.03 Å decrease in the length of the axial bonds compared to neutral imidazole derivatives.

Three five-coordinate iron(III) complexes with neutral axial ligands provide strong evidence for a general decrease in Fe–N$_p$ and in the out-of-plane displacement of the iron atom as a result of the positive charge. This is true for high-spin [Fe(OEP) (2-MeHIm)]$^+$ [16], where Fe–N$_p$ has decreased by ~ 0.03 Å and the displacement by ~ 0.1 Å relative to 2.065 and 0.47 Å-values for Fe(Porph)X derivatives[8].

Two cationic, quantum-admixed intermediate-spin complexes, [Fe(OEP)(3-ClPy)]$^+$ [17] and [Fe(TPP)(THF)]$^+$ [18] also show such decreases in Fe–N$_p$ and displacements relative to [Fe(TPP)(OClO$_3$)]$^{19)}$, the structural standard of reference for five-coordinate intermediate-spin complexes. However, for the intermediate-spin complexes, the prediction of coordination parameters also seems to need the specification of the binding strength of the axial ligand. Shelly et al.[20] have compared the structural parameters of the complexes [Fe(TPP)(OClO$_3$)], [Fe(TPP)(FSbF$_5$)] and [Fe(TPP)(B$_{11}$CH$_{12}$)]. The well-known perchlorate derivative binds to iron(III) via an oxygen atom. The hexafluoroantimonate anion[21] binds to iron via a bridging fluoride and the carborane anion binds to iron via an unsupported Fe–H–B bridge bond. For these anion-ligated iron(III) complexes, it was suggested that the iron porphinato bond parameters were a sensitive guide to the nature of the interaction of the anion. The core coordination parameters of a series of Fe(Porph)X complexes are given in Table I. [Fe(TPP)I][22] is representative[8] of high-spin complexes; the remaining complexes are all intermediate-spin complexes. The authors suggest that the core parameters of the anion-ligated species are the result of two properties of the coordinating anion, the binding strength and the ligand field strength. The decreasing Fe–N$_p$ distance and the displacement culminate in the values for the carborane derivative and were taken as evidence for extremely weak binding of the anion. For comparison, the corresponding values for two intermediate-spin cationic complexes are also given in Table I.

These two cationic, intermediate-spin complexes allow comparison of the change in axial bond distance that accompanies the change in coordination number without change in spin state. Six-coordinate [Fe(OEP)(THF)$_2$]ClO$_4$$^{23)}$ has an axial bond length of 2.187(11) Å and an equatorial Fe–N$_p$ bond length of 1.978(12) Å. Six-coordinate

**Table I.** Porphinato core parameters (Å) for high- and intermediate-spin five-coordinate ferric derivatives$^a$

| Complex | Fe–N$_p$ | Ct′⋯N | Fe⋯Ct′ | Ax$^b$, Fe–Ax | Ref. |
|---|---|---|---|---|---|
| a. Anionic axial ligand | | | | | |
| [Fe(TPP)I] | 2.066(11) | 2.014 | 0.53 | I, 2.554(3) | 22 |
| [Fe(TPP)(OClO$_3$)] | 2.001(5) | 1.981 | 0.30 | O, 2.029(4) | 19 |
| [Fe(TPP)(FSbF$_5$)] | 1.978(3) | 1.974 | 0.15 | F, 2.105(3) | 21 |
| [Fe(TPP)(B$_{11}$CH$_{12}$)] | 1.961(5) | 1.955 | 0.10 | H, 1.82(4) | 20 |
| b. Neutral axial ligand | | | | | |
| [Fe(OEP)(3-ClPy)]$^+$ | 1.979(6) | 1.967 | 0.22 | N, 2.126(5) | 17 |
| [Fe(TPP)(THF)]$^+$ | 1.970(6) | 1.962 | 0.18 | O, 2.066(2) | 18 |

$^a$ Ct′ is the center of the 24-atom core.
$^b$ Ax is the donor atom of the axial ligand.

[Fe(OEP)(3-ClPy)$_2$]ClO$_4$[24)] has an average axial bond length of 2.310(17) Å and an equatorial bond length of 2.005(6) Å. It is readily seen that the axial bond lengths show substantial elongation that occurs as a result of the change in coordination number[25)]. This pattern of anomalously short axial bond distances in the five-coordinate complexes is paralleled in five-coordinate high-spin ferric systems. In the high-spin complexes, the axial bond distances are as short as those expected for low-spin species. We return to this point in the next paragraph. Two other six-coordinate intermediate-spin complexes have been characterized in the author's laboratory. The control of which particular spin state is observed seems to depend on the orientation of the axial ligand plane, a point that will be considered in detail in a later section.

The change in axial bond lengths as a consequence of changes in spin state and coordination number is most cleanly seen in the series of complexes high-spin [Fe(TPP)(NCS)][26)], high-spin [Fe(OEP)(Py)(NCS)] and low-spin [Fe(TPP)(Py)(NCS)][27)]. The pattern of bond distance and displacement changes in these three complexes is illustrated in Fig. 2. An important point to note is that in the six-coordinate iron complexes, there can still be significant displacements of the iron atom out of plane in response to differences in the strength of the axial interactions. In the low-spin complex, the displacement is quite small; these are again seen in the other known mixed axial ligand species: [Fe(TPP)(Py)(CN)][28)], [Fe(TPP)(Py)(N$_3$)][29)] and [Fe(TPP)(NO)(H$_2$O)][30)]. A second set of examples is five- and six-coordinate fluoride species: [Fe(TPP)F][31)] has an Fe–F bond distance of 1.792(3) Å, while the six-coordinate complex, [Fe(TPP)F$_2$]$^-$ [14)], has an axial Fe–F distance of 1.966(2) Å. This distance is a normal high-spin Fe–F bond distance.

A new "pigeonhole" is that of five-coordinate low-spin iron(III) species. There have been three structurally characterized five-coordinate low-spin iron(III) complexes; these demonstrate that one strong field axial ligand can give a sufficiently strong ligand field to yield low-spin species. The complexes characterized are [Fe(OEP)(NO)]$^+$ [30)], [Fe(TPP)(Ph)][32)] and [Fe(TAP)(SH)][33)]. The low-spin nature of the nitrosyl and σ-phenyl complexes might well be expected since the axial ligands are very strong field ligands; the low-spin state of the hydrosulfide complex is somewhat surprising. Although bis(thiolate) complexes are known to be low-spin[34)], all five-coordinate thiolate complexes[35, 36, 37)] are high spin. Further, any difference in ligand field strength between thiolate and hydrosulfide would appear to be rather small. Thus the reason for the differences in spin state

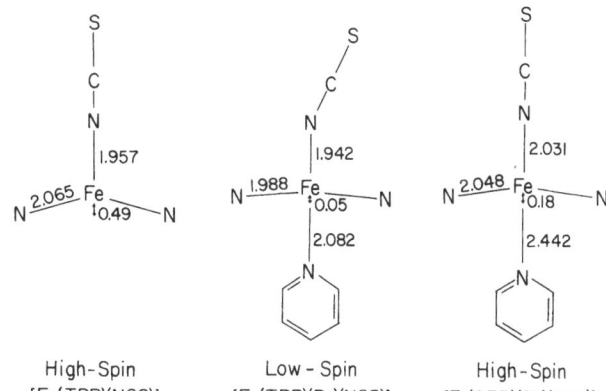

**Fig. 2.** A schematic diagram illustrating the structural differences between the complexes with indicated coordination number and spin state

High-Spin [Fe(TPP)(NCS)]          Low-Spin [Fe(TPP)(Py)(NCS)]          High-Spin [Fe(OEP)(Py)(NCS)]

between $SR^-$ and $SH^-$ complexes is unclear. It is to be noted that the stereochemical features of the five-coordinate thiolate complexes are those of canonical high-spin derivatives.

The stereochemical parameters of these three low-spin species are presented in Table II. It can be seen, as in the five-coordinate intermediate-spin complexes, that there is significant variation in the parameters. The general decreases in Fe–$N_p$ and displacement follow the binding and ligand field strength of the axial ligand. A particularly noteworthy feature is the variation in the out-of-plane displacement of the low-spin iron(III) atom.

Another interesting development in the spin-state/stereochemistry relationships is the characterization of a number of bis(nitrogen donor) iron(III) complexes with novel spin states. One complex, $[Fe(OEP)(3\text{-}ClPy)_2]ClO_4$, has been structurally characterized in three crystalline polymorphs and in all three possible spin states. An intermediate-spin state example has already been briefly mentioned; another crystalline form[38] is a thermal spin-equilibrium between the low- and high-spin states. This form has been characterized[39] by X-ray diffraction at 98 K, the temperature at which the ion is nearly completely low spin; the structure found is that expected[8] for a low-spin complex. The structure found from the data collected at 293 K is intermediate to that expected for either a high- or a low-spin form. A "crystallographic resolution of spin isomers" gives the result depicted in Fig. 3. Complete details of the coordination groups of these two crystalline modifications plus a third[40], a chloroform solvate, are given in Table III. As will be discussed later, these differences appear to result from ligand orientation effects. Two other high-spin iron(III) complexes, $[Fe(OEP)(2\text{-}MeHIm)_2]ClO_4$[41] and $[Fe(TPP)\text{-}(BzHIm)_2]ClO_4$[42], are high spin with Fe–N(Ax) = 2.275 Å and 2.216(5) Å, respectively.

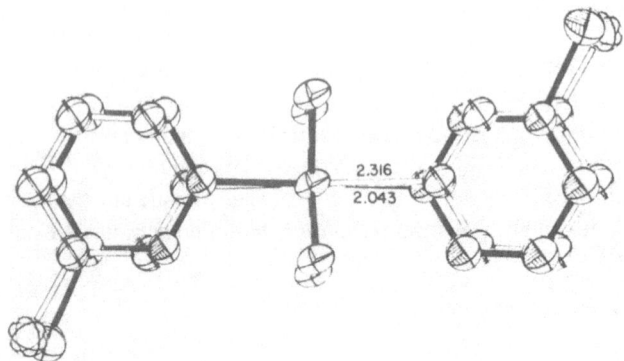

2.316
2.043

**Fig. 3.** A plot showing the "crystallographic resolution of spin isomers". The two positions of the axial 3-chloropyridine ligand corresponding to the two spin states are shown. Reproduced with permission from Ref. 39

**Table II.** Porphinato core parameters (Å) for low-spin five-coordinate ferric derivatives[a]

| Complex | Fe–$N_p$ | Ct′···N | Fe···Ct′ | Ax[b], Fe–Ax | Ref. |
|---|---|---|---|---|---|
| $[Fe(TPP)(Ph)]$ | 1.961(7) | 1.954 | 0.17 | C, 1.955(3) | 32 |
| $[Fe(OEP)(NO)]^+$ | 1.994(1) | 1.973 | 0.29 | N, 1.644(3) | 30 |
| $[Fe(TAP)(SH)]$ | 2.015(2) | 1.988 | 0.33 | S, 2.298(3) | 33 |

[a] All distances in Å.
[b] Ax is axial ligand donor atom.

**Table III.** Comparison of coordination group distances in **tri-** and **mono-**[Fe(OEP)(3-ClPy)$_2$]ClO$_4$ and the chloroform solvate

| Phase, conditions | Fe–N$_p$[a] | Fe–N(3-ClPy)[a] | Ref. |
|---|---|---|---|
| triclinic[b] | | | 39 |
| low-spin form, 98 K | 1.994(6) | 2.031(2) | |
| "average" structure, 293 K | 2.014(4) | 2.194(2) | |
| low-spin form, 293 K | (1.990)[b] | 2.043 | |
| high-spin form, 293 K | (2.045)[b] | 2.316 | |
| monoclinic, 293 K | 2.005(6) | 2.310(17) | 24 |
| CHCl$_3$ solvate | 2.006(8) | 2.304(33) | 40 |

[a] The numbers in parentheses are the estimated standard deviations, calculated on the assumption that all averaged values were drawn from the same population.
[b] Expected value, not experimentally determined from the analysis of the thermal spin mixture.

These complexes are high spin because of the use of the sterically hindered imidazoles as the axial ligands. In the case of [Fe(OEP)(2-MeHIm)$_2$]ClO$_4$[41], it is known that the complex has a temperature-dependent magnetic susceptibility in solution; the high-spin state in the solid is probably the result of ligand orientation effects. Such ligand orientation effects are important in defining the magnetic behavior of a second crystalline form of [Fe(TPP)(Py)(NCS)][43]. This crystalline phase has two crystallographically distinct sites; these sites are also magnetically distinct with one site being a spin-equilibrium site and the second site being a high-spin site under all conditions. These conclusions come from a joint crystallographic and magnetic study over a wide temperature range. (The structure was determined at 96 and 293 K and the magnetic susceptibility from 20 K up.) The phenomenon is consistent with a ligand orientation effect.

There are, in addition, a number of iron(III) complexes that should be briefly mentioned, either because of the design or another feature of the porphinato ligand or because of the presence of a novel axial ligand. Except for the first of these, all of these derivatives are five-coordinate high-spin species. The structure of the chloroiron(III) derivative of a potentially binucleating ligand, FeCl(P–N$_4$), has been determined[44]. This complex has a six-coordinate iron(III) with long axial bonds: Fe–Cl = 2.31(2) Å and Fe–N(Py) = 2.085(6) and Fe–N$_p$ = 2.042(8) Å. The chloride is inside the "pocket" of the four nicotinamide groups. The pyridine ligand originates from another porphyrin ligand so as to generate a polymeric chain. The formation of the polymeric chain also leads to a chain magnetic interaction between iron atoms. The ligand has been prepared as a model for intermetallic coupling in cytochrome oxidase. Fe$_2$(μ-O)FF · H$_2$O[45] is a complex utilizing a "face-to-face" diporphyrin designed to provide binuclear species with a variety of bridging ligands. Structural parameters of the μ-oxo bridged compound are typical of the genre with the possible exception of a smaller Fe–O–Fe (161.1°) angle and the presence of a water molecule that is hydrogen bonded to the μ-oxo ligand. Figure 4 illustrates the bridge of the biporphyrin and the hydrogen bonded water molecule. [Fe(ODM)]$_2$O[46], another μ-oxo iron(III) derivative, has a severely $S_4$-ruffled core as a consequence of the peripheral steric congestion. Parameters of the coordination group are normal. [Fe(TPP)]$_2$(μ-SO$_4$)[47] has a μ-sulfato ligand with each iron bonded to a single oxygen atom of the sulfato ligand. There is no metal ion coupling through the sulfate

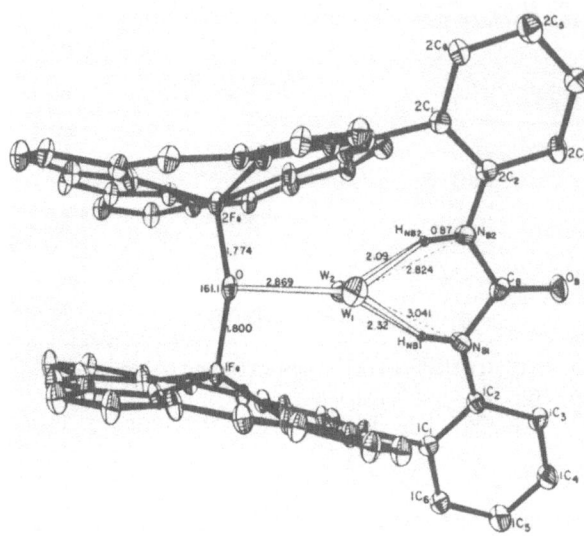

**Fig. 4.** A view of the $F_2(\mu\text{-}O)FF$ · $H_2O$ molecule showing the urea based bridge and the water hydrogen bonded to the $\mu$-oxo ligand and the urea bridge. Reproduced with permission from Ref. 45

bridge. [Fe(TPP)(HSO$_4$)][48] is a typical high-spin iron(III) complex with single axial oxygen donor atom. Fe-O is 1.927 (11) Å. The structure of [Fe(TPP)(OAc)][49] is similar with a monodentate acetate ligand. The length of the axial Fe-O bond is 1.898(4) Å. Another similar complex is a benzenesulfinato[50] derivative which has Fe-O = 1.92(1) Å. Two chloroiron(III) derivatives, a basket handle[51] complex and FeCl($C_2$-Cap)[52] both have typical coordination group parameters. These two ligands were originally synthesized for oxygen binding studies. A hydroxide derivative has also been characterized[53]: the tetrapyrrole ligand used here is a *t*-butylporphodimethene that provides adequate steric hindrance to prevent the condensation reaction and the formation of a binuclear $\mu$-oxo complex. A phenolate bridged iron(III) dimer has been synthesized by using the functionalized 5-(2-hydroxyphenyl)-10,15,20-tritolylporphyrin as the ligand to iron. The crystal structure[54] reveals that the complex is a typical high-spin iron(III) species with a phenolate oxygen donor; the Fe–O distance is 1.847(2) Å.

Finally, a sulfur-bridged iron(III)-copper(II) system has been reported[55] with two [Fe(Tp–ClPP)]$^+$ units that sandwich a [Cu(MNT)$_2$]$^{2-}$ unit. Each iron coordinates to a sulfur of the [Cu(MNT)$_2$]$^{2-}$ complex with Fe1–S = 2.444(2) Å, Fe2–S = 2.549(2) Å. Fe2 additionally forms a weak sixth bond, Fe2–S' = 2.956(2) Å. Fe1 is five-coordinate with Fe–N$_p$ = 1.976(4) Å and Fe···Ct = 0.20 Å, consistent with an intermediate-spin state. Fe2 is quasi-six-coordinate with Fe–N$_p$ = 1.978(2) Å and Fe···Ct = 0.12 Å. The spin state, according to the original authors, is best assigned as intermediate spin. However, the [Fe(TAP)(SH)] result[33] noted earlier, now makes the assignment more tenuous; a low-spin state is now seen as possibility based on structural parameters.

Many fewer iron(II) structures have been reported. The structure of [Fe(TPP)(Py)$_2$] · 2 Py has been briefly communicated[56]: Fe–N$_p$ = 1.993(6) Å and Fe–N(Py) = 2.039(1) Å with iron in-plane. A nitrosoalkane complex [Fe(TPP)(RNO)(amine)][57] has a normal structure for a low-spin six-coordinate complex with Fe–N$_p$ = 1.993(10) Å, an axial Fe–N(nitroso) distance of 1.865(4) Å, and an axial distance to the amine of 2.100(8) Å.

A five-coordinate nitrene complex has been reported[58] where the nitrene ligand is (2,2,6,6-tetramethyl-1-piperidyl)nitrene. [Fe(Tp–ClPP)(nitrene)] has Fe–N(Ax) =

1.809(4) Å with the Fe–N–N unit linear. Fe–$N_p$ = 2.096(2) Å and the iron is 0.49 Å out of the $N_4$ plane and 0.54 Å out of the 24-atom mean plane. These displacements are at the upper end of the values observed for high-spin iron(II) complexes with neutral ligands (with concomitant neutral charge on the complex). Thus in [Fe(TPP)($SC_2H_5$)]$^{-}$ [59], the displacements Fe···Ct and Fe···Ct' are 0.52 and 0.62 Å while for monoimidazole adducts of iron(II) the displacements[2, 60] are 0.4 Å and Fe–$N_p$ is 2.072(4) or 2.086(6) Å. The effect of overall negative charge on the coordination group parameters is also seen in the [Fe(TpivPP)Cl]$^{-}$ [61] complex where Fe–$N_p$ = 2.108(15) Å and the displacements Fe···Ct and Fe···Ct' are 0.53 and 0.59 Å. The axial chloride ligand is in the pocket formed by the pivalamide pickets with an Fe–Cl bond length of 2.301(2) Å. These large coordination distances and displacements are not seen in a related thiolate complex[61], [Fe(TpivPP)($SC_6HF_4$)]$^{-}$, where Fe–$N_p$ = 2.076(20) Å and Fe···Ct = 0.42. However, the thiolate ligand is coordinated to iron on the side opposite the pickets – an overall conformation that probably leads to limits on the displacement of iron because of limited doming of the core. It is to be emphasized that the doming of the core, apparently not observed in this last complex, is a generally observed feature of the high-spin iron(II) porphinato complexes.

The structures of two closely related four-coordinate iron(II) complexes[62] have been reported: Fe(OEP) and Fe(OEC). The average Fe–$N_p$ bond distance is 1.996(16) Å in Fe(OEP). The analogous distance in Fe(OEC) is 1.978(13) Å while the Fe–N(pyrroline) distance is larger at 2.002(4) Å. The effects of having one reduced ring on the coordination group parameters will be discussed subsequently.

The preparation and structures of two reduced iron complexes have been reported[63]. The two complexes are [Fe(TPP)]$^{-}$ and [Fe(TPP)]$^{2-}$ in which the formal oxidation states are Fe$^I$ and Fe$^0$, respectively. The structure determinations established four-coordinate low-spin complexes for both of these species. The pattern of bond distances in these two

**Table IV.** Comparison of averaged bond distances and angles[a, b]

|  | $\langle$Fe(TPP)$L_2\rangle$ | [Fe(TPP)]$^{-}$ | [Fe(TPP)]$^{2-}$ | [Co(TPP)]$^{-}$ |
|---|---|---|---|---|
| **Distances, Å** | | | | |
| M–$N_p$ | 2.000 | 1.980(6) | 1.968(1) | 1.942(3) |
| $N_p$–$C_a$ | 1.384 | 1.401(6) | 1.409(5) | 1.397(3) |
| $C_a$–$C_b$ | 1.439 | 1.429(12) | 1.421(4) | 1.434(3) |
| $C_a$–$C_m$ | 1.393 | 1.385(5) | 1.380(8) | 1.380(3) |
| $C_b$–$C_b$ | 1.341 | 1.338(1) | 1.352(5) | 1.337(8) |
| **Angles, deg.** | | | | |
| MN$C_a$ | 127.2 | 127.7(4) | 128.1(4) | 128.1(2) |
| $C_a$N$C_a$ | 105.2 | 104.6(1) | 103.7(1) | 103.5(3) |
| N$C_a$$C_b$ | 110.1 | 109.8(3) | 110.4(2) | 110.7(2) |
| N$C_a$$C_m$ | 125.6 | 125.2(5) | 125.1(2) | –[c] |
| $C_m$$C_a$$C_b$ | 124.3 | 124.9(7) | 124.4(2) | –[c] |
| $C_a$$C_b$$C_b$ | 107.3 | 107.9(8) | 107.7(8) | 107.4(2) |
| $C_a$$C_m$$C_a$ | 124.0 | 124.0(3) | 123.4(2) | 122.4(4) |

[a] The numbers in parentheses are the estimated standard deviations calculated from the deviation of the averaged value from the population; [b] Data from Refs. 63 and 64; [c] Not reported.

complexes was unexpected. Not only were the $Fe-N_p$ bond distances shorter than expected, a number of bond parameters in the porphinato core also had unusual values. The values are summarized in Table IV. The $N-C_a$ bonds are longer than observed in any other metalloporphyrin. It was pointed out that the lengthening of the $N-C_a$ bonds and the shortening of the $C_a-C_b$ bonds was consistent with the population of the LUMO of $e_g$ symmetry, i.e., a $\pi$-anion species. A subsequent report of the structure of $[Co(TPP)]^-$ [64] showed similar effects on these three bonds. Those investigators suggested an alternative explanation for the bond changes: enhanced back-donation of the $Co^I$ center to the porphyrin $\pi$ system. Both explanations emphasize increased electron density for the porphinato ring at the expense of the metal ion.

## II. $O_2$ Complexes

We include in this section any tetrapyrrole derivative containing a bound $O_2$ unit. $O_2$ complexes of iron porphyrinates are of interest because of the oxygen-carrying and -utilizing functions of the hemoproteins. A major issue has been the nature of the interaction between the bound $O_2$ unit and the metalloporphyrin. Two limiting structures have been suggested: a bent end-on bound structure (Pauling structure) and a structure with both O atoms bound to the metal to form a three-membered ring (Griffith structure). In the event, an exciting development has been the realization of both types of structure – even for the same metal atom (with different formal oxidation states). It is to be noted that $O_2$ metallocycles can be regarded as containing bound peroxide, $O_2^{2-}$, while the end-on complexes can be considered as bound superoxide, $O_2^-$. It can be noted that there is a formal aspect to this description – but most investigators find the formalism useful in describing the general pattern of electron density in the complexes.

The first isolated crystalline $O_2$ metalloporphyrin complex was prepared by Collman et al.[65]. The synthetic strategy applied was to synthesize a porphyrin with a binding pocket for $O_2$. The binding pocket is used to inhibit the side-reaction that ultimately leads to the irreversible formation of $\mu$-oxo iron(III) species from the iron(II) starting complex. This synthetic strategy has been used to prepare two kinds of crystalline adducts: those with a nitrogen base as the sixth ligand[66, 67], i.e., models for the oxygen carrying proteins, myoglobin and hemoglobin, and derivatives with an axial thiolate[68], i.e., models for the first oxygen adduct in the catalytic cycle of cytochrome P-450. Structural data are listed in Table V. It should be noted that there is an understandable loss of precision in the structural parameters of these end-on complexes owing to the need to use the elaborate porphyrin ligand. Thus the structural differences of the first two entries of Table V are only marginally different based on the formal uncertainties. In all three of the end-on complexes, there is disorder in the position of the $O_2$ ligand. In the thiolate complex, Weiss et al. have strong circumstantial evidence to suggest that the disorder is dynamic, i.e., the oxygen molecule is rotating about the Fe–O bond in the solid state. Weiss and coworkers have also been able to crystallize an ordered example by using another counterion; details of this result will be forthcoming.

Table V also presents the structural data for the other class of oxygen adducts. As noted earlier, these complexes can be regarded as peroxo complexes. Indeed the titanium[69] and molybdenum[70] complexes can be prepared by reaction with peroxides. An interesting feature of these complexes is the eclipsing of the $M-O_2$ plane with respect

**Table V.** Bond parameters of metalloporphyrins containing a bound $O_2$

### a. End-on complexes

| Complex | M–N$_p$[a] | M–Ax[a] | M–O[a] | M–O–O[b] | O–O[a] | Ref. |
|---|---|---|---|---|---|---|
| [Fe(TpivPP)(1-MeIm)(O$_2$)] | 1.979(13) | 2.068(18) | 1.745(18) | 129(2) 133(2) | 1.17(4) 1.15(4) | 66 |
| [Fe(TpivPP)(2-MeHIm)(O$_2$)] | 1.996(4) | 2.107(4) | 1.898(7) | 129.0(12) 128.5(18) | 1.205(6) 1.232(22) | 67 |
| [Fe(TpivPP)(SC$_6$HF$_4$)(O$_2$)]$^-$ | 1.990(6) | 2.369(2) | 1.818(8) | 128(2) | 1.14(5) | 68 |

### b. Peroxo Complexes

| Complex | M–N$_p$[a] | M–O[a] | O–O[a] | M$\cdots$Ct[a] | Orientation[c] | Ref. |
|---|---|---|---|---|---|---|
| [Mo(TTP)(O$_2$)$_2$] | 2.096(4) | 1.958(4) | 1.399(6) | 0 | Eclipsed | 70 |
| [Ti(OEP)(O$_2$)] | 2.109(23)[d] | 1.825(3) | 1.445(5) | 0.62 | Eclipsed | 69 |
| [Mn(TPP)(O$_2$)]$^-$ | 2.184(23)[e] | 1.895(9) | 1.421(5) | 0.76[f] | Eclipsed | 75 |
| [Fe(TPP)O$_2$]$^-$ | – | 1.80(3) | – | 0.3 | – | 71 |

[a] Values in Å; [b] Values in degrees; [c] Orientation with respect to N–M–N coordinate plane; [d] Two sets, Ti–N ($\parallel$ to TiO$_2$ plane) 2.128(5); Ti–N ($\perp$ to TiO$_2$ plane) 2.090(4) Å; [e] Two sets, Mn–N ($\parallel$ to MnO$_2$ plane) 2.168(23); Mn–N ($\perp$ to MnO$_2$ plane) 2.200(2); [f] 0.91 Å from 24-atom plane

to a coordinate plane defined by the metal and an opposite pair of nitrogen atoms. The iron complex, characterized by EXAFS spectroscopy[71] appears to have a triangulo peroxidic structure. The complex can be described as a high-spin iron(III) species whose structure was originally suggested by Valentine[72a]. Interestingly, the same compound can be prepared by reaction of superoxide with iron(II)[72], reaction of molecular $O_2$ with the formal Fe$^I$ complex[73] or by the one electron reduction[74] of the iron(II) oxygen adduct. The structure of this iron species is important because of its relevance to the intermediate formed by the one electron reduction of oxygen adduct of cytochrome P-450. Valentine and coworkers[75] have recently characterized an analogous manganese derivative.

## III. High and Low Oxidation States

The porphinato ligand (and related tetrapyrroles) does not appear to be exceptional in stabilizing unusually low or high oxidation states of metal ions. Moreover, the macrocyclic ligand is itself electroactive, undergoing both reductions and oxidations to form π-anions and π-cations, respectively. This feature can give rise to ambiguity in the assignment of the oxidation state of the metal, although the question of whether the redox process is ring-based or metal-based can usually be answered especially for oxidation processes. Reed[73] has pointed out that a number of species that were formulated as iron(IV) complexes are better considered to be iron(III) complexes of π-cation radicals.

The species with the lowest known oxidation states of the metal (M$^I$ and M$^0$) have already been described under iron derivatives. All other species are relatively less reduced; this section will include a number of species in which the metal oxidation state is low compared to the species normally isolated.

Marchon and coworkers[76] have prepared a methoxytitanium(III) complex. An unusual feature of the complex is a nearly linear Ti–O–CH$_3$ bond (171(1)°) and a short Ti–O bond (1.77(1) Å). These features were associated with strong π-bonding to titanium because of the electron deficient nature of the Ti$^{III}$ ion. The Ti$^{III}$–N$_p$ bonds are as long as the Ti$^{IV}$–N$_p$ bonds in [Ti(OEP)O][69]. Guilard and coworkers[77] have reported the structure of two six-coordinate vanadium(II) complexes (the axial ligands are bis-tetrahydrofuran or bis-dimethylphenylphosphine). These have normal trans geometries with modestly long equatorial bonds: V–N$_p$ = 2.048(4) Å. Axial bonds are V–O = 2.174(4) or V–P = 2.523(1) Å. The common oxidation state for these two metals is V$^{IV}$ and Ti$^{IV}$.

Weiss and coworkers have characterized a number of molybdenum complexes in which the oxidation state of Mo is lower than the normal value of + 5. First is a series of Mo$^{IV}$ complexes: five-coordinate [Mo(TTP)O][78] which has Mo–N$_p$ = 2.111(5) Å and Mo···Ct = 0.64 Å. The remaining three complexes are six-coordinate species with trans axial ligands. [Mo(TTP)Cl$_2$][78], [Mo(TTP)(N$_2$C$_6$H$_5$)$_2$][79], and [Mo(TPP)(σ-C$_6$H$_5$)Cl][80] have, respectively, Mo–N$_p$ = 2.074(23), 2.071(15), and 2.070(7) Å.

A number of Mo$^{II}$ complexes have also been characterized. The compounds are: [Mo(TTP)(PhC≡CPh)][81], [Mo(TTP)(NO)(CH$_3$OH)] and [Mo(TTP)(NO)$_2$][82], [Mo(TT-P)(Py)$_2$] and [Mo(TTP)(CO)$_2$][83]. Three of these species, the dinitrosyl, the dicarbonyl and the π-bonded diphenylacetylene complex have "cis" geometry, i.e., the axial ligands are on the same side of the porphyrin plane. The remaining two complexes have the more normal trans geometry of axial ligands. An unusual feature of the dinitrosyl and acetylene complexes is that the axial ligand plane is eclipsed with respect to an opposite pair of porphinato nitrogen atoms. The dicarbonyl has the staggered geometry that is expected on the basis of minimizing nonbonded contacts. The Mo–N$_p$ bonds form two distinct classes in the eclipsed geometry structures. Theoretical calculations on the pyridine and carbonyl complexes[83] support the idea that the cis geometry is the result of π-bonding. The cis carbonyl and acetylene complexes are diamagnetic while the pyridine complex is paramagnetic, further emphasizing the differences in the electronic structures. A summary of the geometrical features of the complexes is given in Table VI.

**Table VI.** Structural parameters of Mo$^{II}$ complexes[a]

| Complex | Mo–N$_p$ | Ax, Mo–Ax | Mo···Ct | Mo···Ct' | Ref. |
|---|---|---|---|---|---|
| [Mo(TTP)(PhC≡CPh)] | 2.149(2)[b] 2.105(1)[c] | C, 1.974(13) | 0.63 | 0.73 | 81 |
| [Mo(TTP)(NO)(CH$_3$OH)] | 2.091(4) | N, 1.746(6) | –[d] | 0.28 | 82 |
| [Mo(TTP)(NO)$_2$] | 2.195(13)[b] 2.135(7)[c] | N, 1.70(1) | –[d] | 0.99 | 82 |
| [Mo(TPP)(CO)$_2$] | 2.143(24) | C, 1.86(2) | 0.62 | 0.76 | 83 |
| [Mo(TPP)(Py)$_2$] | 2.071(6) | N, 2.215(3) | 0 | 0 | 83 |

[a] All distances in Å;   [b] Mo–N bonds parallel to axial plane;   [c] Mo–N bonds perpendicular to axial plane;   [d] Not reported

A number of high-valent metalloporphyrin complexes have been reported. Again high-valency is taken relative to normal species. Many of these species are stabilized by the presence of oxygen donors in the form of oxo or alkoxo ligands. One such set is a series of binuclear μ-oxo complexes where the metals (Ru, Os, Mn) have the formal oxidation state $+4$. All complexes of this class are six-coordinate and include [Ru(OEP)(OH)]$_2$O[84], [Ru(TPP)(OC$_6$H$_4$CH$_3$)]$_2$O[85], [Ru(OEP)Cl]$_2$O[86], [Os(OEP)(OCH$_3$)]$_2$O[87], and [Mn(TPP)(N$_3$)]$_2$O[88]. All but the last of these species are diamagnetic d$^4$ species; the Mn$^{IV}$ complex has a reduced magnetic moment from antiferromagnetic coupling through the μ-oxo bridge. Two other bridged complexes have formal states of iron(III 1/2) and iron(IV); the five-coordinate complexes are [Fe(TPP)]$_2$N[89] and [Fe(TPP)]$_2$C[90].

Mononuclear high oxidation state species are rarer. Hill and coworkers have reported the structures of two six-coordinate Mn$^{IV}$ complexes: [Mn(TPP)(OCH$_3$)$_2$][91] and [Mn(TPP)(NCO)$_2$][92]. The equatorial Mn–N$_p$ bond distances, 2.012(27) and 1.970(2) Å, respectively, are slightly shorter than those of Mn$^{III}$ complexes. However, the axial bond distances, Mn–O = 1.839(2) and Mn–N = 1.926(11) Å, are substantially shorter than those of six-coordinate Mn$^{III}$ complexes. These distances are entirely consistent with those expected for Mn$^{IV}$ d$^3$ complexes (S = 3/2).

All remaining mononuclear species have the highly charged ligands oxo or nitrido to balance the high formal charge on the metal; all save [Mo(TPP)(O)$_2$][93] have a single axial ligand. The molybdenum(VI) complex has two oxo ligands cis to each other. The O···O separation of 2.547 Å clearly shows that the ligands are oxo and not a single peroxo ligand. The O–Mo–O plane is essentially perpendicular to the porphyrin plane and the plane is staggered with respect to the four porphinato nitrogen atoms. There is no apparent bonding reason for the observation of two types of Mo–N$_p$ bonds that are given in Table VII. The complete structural parameters are reported in Table VII along with the values for four d$^2$ complexes of nitridoMn$^V$ [94, 95] and oxoCr$^{IV}$ [96]. The structure of a d$^1$ Cr$^V$ nitrido complex[97] is also given in Table VII. The M–N$_p$ bond distances are approximately the same length as those of low-spin Mn$^{II}$, Mn$^{III}$ and Cr$^{II}$. Thus the increased formal charge in these highly covalent species has only a small effect on the complexing bond lengths to porphyrin. Obviously, the nitrido and oxo ligand transfer so much electron density to metal via σ and π donation that the "effective oxidation state"

**Table VII.** Structural parameters of high-oxidation state metalloporphyrins[a]

| Complex | M–N$_p$ | M–Ax | M···Ct | M···Ct' | Ref. |
|---|---|---|---|---|---|
| [Mo(TTP)(O)$_2$][b] | 2.246(1) | 1.744(9) | 0.97 | 1.10 | 93 |
| | 2.157(11) | 1.709(9) | | | |
| [Mn(OEPMe$_2$)N] | 2.006(4) | 1.512(2) | 0.43 | 0.60 | 95 |
| [Mn(TAP)N] | 2.021(31) | 1.515(3) | 0.39 | –[c] | 94 |
| [Cr(TTP)O] | 2.032(7) | 1.572(6) | 0.47 | –[c] | 96a |
| [Cr(TPP)O] | 2.036(9) | 1.62(2) | 0.49 | –[c] | 96b |
| [Cr(TTP)N] | 2.042(9) | 1.565(6) | 0.42 | –[c] | 97 |

[a] All distances in Å; [b] Two apparent sets of Mo–N$_p$ bonds; [c] Not reported

of the metal is much lower than the "formal oxidation number" and hence the ionic radius of metal does not vary much. Despite the strong interest in the structure of ferryl (Fe=O) derivatives, an intermediate in the catalytic cycle of cytochrome P-450 and related systems, no crystals have been obtained.

## IV. Tetrapyrroles With N-Substituents

There are a number of metalloderivatives that have additional substituents bonded to one or more of the four donor nitrogen atoms. These species can be divided into a number of groups as shown below. One group, illustrated as type *A*, has an alkyl or aryl substituent to one nitrogen atom. A second type *B*, has a group inserted into a M–N bond. Type *C*, has two adjacent nitrogen atoms bridged by a carbon atom. There is only one example of type *D* and type *E*. These examples will be considered along with some miscellaneous structures.

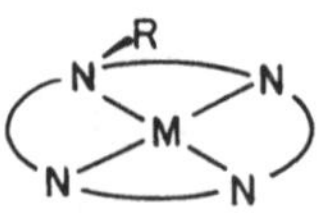

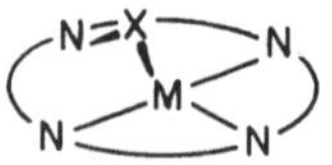

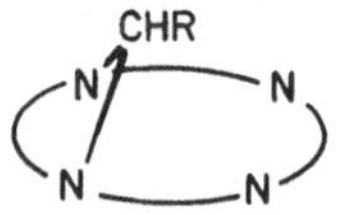

## 1. N-Aryl and N-Alkyl Compounds

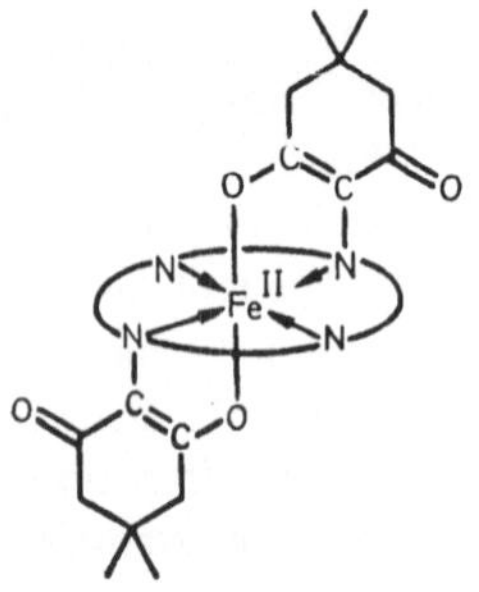

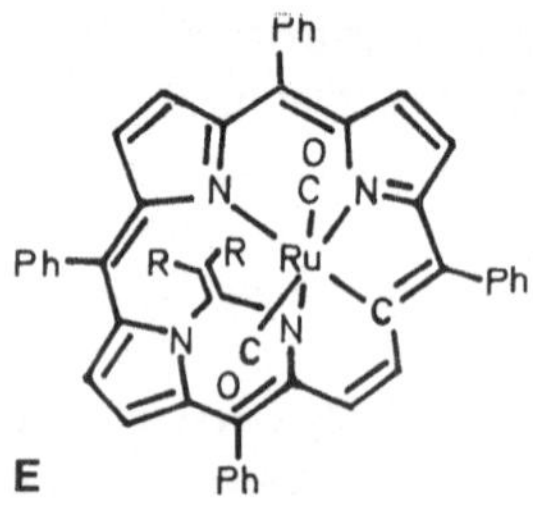

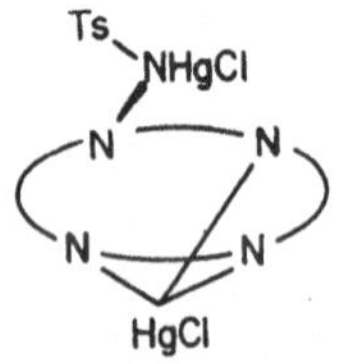

These compounds can be considered as resulting from the replacement of one hydrogen atom of the free base with an aryl or alkyl group. This ligand loses the remaining proton when coordinated to a metal ion. The ligand is thus a uninegative species. Generally, the complexes have a single axial anionic ligand to complete the charge balance. Lavallee and Anderson have characterized a substantial series of N-methyltetraphenylporphyrin derivatives. Species in this series are the cobalt(II)[98], manganese(II)[99], zinc(II)[100], iron(II)[101]. and a related free base[102] compound. These workers have also characterized an N-phenyl zinc(II) derivative[103]. In addition, the structure of a cobalt(II) derivative of N-ethylacetate octaethylporphyrin has been described[104] as well as the nonmetallo derivative[105]. A noteworthy feature of all these N-substituted species is the increased stability of the divalent oxidation state. Commonly observed structural features include a severely canted N–R substituted pyrrole ring and three classes of M–N bond distances

(M–NR, M–N adjacent and M–N trans to the N-substituent). The N-substituent is found on the side of the plane opposite the axial ligand and tends to block the sixth coordination site. Stereochemical parameters are summarized in Table VIII. A related nonmetalled complex is a N,N'-dimethyletioporphyrin I species[106]. The two methyl groups are on adjacent pyrrole rings but on opposite sides of the porphinato plane.

## 2. Insertion Products

These derivatives (Type $B$) are at least formally prepared by the insertion of a fragment into a M–N bond to yield a new M–X–N unit. Such species have been suggested as possible intermediates in the insertion of an oxygen atom into a C–H bond by cytochrome P-450. Several of these derivatives have formally had a carbene fragment inserted into a M–N bond. Such derivatives include a $Ni^{II}$ derivative[107] and a cobalt(III) species[108] that has undergone two such insertion reactions. Other species represent the formal reaction of a vinylidene with iron(III)[109, 110] (two different crystalline forms). This iron derivative has an intermediate-spin state. Other complexes result from the insertion of a nitrene[111] or an oxene[112]. This last derivative can also be considered to be a porphyrin N-oxide derivative and the structure of a free base species of a porphyrin N-oxide has also been reported[113]. Appropriate stereochemical parameters for the members of this class are found in Table IX.

**Table VIII.** Stereochemical parameters of N-substituted metalloderivatives[a]

| Complex | M–NR | M–N(adj) | M–N(opp) | M–Ax | Disp[b] | Ref. |
|---|---|---|---|---|---|---|
| [Co(21-EtO–OEP)Cl] | 2.455(5) | 2.068(6) | 1.992(5) | 2.271(2) | 0.61 | 104 |
| [Co(N–CH₃TPP)Cl] | 2.381(5) | 2.063(5) | 2.016(4) | 2.243(2) | 0.55 | 98 |
| [Mn(N–CH₃TPP)Cl] | 2.368(5) | 2.156(1) | 2.118(5) | 2.295(3) | 0.69 | 99 |
| [Fe(N–CH₃TPP)Cl] | 2.329(2) | 2.117(2) | 2.082(2) | 2.244(1) | 0.62 | 101 |
| [Zn(N–CH₃TPP)Cl] | 2.530(7) | 2.085(6) | 2.018(9) | 2.232(3) | 0.65 | 100 |
| [Zn(N–PhTPP)Cl] | 2.499 | 2.094(11) | 2.015 | 2.24 | 0.67 | 103 |

[a] All distances in angstroms; [b] Disp – displacement of metal ion from plane defined by the three unsubstituted nitrogen atoms

**Table IX.** Stereochemical parameters of M–X–N derivatives[a]

| Complex | M–N$_p$ | X | M–X | M⋯Ct | M–Ax | Ref. |
|---|---|---|---|---|---|---|
| Ni(R)TPP[b] | 1.916(10) | C | 1.905(4) | 0.19 | – | 107 |
| Co(R)₂OEP[b, c] | 1.92(2) | C | 1.99(1) | – | 2.05(2)[d] | 108 |
| [Fe(R)TTPCl][e] | 1.993(9) | C | 1.921(5) | 0.31[f] | 2.299(1) | 109, 110 |
| Ni(R)TPP[g] | 1.908(21) | N | 1.830(4) | 0.20 | – | 111 |
| Ni(O)OEP[h] | 1.917(15) | O | 1.788(4) | 0.05[f] | – | 112 |

[a] All distances in angstroms; [b] Carbene fragment R = Ethoxycarbonylcarbene; [c] Two carbene fragments inserted; [d] Ax = bidentate nitrate, average Co–O distance reported; [e] R = Carbene = (p-ClC₆H₄)₂C=C; [f] Displacement from 3 unreacted pyrrole rings; [g] R = nitrene = N-tosyl-amino; [h] O = Oxene or porphyrin N-oxide

## 3. N,N'-Bridged Species

The (adjacent) N,N'-bridged porphyrin species are derivatives that do not coordinate to metals via all four pyrrolic nitrogen atoms. The structures of two palladium(II) derivatives have been reported[114, 115]. The carbon substituent bonded to the two nitrogen atoms is benzyloxymethylene or ethoxycarbonylmethylene, respectively. The palladium ion forms square planar complexes in which the metal is coordinated to two pyrrolic nitrogen atoms (the unsubstituted ones) and two halide ions (but see below). Thus two pairs of adjacent nitrogen atoms are bridged: one pair by carbon, the second by a metal ion. The square coordination plane ($N_2PdX_2$) is tilted by about 65° from the plane of the four nitrogen atoms. The geometry requires that the metal atom be substantially out of the porphyrin plane (1.46 or 1.37 Å). The other nitrogen substituent is on the other side of the porphyrin plane; the tetrahedral geometry of the N,N'-bridging carbon atom "pushes" a hydrogen atom at the palladium ion. The intramolecular Pd-H distances are 2.58 or 2.47 Å. Thus, the coordination polyhedron around Pd might be described as square-pyramidal. The Pd–N and Pd–Cl or Pd–Br distances are normal. The complexes can be regarded as sitting-atop complexes that are postulated as intermediates in the metal insertion reactions. The structure of the related free base has also been reported[116].

A related substance, another sitting-atop model, is a bischloromercury(II) complex[117] of N-tosylaminooctaethylporphyrin. The two $Hg^{II}$ ions are on opposite sides of the porphyrin plane and are in quite different coordination environments. One $Hg^{II}$ ion is coordinated to the tosylamino nitrogen and a chloride ion in a linear fashion. The second $Hg^{II}$ ion is coordinated to the three other porphyrin nitrogen atoms and one chloride ion. (See structure F).

## 4. Miscellaneous Species

A most remarkable compound with N-substituents has been obtained by reaction of iodonium ylides with an $Fe^{II}$ porphyrinate. The product is a bis(metallocycle) of type D. The iron(II) ion is high spin. The structure[118] has an in-plane iron(II) with Fe–N(alkyl) = 2.242(5) Å and Fe–N = 2.082(5) Å. The latter bond distance is slightly larger than that of other six-coordinate high-spin iron(II) species.

Balch and coworkers[119] have reported that triruthenium dodecacarbonyl reacts with a N,N'-vinyl bridged porphyrin to yield structure E. Ru–N bond distances range from 2.068 to 2.257 Å.

The structures of two crystalline forms of monoprotonated octaethylporphyrin (monocation) have now been reported[120] in detail. One pyrrole ring is tipped out of the plane of the other three in order to allow the simultaneous positioning of three hydrogen atoms at the center of the ring.

# V. Electron Density Studies

Experimental electron density studies are a recent development in the structural characterization of metallotetrapyrroles. It is hoped that these studies will provide an additional

**Table X.** Experimental d-orbital populations from accurate x-ray diffraction experiments

| Complex | Orbital population | | | | | |
| | $d_{x^2-y^2}$ | $d_{z^2}$ | $(d_{xz}, d_{yz})$ | $d_{xy}$ | $d^n$, spin state | Ref. |
| --- | --- | --- | --- | --- | --- | --- |
| Co(TPP) | 1.0(2) | 1.0(2) | 3.7(3) | 1.3(2) | $d^7$, L.S. | 121 |
| Fe(Pc) | 0.70(7) | 0.93(6) | 2.12(7) | 1.68(10) | $d^6$, I.S. | 122 |
| Fe(TPP) | 0.43(13) | 1.70(11) | 3.54(14) | 1.50 | $d^6$, I.S. | 123 |
| [Fe(TPP)(OCH₃)]ᵃ | 1.15(6) | 1.15(5) | 1.98(6) | 1.21(6) | $d^5$, H.S. | 124 |
| | 1.04 | 1.07(6) | 1.81(8) | 1.08(6) | | |
| [Fe(TPP)(THF)₂]ᵃ | 1.31 | 0.88 | 2.72 | 1.03 | $d^6$, H.S. | 125 |
| | 1.42 | 1.04 | 2.52 | 0.93 | | |

ᵃ Two different refinements

way to examine the electronic structure and bonding. The first such study to be reported was the electron density determination for Co(TPP)[121]. This study demonstrated the feasibility of determining, experimentally, the populations of the d-orbitals. At the risk of oversimplifying difficultly obtained results, a summary of d-orbital populations is given in Table X. The results for Fe(Pc)[122] and Fe(TPP)[123] clearly show that the detailed ground state for Fe(TPP) is different than that for Fe(Pc) even though both are nominal intermediate-spin species. For Fe(TPP), the $^3A_{2g}$ state is the predominant state, although other states must be mixed in as well. A high-spin iron(III) system, [Fe(TPP)(OCH₃)][124], and a high-spin iron(II) complex, [Fe(TPP)(THF)₂][125], are also summarized in Table X. It can be seen that the $d_{x^2-y^2}$ orbital is always populated to a larger extent than that required by the formal electronic configuration. It is presumed that this is the result of strong σ-bonding from the porphinato ligand. It is to be hoped that the method will allow more detailed descriptions of bonding if the charge density determination is coupled with the determination of spin densities.

## VI. π-Cation Radical Complexes

The structures of a number of π-cation radical complexes have been determined. As noted earlier, these are species in which the macrocyclic ring rather than the metal has been oxidized. Almost all of these species can be considered as high-valent species since they exist in a total oxidation level above that found under aerobic conditions. A summary of the structural parameters of the metal complexes of known structure is given in Table XI. The structural parameters of the [Mg(TPP)(OClO₃)][126] and [Zn(TPP)(OClO₃)][127] are those typical of normal five-coordinate zinc and magnesium porphyrinates. The structures of five-coordinate [Fe(TPP)Cl]⁺ and six-coordinate [Fe(TPP)(OClO₃)₂][128] are exactly that expected for the analogous five- and six-coordinate high-spin iron(III) porphyrinates. The same is true of the [Cu(TPP)]⁺ radical cation derivative[129]. Thus, one question can be answered: ring oxidation does not appear to change the geometric aspects of metal coordination.

**Table XI.** Stereochemical parameters of π-cation complexes[a]

| Complex | M–N$_p$ | Ax, M–Ax | M···Ct' | Ref. |
|---------|---------|----------|---------|------|
| [Mg(TPP)(OClO$_3$)] | 2.096(14) | O, 2.012(5) | 0.52 | 126 |
| [Fe(TTP)Cl]$^+$ | 2.07(2) | Cl, 2.168(5) | 0.37 | 128 |
| [Fe(TPP)(OClO$_3$)$_2$] | 2.045(10) | O, 2.13(1) | 0 | 128 |
| [Cu(TPP)]$^+$ | 1.988(4) | – | 0.06 | 129 |
| [Zn(TPP)(OClO$_3$)] | 2.076(9) | O, 2.079(8) | 0.34 | 127 |

[a] All values in Å

When the coordinated metal ion is paramagnetic, there is the possibility of magnetic interaction between the metal ion and the paramagnetic ligand. Three possible descriptions can be offered: antiferromagnetic coupling, ferromagnetic coupling, and lastly non-interacting systems. A second question thus concerns the possible relationship of structure and the magnetic properties of the metalloderivatives. It has been suggested[128, 129] that antiferromagnetic coupling can result only when the porphinato complex deviates significantly from $D_{4h}$ symmetry. Only in such cases are the metal d and ligand a$_{1u}$ and a$_{2u}$ orbitals not orthogonal. The structures of diamagnetic [Cu(TPP)]$^+$ [129] and the reduced moment of [Fe(TTP)Cl]$^+$ [128] are consistent with this idea of nonplanar cores and antiferromagnetic coupling. The distinction between a noninteracting or a ferromagnetically coupled system is more difficult to make, but an effectively planar porphinato core has been claimed as necessary for either. Magnetic susceptibility data for [Fe(TPP)(OClO$_3$)$_2$][128] are consistent with ferromagnetic coupling.

A final question concerns the unusual core conformation of all of the four- and five-coordinate π-cation radical species listed in Table XI. All these species have an unusual saddle-shaped conformation. Possible reasons for this are to be described in the conformation section.

## VII. Reduced Ring Tetrapyrroles

There has been substantial recent interest in the structures of reduced porphyrin derivatives. Reduced derivatives mean the reduction of the porphyrin system by the addition of one or more moles of hydrogen. This chemical change can have important effects on physical properties. However, in this review, the effect of such modified macrocycles on the coordination parameters of the metalloderivatives is the major question. Characterized species include three levels of reduction, di-, tetra- and hexahydroporphyrins. Examples of the dihydroporphyrins are chlorins (G) and porphodimethenes (H). Tetrahydroporphyrins include bacteriochlorins (I) and isobacteriochlorins (J). Characterized hexahydroporphyrins are the pyrrocorphins (K) and hexahydroporphyrins (L).

We first simply note that there are i) a number of free base compounds of known structure[130–136] and ii) a number of magnesium derivatives of known structure[137–140] that are related to the naturally occuring chlorophyls.

One effect of reducing one or more pyrrole rings is to increase the metal-nitrogen bond distance to the reduced ring(s) if everything else remains equal. This is cleanly seen in the structure of [Zn(TPC)(Py)][141]. In this complex, the Zn–N$_p$ bond distances average

G

H

I

J

K

L

to 2.065(6) Å, a normal value for zinc porphyrinates, while the Zn–N(pyrroline) distance is 2.130(1) Å. A similar pattern, $Zn–N_p$ = 2.070(23) Å and Zn–N(pyrroline) = 2.102(21) Å, is seen in the isobacteriochlorin complex, $[Zn(TPiBC)(Py)]$[142]. There is a real possibility that the $Zn–N_p$ distances are slightly overestimated and the Zn–N(pyrroline) distances are slightly underestimated owing to the presence of rotational disorder which would occasionally interchange a pyrrole ring with a pyrroline ring.

However, everything does not remain equal. Consideration of all structure determinations suggest that reduced derivatives have substantial conformational flexibility. Indeed, reduced porphyrins species are probably more readily deformed in a direction perpendicular to the mean plane than porphyrin complexes. Note, however, that this is a permissive statement and *not* a requirement; reduced porphyrin species with planar cores are possible and observed. This flexibility does have one significant consequence for the reduced porphyrin species; it permits shortening of *all* M–N bond distances if the species undergoes an $S_4$-ruffling of the core. A quantitative treatment of the effect has been given by Hoard[11]. Very roughly, a proper $S_4$-ruffling can lead to a decrease in the M–N bond distances by 0.02–0.03 Å. Thus planar Fe(OEP) has an average bond distance of 1.996(16) Å, while the reduced derivative[62], Fe(OEC), has a ruffled core and $Fe–N_p$ = 1.978(13) Å and Fe–N(pyrroline) = 2.002(4) Å. Thus, we can expect that the Fe–N(pyrroline) distance would be 2.02–2.03 Å in a planar form. Similarly, in the series Ni(TMP), Ni(TMC)[143] and Ni(TMiBC)[144] the observed $Ni–N_p$ bond distance decreases in the series from 1.953(14) Å, 1.922(6) Å, and 1.917(7) Å, respectively. This distance correlates with the amount of $S_4$ ruffling in the series (starting with almost planar Ni(TMP)). The values of the Ni–N distances to the reduced rings are 1.936(1) Å for Ni(TMC) and 1.926(6) Å for Ni(TMiBC).

These data for the isobacteriochlorins also suggest one other possible trend in the coordination parameters, namely that as the number of reduced rings is increased the *differences* in the M–N bond distances to the two types of rings will decrease. Data presented by Kratky[145] for a number of tetra- and hexahydrooctaethylporphyrin derivatives are in accord with this idea. Unfortunately, most of the metallo species are four-coordinate nickel complexes. These low-spin square-planar complexes are expected to have short M–N bond distances. A complete test of the effect of the number of reduced rings on M–N bond distances will require additional structural data for metallo species with intrinsically larger M–N bond distances.

Kratky and coworkers[145] have summarized the structures of thirteen reduced porphyrin species including ten metallo derivatives of which eight are four-coordinate nickel(II) species. Only selected stereochemical parameters have been presented. A summary of some available information is given in Table XII. It is seen that these species have very large $S_4$ rufflings of the macrocyclic cores. A probable effect of this ruffling is to allow for better equivalence of the M–N bond distances. It is to be noted that almost all of the Ni–N bonds are shorter than the 1.93 Å-value found[146, 147] for ruffled $Ni^{II}$ porphyrinates. Hence the bond distances in all reduced derivatives tend more towards the unconstrained Ni–N bond distance. This is in accord with the notion of greater conformational flexibility for the reduced tetrapyrrole derivatives. However, it is again to be emphasized that nonplanarity is not a requirement of reduced tetrapyrroles.

The structure of two related species having β-oxo substituents have been reported. These species are formally analogous to hydroporphyrins. These species are of interest because it appears that oxo-porphyrins are the chromophoric group in certain "green" hemoproteins. Stolzenberg[148] has reported the structure of Ni(OxoOEP), a species formally equivalent to a chlorin. Interestingly, there are two crystallographically distinct molecules of Ni(OxoOEP) in the solid state: one with an $S_4$-ruffled core and a second with a nearly planar core. This again points out the permissive nature of the conformations of the reduced porphyrin species. The structure of a species analogous to an isobacteriochlorin, Cu(DioxoOEP)[149], has also been reported. Structural data, equivalent to that of Kratky, is summarized at the bottom of Table XII for these two β-oxo species.

**Table XII.** Summary of stereochemical parameters of reduced porphyrin species[a]

| Complex | Configuration[b] | M–N$_p$ | M–N(red) | $d_m$[c] | No. Red. Rings | Ref. |
|---|---|---|---|---|---|---|
| Ni(PCOEP) | ccccc | 1.913(8) | 1.908(8) | 0.74 | 3 | 145 |
| Ni(PCOEP) | tctcc | 1.925(8) | 1.931(6) | 0.66 | 3 | 145 |
| Ni(PCOEP) | tcttc | 1.938(12) | 1.927(12) | 0.66 | 3 | 145 |
| [Co(PCOEP)(Py)] | ccccc | 1.950(7) | 1.982(10) | 0.51 | 3 | 145 |
| Cu(PCOEP) | tctct | 1.990(10) | 2.045(19) | 0.26 | 3 | 145 |
| Ni(OEPiBC) | ttt | 1.915(3) | 1.920(4) | 0.64 | 2 | 145 |
| Ni(OEPiBC) | tct | 1.94(3) | 1.94(4) | 0.64 | 2 | 145 |
| Ni(OEPBC) | ccc | 1.920(2) | 1.922(3) | 0.61 | 2 | 145 |
| Ni(HHOEP) | tttt | 1.91(3) | 1.91(3) | 0.72 | 2 | 145 |
| Ni(HHOEP) | ttct | 1.925(7) | 1.94(4) | 0.68 | 2 | 145 |
| Ni(OxoOEP)[d] |  | 1.967(7) | 1.994(18) | 0.05 | 1 | 148 |
|  |  | 1.953(11) | 2.030(29) | 0.36 | 1 |  |
| Cu(DioxoOEP) |  | 1.999(10) | 2.044(13) | 0.09 | 2 | 149 |

[a] All values in Å;   [b] The configurations describes the relative orientation, in pairs, of the groups on the reduced rings starting with the first substituent on Ring A;   [c] $d_m$ is the average absolute displacement of the *meso*-carbon atoms from the mean plane of the four nitrogen atoms. All cores have $S_4$-ruffled geometry;   [d] Two independent molecules

Finally, it should be noted that the expected increase in the $C_b$–$C_b$ and $C_a$–$C_b$ bond distances of the five-membered ring are observed upon reduction. It is expected and generally observed that the pyrroline ring is no longer planar.

## VIII. Novel Systems

In this section, we describe the stereochemistry of a small group of species with "nontraditional" or unexpected structures. Classification of these species has a certain amount of arbitrariness.

Collman et al.[150] have described the preparation and characterization of a binuclear porphyrin complex containing a Ru–Ru metal-metal bond of order two. The Ru–Ru bond length in $[Ru(OEP)]_2$ is 2.408(1) Å. The average Ru–$N_p$ bond distance is 2.050(5) Å. The two porphinato planes are twisted 24° with respect to each other. This derivative is the only one with metal-metal bonded porphyrin units structurally characterized.

In addition, two new derivatives containing metal-metal bonded species in which the second metal is an "axial" ligand have been characterized. Both species have tin porphyrinates at their heart. The structure of a species containing a chain of four metal atoms has been reported[151]. The molecule is $(TPP)Sn–Mn(CO)_4–Hg–Mn(CO)_5$. The "five-coordinate" tin ion is displaced 0.85 Å from the porphyrin plane and the average Sn–$N_p$ bond distance is 2.183(15) Å. These parameters are entirely consistent with a formal oxidation state of + 2 for the tin. The Sn–Mn bond distance is 2.554(3) Å. The structure of $(OEP)Sn–Fe(CO)_4$ has also been reported[152]. The structure is very similar with an average Sn–$N_p$ distance of 2.187 Å and Fe–Sn = 2.492 Å. A somewhat related complex has two –C–Re(CO)$_3$ groups as axial ligands to $[Sn^{IV}(TPP)]$[153]. The Sn–C distances are 2.14 Å and the Re–C distance is 1.75 Å. The Sn–C–Re angle is 138.5(8)°. These values are consistent with the formulation Sn–C≡Re. The bond parameters to the in-plane tin atom are commensurate with tin(IV).

The structure of seven-coordinate $[Nb_2(TPP)_2O_3]$[154] and $[Nb(TPP)(OAc)(O)]$[155] and six-coordinate $[Nb(OEP)(F)(O)]$[156] are all found to have all axial ligand on the same side of the porphyrin plane in a 4 + 3 or 4 + 2 arrangement. The niobium(V) ion is displaced out of the porphyrin nitrogen atom plane by 0.9–1.0 Å and the Nb–$N_p$ distances are 2.21 to 2.24 Å.

The first six-coordinate zinc(II) porphyrinate[157] complex has been reported. $[Zn(TPP)(THF)_2]$ has two axial THF ligands at Zn–O = 2.380(2) Å. The Zn–$N_p$ distance is 2.057(1) Å. As described later, we think that this compound is stabilized by solid state factors. We[158] have also recently found a six-coordinate zinc(II) complex, again solid state factor appear responsible for the unusual coordination number.

Buchler, Weiss and coworkers have recently described the structures of lanthanide porphyrins. The structures of cerium(IV) "doubledecker" and biscerium(III) "tripledecker" molecules have been described[159]. $[Ce(OEP)_2]$ and $[Ce_2(OEP)_3]$ have cerium ions with approximate square-antiprismatic geometry in both complexes. In the doubledecker complex, the Ce–$N_p$ bond length is 2.475(1) Å. The two porphyrin rings are rotated by about 43° with respect to each other. A quite similar structure is seen for $[Eu(OEP)_2]$[160]. In this structure, Eu–$N_p$ is slightly longer at 2.510(2) Å. The two rings have a mean plane separation of 3.40 Å. In the tripledecker [$Ce_2(OEP)_3$], the Ce–$N_p$ bond distance to the external rings is 2.501(2) Å, and the Ce–$N_p$ bond distance to the internal (center) ring nitrogen atoms is 2.758(2) Å. The porphinato cores are separated by 3.54 Å in the tripledecker. In the tripledecker complex, the central cerium(III) ion sits at an inversion center and hence the two outer rings have the same relative orientation to each other. The central ring is twisted by 24.5° from the two outer rings. The $Ce^{III}$ ions are displaced by 1.39 Å from the external ring and 1.88 Å from the internal (central) ring. The central ring is, of course, shared between two metal ions.

Hitchcock has described[161] a bis(NiOEP) derivative where the two porphyrin rings are joined by a *meso,meso'*-ethylene bridge. The notable point is a lack of interaction between the two rings. The two rings are required to be parallel to each other by a center of symmetry located at the center of the ethylene bridge. Metal coordination parameters are normal.

# IX. Other Species

Included in this section is a miscellany of species, of recent origin, that warrant at least brief descriptions.

## 1. Main Group Derivatives

A most unusual main group derivative is a phosphorous(V)[162] derivative which is probably the least metallic of the metallo species. The structure of the dihydroxo derivative, [$P(TPP)(OH)_2$]OH[163], has been determined. The average P–$N_p$ bond distance is 1.891(19) Å and the average axial P–O(OH) is 1.59(7) Å. In order to accommodate the very small $P^{(V)}$ cation, the porphyrin core is exceedingly $S_4$-ruffled with deviations of the meso carbon atoms of 0.80 from the mean plane.

The structures of a number of five-coordinate thallium(III)[164-166] complexes lead to an unprecedented observation in metalloporphyrin species: substantial variation in the coordination group geometry as a function of the axial ligand. (For transition metal complexes such changes are never observed unless there is a concommitant change in spin state.) Table XIII summarizes coordination group parameters of a number of thallium complexes. Even though the norbornenyl structure is not of high precision, it is seen that that derivative, as well as the methyl species, lead to larger Tl–$N_p$ bonds and substantially larger displacements of the $Tl^{III}$ ion from the plane of the macrocycle

**Table XIII.** Stereochemical parameters of thallium and indium complexes[a]

| Complex | M–$N_p$ | M–Ax | M⋯Ct | M⋯Ct' | Ref. |
|---|---|---|---|---|---|
| [Tl(OEP)Cl] | 2.212(6) | 2.449(3) | 0.69 | 0.75 | 164 |
| [Tl(TPP)Cl] | 2.210(4) | 2.420(4) | 0.62 | 0.74 | 165 |
| [Tl(TPP)(CH$_3$)] | 2.291(5) | 2.147(12) | 0.84 | 0.98 | 165 |
| [Tl(TPP)(NOR)] | 2.292(131) | 2.089(10) | 0.99 | 1.10 | 166 |
| [In(TTP)Cl] | 2.156(6) | 2.369(2) | 0.61 | 0.71 | 167 |
| [In(TPP)(CH$_3$)] | 2.205(6) | 2.13(1) | 0.78 | 0.92 | 168 |

[a] All distances in Å

compared to the two chloride complexes. Similar observations are applicable to analogous indium(III) species[167, 168] also listed in Table XIII. The only other characterized $In^{III}$ species is a polymer $[In(TPP)(O_3SCH_3)]_n$[169]. This complex has a six-coordinate metal ion; the axial methylsulfonate ligand bridges two porphinatothallium units to form an –In–O–In–O– chain. The average In–$N_p$ bond distance is 2.107(22) Å.

The structure of $[Cd(TPP)(Pip)]$ has been described[170]. The large $Cd^{II}$ ion is displaced 0.65 Å from the porphyrin mean plane. The average Cd–$N_p$ bond distance is 2.204(14) Å. The axial Cd–N bond distance is 2.323 Å. This structure is quite different from that reported[171], in preliminary form, for a bis(dioxane) adduct of Cd(TPP). We believe that this structure is better described as a disordered structure in which the $Cd^{II}$ ion is displaced from the center of the ring on either side of the porphyrin plane, much like that found for four-coordinate Mn(TPP)[172]. This "six-coordinate" $Cd^{II}$ structure is regarded by us as another example of the effects of crystal packing to yield crystals of otherwise unstable species.

A number of six-coordinate magnesium porphyrinates have been recently characterized. The Mg–$N_p$ distances are all nearly equal with the average distance to in-plane $Mg^{II}$ = 2.072(5) Å. The axial distances do vary with Mg–N(Im)[173] = 2.297(8) Å, Mg–N(py)[174] = 2.389(2) Å, Mg–N(4-MePy)[173] = 2.386(2) Å, and Mg–N(Pip)[173] = 2.419(4) Å.

## 2. Transition Metal Complexes

In this section, we briefly note the results of a variety of transition metal porphyrinate structures. The structure of two low-spin $Mn^{II}$ porphyrinates, each having the noninnocent axial ligand NO, have been reported[175]. The complexes are $[Mn(TTP)(NO)]$ and $[Mn(TPP)(Pip)(NO)]$ and thus allow for a comparison of the effects of increased coordination number: Mn–$N_p$ = 2.004(5) and 2.027(3) Å, respectively. A novel $Mn^{III}$ complex has been prepared and characterized[176]. $[Mn(TPP)(Im)]_n$ is a chain polymer with two magnetically and structurally distinct $Mn^{III}$ ions: one high spin and one low spin. The axial bond to the imidazolate ligand of the low-spin species is 2.186(5) Å and that to the high-spin $Mn^{III}$ ion is 2.280(4) Å. This is the only structurally characterized low-spin $Mn^{III}$ species as even $[Mn(TPP)(CN)]$[28] is found to be high spin. The structures of three bis-oxygen ligated $Mn^{III}$ complexes show no variation in Mn–$N_p$ (average 2.003(9) Å) and only very minor variation in the Mn–O bond lengths. The compounds (axial distance) are $[Mn(TPP)(MeOH)_2]ClO_4$[177] (2.261(13) Å), $[Mn(TPP)(DMF)_2]ClO_4$[178] (2.217(4) Å), and $[Mn(TPP)(2,6-LutNO)_2]ClO_4$[179] (2.263(4) Å). The structures are thus typical of high-spin $d^4$ manganese(III) complexes.

There have been a number of species whose structures have been determined that have typical organometallic ligands. The structures of $[Rh(OEP)(CHO)]$[180] and $[Rh(Etio\ I)(PhCO)]$[181] have relatively short Rh–C bonds of 1.896(6) and 1.963(7) Å, respectively. The benzoyl complex has the $Rh^{III}$ ion displaced 0.09 Å out of the four nitrogen plane, *away* from the axial ligand towards the vacant coordination site. This is a most unusual stereochemical feature for a five-coordinate complex. Dolphin, James and coworkers have reported the structure of $Ru^{II}$[182] and $Ru^{III}$[183] complexes with axial phosphine ligands. Although the $Ru^{II}$ species has potential bidentate phosphine ligands, there is nothing unusual about the coordination geometry.

The structures of several new cobalt(III) porphyrinate derivatives have also been reported. The displacement of the cobalt ion in five-coordinate $[Co(TPP)(CH_2CHO)]$[184] is found to be 0.14 Å from the four nitrogen plane and 0.10 Å from the 24-atom plane. This complex is thus an example of the relatively rare "reverse domed" core. Other features are completely typical of cobalt(III). An EXAFS study[185] of $[Co(TPP)(Br)]$ has shown that the axial bond appears to be anomalously short with a value, Co–Br = 2.28 Å and a very small (0.05 ± 0.07 Å) displacement of the cobalt. This had also been noted in an unpublished structure from this laboratory[186]. The structure of three six-coordinate cobalt(III) species finish the transition metal complexes. $[Co(TPP)(OH_2)_2]ClO_4$[187] has an axial Co–O bond distance of 1.936(5) Å and a Co–$N_p$ bond average of 1.964(6) Å. Similar Co–$N_p$ distances are seen in two bis(thiolate)cobalt(III) complexes reported by Weiss[188]. The distances in two different crystalline forms of the complex, $[Co(TPP)(SC_6HF_4)_2]^-$, are 1.977(8) Å and 1.973(11) Å. The small differences in the bond lengths of these three species are readily correlated with the ruffling of the core, with the smallest Co–$N_p$ distance seen in the most ruffled species. The axial Co–S bond length in the two cobalt cytochrome P-450 models of Weiss is 2.343(9) Å. Any effect resulting from the use of the strong electron-withdrawing fluoro-substituted thiolate ligand on the structure is regarded as small.

## C. The Geometry of π-π Interactions in Porphyrin Derivatives

Our interest in the geometry of π-π interactions between porphyrin rings in the solid state was kindled by the unexpected observation of a tight "dimer" between molecules of five-coordinate low-spin $[Fe(OEP)(NO)]^{+}$[30] in the solid state. The interaction is depicted in two views shown in Figs. 5 and 6. The close separation of the two porphinato cores (3.362 Å) and the strong overlap of the two rings is noteworthy. Our interest was intensified by a second result: the X-ray structure determination of $[Fe(OEP)(2-MeHIm)]^{+}$[16]. Two crystalline forms of this five-coordinate high-spin iron(III) complex have been investigated; both are found to have π-π dimers in the solid state with similar structures. Indeed, one modification is virtually quantitatively identical to that of

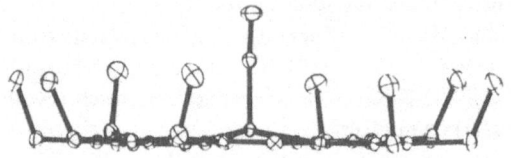

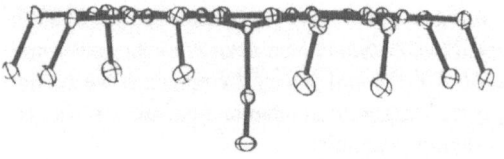

**Fig. 5.** An edge-on view of the π-π dimer formed by the $[Fe(OEP)(NO)]^{+}$ ion. Reproduced with permission from Ref. 30

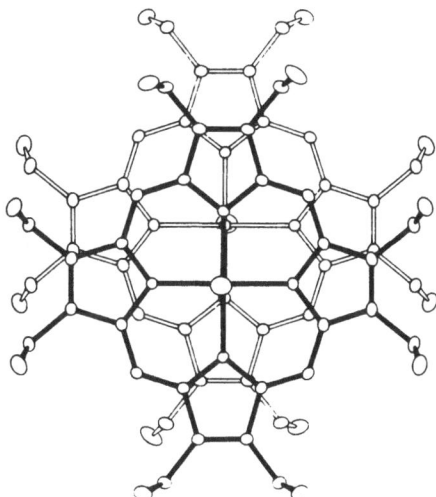

**Fig. 6.** A computer-drawn model of the two $[Fe(OEP)(NO)]^+$ ions related by an inversion center showing the overlap of the two cores. The planes of the two parallel cores are also parallel to the plane of the paper, and the heavy-lined core is closest to the viewer. Reproduced with permission from Ref. 30

$[Fe(OEP)(NO)]^+$. Furthermore, a detailed analysis[189] of the Mossbauer spectra and temperature-dependent magnetic susceptibilities demonstrated that the two iron(III) ions were coupled through the $\pi$ systems rather than simply by a through space interaction.

This similarity of the $\pi$-$\pi$ interactions in $[Fe(OEP)(2\text{-MeHIm})]^+$ and $[Fe(OEP)(NO)]^+$ suggested that common geometric patterns might be found and led us to systematically evaluate the literature for solid-state interactions between porphyrin rings. Calculations using the original crystallographic coordinates were generally required in order to characterize the species in a completely uniform manner. We have examined all appropriate structures known to us for the possibility of such interactions. We considered any porphyrin or reduced porphyrin molecule with only alkyl peripheral substituents. For the metallo derivatives, only species with no axial ligands or ligand(s) on one side of the porphyrin plane were considered. In other words, we considered any porphyrin with at least one unhindered face. Almost all of the derivatives satisfying these criteria display an apparent aromatic interaction between pyrrole rings of the porphyrin derivative. Two limiting types of structures have been found: one in which porphyrin rings interact in an extended stacked arrangement (aggregates) and a second in which porphyrin rings interact in pairs. It is presumed that these solid state phenomena bear some relation to self-association under less constraining environments.

The results of this survey are shown in Table XIV. The criterion for inclusion in this table was a possible $\pi$-$\pi$ interaction as judged by a mean plane separation of less than 4.0 Å and at least a modest overlap of the two macrocyclic rings. The entries of Table XIV are grouped by type: free base derivatives and then the metallo derivatives in alphabetical order. The geometric parameters used in Table XIV can be understood with the aid of Fig. 7. The mean plane separation (M.P.S.) is the perpendicular distance between the pair of best planes of the 24-atom cores of the macrocyclic rings. Ct represents the center of the hole defined by the four nitrogen atoms of the macrocycle. Ct···Ct and M···M represent the distances between these respective positions in the two rings. L.S. represents the lateral shift of the "slipped" arrangement of the two rings and is independent of the relative orientation of the rings. L.S. in this table is defined as

**Table XIV.** Summary of mean plane separations, lateral shifts and slip angles

| Compound | Type | M.P.S.[a] | Ct–Ct[a] | M–M[a] | L.S.[a] | S.A.[b] (Ct) | S.A.[b] (M) | Dihed. angle[b] | Group | Ref. |
|---|---|---|---|---|---|---|---|---|---|---|
| 21-EtO–HOEP | Dimer | 3.41 | 3.738 | – | 1.53 | 24.2 | – | 0 | S | 105 |
| 25-EtO–H2OEP | Dimer | 3.46 | 3.592 | – | 0.97 | 15.6 | – | 0 | S | 115 |
| 4-Bu-5-Et MeBPheo d | Agg-A | 3.66 | 3.859 | – | 1.22 | 18.5 | – | 1.4 | S | 133 |
| 4-Bu-5-Et MeBPheo d | Agg-B | 3.74 | 5.514 | – | 4.05 | 47.3 | – | 1.4 | W | 133 |
| 4,5DiEt MeBPheo d | Agg-A | 3.51 | 3.852 | – | 1.59 | 24.3 | – | 1.7 | S | 133 |
| 4,5DiEt MeBPheo d | Agg-B | 3.72 | 5.358 | – | 3.86 | 46.0 | – | 1.7 | I | 133 |
| Siro-iBC | Agg-A | 3.65 | 5.229 | – | 3.74 | 45.7 | – | 0 | I | 132 |
| Siro-iBC | Agg-B | 3.44 | 7.214 | – | 6.34 | 61.5 | – | 0 | W | 132 |
| Formylporphyrin | Agg | 3.41 | 6.677 | – | 5.74 | 59.3 | – | 0 | W | 204 |
| MeBPheo a | Agg | 3.50 | 8.107 | – | 7.31 | 64.4 | – | 0 | W | 134 |
| H2Meso IX DME | Agg | 3.41 | 5.971 | – | 4.90 | 55.1 | – | 0 | W | 205 |
| MePheo a | Agg | 3.48 | 8.035 | – | 7.24 | 64.4 | – | 0 | W | 131 |
| NO2Proto IX DME | Dimer | 3.39 | 3.711 | – | 1.51 | 24.0 | – | 0 | S | 206 |
| H2OEP | Agg | 3.42 | 7.483 | – | 6.66 | 62.8 | – | 0 | W | 207 |
| H3OEP+ I3− (tricl) | Dimer | 3.38 | 3.488 | – | 0.86 | 14.3 | – | 0 | S | 233 |
| H3OEP+ I3− (ortho) | Dimer | 3.41 | 3.563 | – | 1.04 | 16.9 | – | 0 | S | 233 |
| H3OEP+ [Re2(CO)6Cl3]− | Dimer | 3.46 | 4.673 | – | 3.14 | 42.2 | – | 0 | I | 234 |
| 5-Bis(EtO)H2OEP | Dimer | 4.59 | 7.262 | – | 5.63 | 50.8 | – | 0 | W | 208 |
| Phyllochlorinester | Agg | 3.20 | 8.194 | – | 7.54 | 67.0 | – | 8.9 | W | 130 |
| Porphine | Dimer | 3.41 | 3.776 | – | 1.63 | 25.5 | – | 0 | S | 209 |
| H2Proto IX DME | Agg | 3.42 | 6.079 | – | 5.02 | 55.8 | – | 0 | W | 210 |
| H2TPrP | Agg | 3.37 | 5.078 | – | 3.80 | 48.4 | – | 0 | I | 211 |
| Co(OEP)(1-MeIm) | Dimer | 3.87 | 8.207 | 8.334 | 7.24 | 61.8 | 62.3 | 0 | W | 212 |
| Cu(5-F-10-BBV-OEP) | Dimer | 3.86 | 5.323 | 5.299 | 3.67 | 43.5 | 43.2 | 0 | I | 213 |
| Cu(DioxoOEP) | Dimer | 3.43 | 6.146 | 6.122 | 5.10 | 56.1 | 55.9 | 0 | W | 149 |
| Cu(TPrP) | Agg | 3.38 | 5.010 | 5.010 | 3.70 | 47.6 | 47.6 | 0 | I | 214 |
| Cu2(DP-7) | Agg | 3.45 | 4.67 | 4.69 | 3.15 | 42.4 | 42.6 | 0 | I | 215 |

| | | | | | | | | | | |
|---|---|---|---|---|---|---|---|---|---|---|
| Cu$_2$(DP-7) | Requ | 3.52 | 5.23 | 5.21 | 3.87 | 47.7 | 47.5 | 0 | I | 215 |
| Cu$_2$(DP–B) | Requ | 3.55 | 3.861 | 3.807 | 1.52 | 23.2 | 21.2 | 4.4 | S | 193 |
| Cu$_2$(FTF) | Requ | 3.87 | 6.341 | 6.331 | 5.02 | 52.4 | 52.3 | 0 | W | 216 |
| Cu$_2$(FTF) | Agg | 4.06 | 8.756 | 8.760 | 7.76 | 62.4 | 62.4 | 6.3 | W | 216 |
| Fe(Proto IX)Cl | Dimer | 3.48 | 6.075 | 6.665 | 4.98 | 55.0 | 58.5 | 0 | W | 217 |
| Fe(Meso IX DME)(OMe) | Dimer | 3.51 | 4.843 | 5.549 | 3.33 | 43.5 | 50.7 | 0 | I | 218 |
| Fe(OEC) | Agg | 3.55 | 4.879 | 4.886 | 3.35 | 43.3 | 43.4 | 0 | I | 62 |
| Fe(OEP) | Agg | 3.42 | 4.812 | 4.812 | 3.39 | 44.7 | 44.7 | 0 | I | 62 |
| Fe(OEP)(2-MeHIm)$^+$ | Dimer | 3.31 | 3.635 | 4.280 | 1.50 | 24.3 | 39.3 | 0 | S | 16 |
| Fe(OEP)(2-MeHIm)$^+$ C2/c | Dimer | 3.42 | 4.039 | 4.602 | 2.15 | 32.1 | 42.0 | 0.9 | S | 16 |
| Fe(OEP)(3-ClPy)$^+$ | Dimer | 3.51 | 4.822 | 5.137 | 3.31 | 43.4 | 47.0 | 4.6 | I | 17 |
| Fe(OEP)(SPh) | Dimer | 3.49 | 8.221 | 8.670 | 7.45 | 64.9 | 66.3 | 0 | W | 219 |
| Fe(OEP)(CS) | Dimer | 3.50 | 4.961 | 5.286 | 3.52 | 45.1 | 48.5 | 0 | I | 220 |
| Fe(OEP)(NCS) | Dimer | 3.46 | 4.813 | 5.475 | 3.35 | 44.0 | 50.8 | 0 | I | 307 |
| Fe(OEP)(NO)$^+$ | Dimer | 3.36 | 3.652 | 4.238 | 1.43 | 23.0 | 37.5 | 0 | S | 30 |
| Fe(OEP)(OClO$_3$) | Dimer | 3.51 | 4.897 | 5.259 | 3.42 | 44.2 | 48.2 | 0 | I | 221 |
| Fe(Proto IX DME)(SPhNO$_2$) | Dimer | 3.44 | 4.688 | 5.354 | 3.19 | 42.8 | 50.0 | 0 | I | 37 |
| Mg(EtChl a) | Agg | 3.63 | 8.859 | 8.859 | 8.08 | 65.8 | 65.8 | 0 | W | 139 |
| Mg(EtChl b) | Agg | 3.52 | 8.760 | 8.760 | 8.02 | 66.3 | 66.3 | 0 | W | 140 |
| Mg(MeChl a) | Agg | 3.55 | 8.470 | 8.470 | 7.69 | 65.2 | 65.2 | 0 | W | 138 |
| Mg(MePyrChl a) | Agg | 3.50 | 8.420 | 8.420 | 7.66 | 65.4 | 65.4 | 0 | W | 137 |
| Mn(OEP)(2-MeHIm)$^+$ | Dimer | 3.43 | 3.756 | 4.144 | 1.52 | 23.9 | 34.1 | 0 | S | 194 |
| Nb(OEP)(O)(F) | Dimer | 3.60 | 8.204 | 9.208 | 7.37 | 64.0 | 67.0 | 0 | W | 222 |
| Ni(Deut IX DME) | Dimer | 3.45 | 3.814 | 3.770 | 1.52 | 25.3 | 23.8 | 0 | S | 223 |
| Ni(DPEP) | Dimer | 3.49 | 4.877 | 4.864 | 3.40 | 44.2 | 44.1 | 0 | I | 224 |
| Ni(OEP) (tricl) | Agg | 3.48 | 7.617 | 7.617 | 6.78 | 62.8 | 62.8 | 0 | W | 225 |
| Ni(OEP) (tricl II) | Agg | 3.45 | 4.814 | 4.814 | 3.36 | 44.2 | 44.2 | 0 | I | 198 |
| Ni(OEP) (tetrag) | Agg | 3.46 | 8.228 | 8.228 | 7.46 | 65.1 | 65.1 | 0 | W | 146 |
| Ni(OxoOEP) | Dimer | 3.31 | 6.473 | 6.467 | 5.56 | 59.2 | 59.2 | 0 | W | 148 |
| Ni(TMC) | Agg | 3.71 | 5.093 | 5.091 | 3.49 | 43.3 | 44.3 | 16.1 | I | 143 |

**Table XIV** (Continued)

| Compound | Type | M.P.S.[a] | Ct-Ct[a] | M-M[a] | L.S.[a] | S.A.[b] (Ct) | S.A.[b] (M) | Dihed. angle[b] | Group | Ref. |
|---|---|---|---|---|---|---|---|---|---|---|
| Ni(TMP) | Agg | 3.35 | 5.664 | 5.664 | 4.57 | 53.8 | 53.8 | 0 | W | 143 |
| Ni(TMiBC)A | Agg | 3.79 | 4.673 | 4.675 | 2.73 | 35.8 | 35.8 | 15.0 | I | 144 |
| Ni(TMiBC)B | Agg | 3.76 | 5.109 | 5.115 | 3.46 | 42.7 | 42.8 | 21.4 | I | 144 |
| Ni$_2$(DP–A) | Requ | 3.90 | 4.593 | 4.566 | 2.43 | 31.9 | 31.3 | 3.9 | I | 193 |
| Pb(TPrP) | Dimer | 3.48 | 5.468 | 7.291 | 4.22 | 50.5 | 61.5 | 0 | I | 226 |
| Rh(Etio I)(PhCO) | Dimer | 3.41 | 4.690 | 4.843 | 3.22 | 43.4 | 45.2 | 0 | I | 227 |
| Rh(OEP)(CH$_3$) | Dimer | 3.44 | 7.166 | 7.224 | 6.28 | 61.3 | 61.5 | 0 | W | 228 |
| [Ru(OEP)]$_2$ | Requ | 3.28 | 3.005 | 2.408 | 0.0 | 0.0 | 0.0 | 0.2 | | 150 |
| Ti(OEP)(O$_2$) | Dimer | 3.62 | 7.243 | 7.959 | 6.27 | 60.0 | 62.9 | 0 | W | 69 |
| Ti(OEP)(O) | Dimer | 3.52 | 7.091 | 7.738 | 6.16 | 60.3 | 63.0 | 0 | W | 69 |
| V(O)(DPEP) | Dimer | 3.54 | 5.089 | 5.820 | 3.66 | 46.0 | 52.6 | 0 | I | 229 |
| V(O)(Etio I) | Dimer | 3.53 | 4.975 | 5.727 | 3.50 | 44.8 | 51.9 | 0 | I | 190 |
| V(O)(Etio I) H$_2$Quin | Dimer | 3.56 | 4.909 | 5.726 | 3.38 | 43.5 | 51.5 | 0 | I | 190 |
| V(O)(OEP) | Dimer | 3.50 | 7.045 | 7.678 | 6.11 | 60.2 | 62.9 | 0 | W | 230 |
| Zn(5-OxoProtoIXDME)Cl | Dimer | 3.60 | 5.484 | 6.316 | 4.14 | 49.0 | 55.3 | 0 | I | 231 |
| Zn(5-NO$_2$OEP) | Agg-A | 3.43 | 4.78 | 4.75 | 3.29 | 44.1 | 43.8 | 0 | I | 232 |
| Zn(5-NO$_2$OEP) | Agg-B | 3.56 | 4.95 | 4.97 | 3.47 | 44.0 | 44.2 | 0 | I | 232 |

[a] Value in Å;    [b] Value in degrees

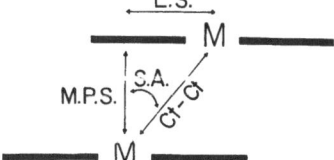

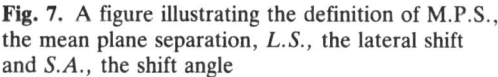

**Fig. 7.** A figure illustrating the definition of M.P.S., the mean plane separation, *L.S.*, the lateral shift and *S.A.*, the shift angle

[sin(S.A.) × Ct–Ct] where S.A. is the "slip angle" between the ring centers. The slip angle between the two metal atoms is also given. The two slip angles will obviously differ when the metal ion is not centered in the ring. The dihedral angle is the angle between the pair of ring planes. In the solid-state structures, the rings (in pairs or in aggregates) are commonly related by an inversion center. This leads to rings that have precisely parallel mean planes and are represented in the tables by a dihedral angle of "0". The inversion symmetry operator also leads to another important geometric requirement that is seen in Fig. 6: the vectors joining opposite nitrogen atoms in one ring must be parallel to those in the second ring. The type of interaction found in the solid state is given by the following symbols: Agg for a symmetric linear chain aggregate, Agg-A for the tighter and Agg-B for the longer interaction in unsymmetrical linear aggregates, Dimer for derivatives that have an interaction that is best regarded as "dimeric", and Requ for an interaction that is required by the molecular connectivity of the molecule. For comparison, also tabulated are the values for $[Ru(OEP)]_2$[150], a binuclear complex with a Ru–Ru metal bond. The metal-metal bond (length = 2.408(1) Å) requires that the two porphyrin cores be essentially directly above each other and to have a rather small interplanar spacing. (The mean plane spacing increases because of significant doming of both cores.)

An immediate concern is to bring some order and understanding to the large number of data found in Table XIV. Two issues need to be addressed. The first is whether any or all of the observed geometrical features simply represent the consequences of crystal packing in the solid and not any real "π-π" interaction between macrocycles. We believe that many of the entries of Table XIV represent (π-π) interactions that determine, at least in part, the crystal packing rather than the opposite. Firstly, it is to be noted that there are appropriately considered derivatives that do not merit inclusion in the table, thus demonstrating that efficient solid-state packing can be achieved without requiring a π-π type of interaction. Secondly, there are "duplicate" derivatives reported in Table XIV. These show virtually identical π-π interactions. We indicate two such examples: 1) the two different solvated crystalline forms of $[Fe(OEP)(2-MeHIm)]^+$ [16] have very similar dimeric structures and appear to persist in solution. 2) The overlaps (detailed in Figs. 8 and 9) of the cores of two different VO(Etio I) derivatives[190] are seen to have nearly identical overlaps that differ only in the relative orientation of the ethyl and methyl substituents. Thirdly, similar dimeric structures have been deduced from solution measurements by NMR[191] and EPR[192] techniques, although the lateral shifts and interplanar spacings are larger than the smallest such values given in Table XIV. Fourthly, a detailed Mossbauer and magnetic susceptibility study[189] of the two forms of $[Fe(OEP)(2-MeHIm)]^+$ shows properties that can be fit only by assuming electronic coupling between the iron(III) centers. The magnitude of the coupling (J = −0.80 and −0.85 cm$^{-1}$) is about 20 times larger than expected from the iron-iron separation in the dimers. This result was interpreted in terms of the overlap of the π systems of the subunits.

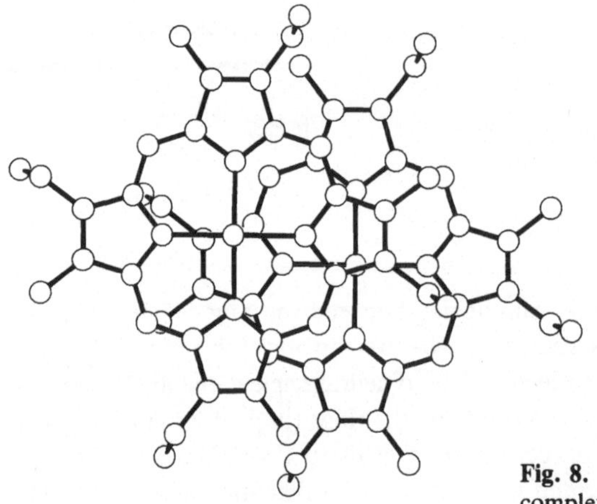

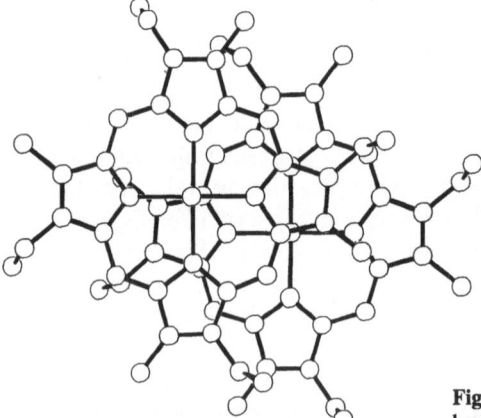

V(O)(Etio I)

**Fig. 8.** A view of the overlap in the V(O)(Etio I) complex. The figure was drawn from coordinates presented in Ref. 190

V(O)(Etio I) H₂Quin

**Fig. 9.** A view of the overlap in the V(O)(Etio I) hydroquinone solvate complex. The figure was drawn from coordinates presented in Ref. 190

Consistent with this interpretation, the species exhibiting the larger spin coupling is the derivative with the smaller M.P.S. and L.S. Fifthly, the structures of some diporphyrin species suggest real driving forces to achieve the kind of π-π interactions detailed in Table XIV. Chang et al.[193] have characterized two diporphyrin species connected by rigid pillars (anthracene and biphenylene) that were designed to maintain a fixed separation between the two porphyrin planes. These derivatives have structures much more like those of the completely unconstrained derivatives constructed from monomeric units than was expected by the investigators. Finally, we[17, 194] have apparently made use of π-π interactions to prepare synthetically difficult species. With this issue resolved, we turn to the second issue, an examination of any underlying pattern(s) for the species given in Table XIV. It is important to note that all species have "slipped" arrangements and there are no π-π species that have the two cores directly above each other.

We believe that the lateral shifts are a useful semiquantitative way to consider the data of Table XIV. In Fig. 10 we plot the lateral shift vs. the mean plane separation for the entries of Table XIV. The plot suggests, without excessive arbitrariness, that the π-π interactions fall into three categories: Group S with small lateral shifts, Group I with moderate lateral shifts and Group W, the remaining members. We use S, I, and W to suggest strong, intermediate, and weak interactions. We believe it clear that Groups S and I represent real interactions, the status of Group W is less certain. The geometric features of Group W may be determined, completely or in part, by the necessity of achieving efficient packing in the lattice. However, it seems highly unlikely that the geometries of Groups S and I are determined by such effects. It is interesting to note that the observed overlap geometry[195] of the "special pair" in the reaction center of *Rhodopseudomonas Viridis* would lead to its classification in Group W.

The Groups S and I are each characterized by a narrow range of values for the lateral shift. These are ~ 1.50 Å and ~ 3.50 Å, respectively. Although the lateral shift is not a definitive descriptor of the interring geometry[196], the pattern of lateral shifts does suggest that the most stable π-π interactions are confined to a narrow range of overlap areas. Classification of the members into the three types are listed in the next to last column of Table XIV. An inspection of the table shows that the populations of all three groups are drawn from the population of individual species of Table XIV in a fairly uniform manner.

A more detailed geometrical analysis has also been carried out to search for the possibility of preferred and specific relative orientations of the two macrocyclic rings. This analysis is also much more numerically exact than that given in Table XIV. In this analysis, the coordinate system is defined in terms of the macrocyclic entities. Two vectors, $u$ and $v$, are defined by the opposite pairs of nitrogen atoms of the tetrapyrrole. These two vectors are then used to define an orthogonal coordinate system x, y, z where

$$x = u$$
$$y = (u \times v) \times u$$
$$z = (u \times v)$$

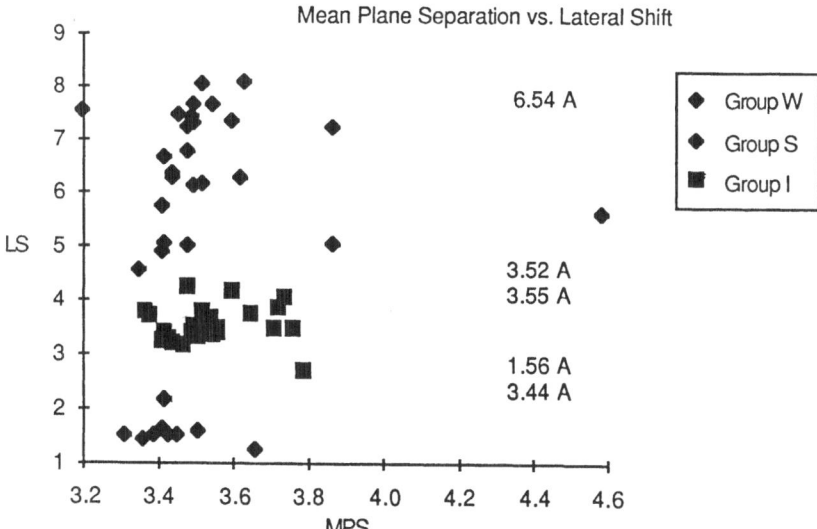

Fig. 10. A plot of the lateral shift vs. mean plane separation for the species listed in Table XIV

**Table XV.** Summary of $x$, $y$, $z$ displacements between Ct's

| Compound | Type | Ct–Ct[a] | M–M[a] | x[a] | y[a] | z[a] | L.S.[a] | Dihed. angle[b] | Group | Ref. |
|---|---|---|---|---|---|---|---|---|---|---|
| 21-EtO-HOEP | Dimer | 3.738 | – | 1.89 | 0.23 | 3.22 | 1.91 | 0 | S | 105 |
| 25-EtO-H2OEP | Dimer | 3.592 | – | 1.15 | 0.35 | 3.39 | 1.20 | 0 | S | 115 |
| 4-Bu-5-Et MeBPheo d | Agg-A | 3.859 | – | 1.12 | 0.78 | -3.61 | 1.37 | 1.4 | S | 133 |
| 4-Bu-5-Et MeBPheo d | Agg-B | 5.514 | – | 3.99 | 0.41 | 3.78 | 4.01 | 1.4 | W | 133 |
| 4,5DiEt MeBPheo d | Agg-A | 3.852 | – | 1.84 | -0.20 | 3.38 | 1.85 | 1.7 | S | 133 |
| 4,5DiEt MeBPheo d | Agg-B | 5.358 | – | 0.34 | 3.82 | -3.74 | 3.84 | 1.7 | I | 133 |
| Siro-iBC | Agg-A | 5.229 | – | -3.80 | -0.28 | -3.58 | 3.81 | 0 | I | 132 |
| Siro-iBC | Agg-B | 7.214 | – | 6.09 | 1.76 | 3.44 | 6.34 | 0 | W | 132 |
| Formylporphyrin | Agg | 6.677 | – | 5.46 | 1.72 | 3.43 | 5.73 | 0 | W | 204 |
| MeBPheo a | Agg | 8.107 | – | 7.36 | 0.42 | 3.37 | 7.37 | 0 | W | 134 |
| H2Meso IX DME | Agg | 5.971 | – | 4.84 | 0.43 | 3.48 | 4.86 | 0 | W | 205 |
| MePheo a | Agg | 8.035 | – | 7.29 | 0.39 | 3.36 | 7.30 | 0 | W | 131 |
| NO2Proto IX DME | Dimer | 3.711 | – | 1.63 | 0.12 | 3.33 | 1.64 | 0 | S | 206 |
| H2OEP | Agg | 7.483 | – | 6.25 | 2.11 | 3.53 | 6.60 | 0 | W | 207 |
| H3OEP+ I3- (tricl) | Dimer | 3.488 | – | 1.41 | 0.13 | 3.19 | 1.42 | 0 | S | 233 |
| H3OEP+ I5- (ortho) | Dimer | 3.563 | – | 1.37 | 0.00 | 3.29 | 1.37 | 0 | S | 233 |
| H3OEP+ [Re2(CO)6Cl3]- | Dimer | 4.673 | – | 3.11 | 0.13 | 3.49 | 3.11 | 0 | I | 234 |
| 5-Bis(EtO)H2OEP | Dimer | 7.262 | – | 5.38 | 1.79 | 4.54 | 5.67 | 0 | W | 208 |
| Phyllochlorinester | Agg | 8.194 | – | 7.15 | 0.63 | 3.96 | 7.18 | 8.9 | W | 130 |
| Porphine | Dimer | 3.776 | – | 1.52 | 0.49 | 3.42 | 1.59 | 0 | S | 209 |
| H2Proto IX DME | Agg | 6.079 | – | 4.98 | 0.48 | 3.45 | 5.00 | 0 | W | 210 |
| H2TPrP | Agg | 5.078 | – | 3.52 | 1.40 | 3.38 | 3.79 | 0 | I | 211 |
| Co(OEP)(1-MeIm) | Dimer | 8.207 | 8.334 | 6.88 | 1.91 | 4.04 | 7.14 | 0 | W | 212 |
| Cu(5-F-10-BBV-OEP) | Dimer | 5.323 | 5.299 | 3.63 | 1.00 | 3.76 | 3.77 | 0 | I | 213 |
| Cu(DioxoOEP) | Dimer | 6.146 | 6.122 | 4.97 | 0.74 | 3.54 | 5.02 | 0 | W | 149 |
| Cu(TPrP) | Agg | 5.010 | 5.010 | 3.42 | 1.36 | 3.40 | 3.68 | 0 | I | 214 |
| Cu2(DP-7) | Agg | 4.67 | 4.69 | -3.02 | 1.15 | -3.36 | 3.24 | 0 | I | 215 |

| Compound | Type | | | | | | | | | Ref. |
|---|---|---|---|---|---|---|---|---|---|---|
| Cu₂(DP-7) | Requ | 5.23 | 5.21 | 3.18 | 2.16 | 3.55 | 3.84 | 0 | I | 215 |
| Cu₂(DP-B) | Requ | 3.861 | 3.807 | 1.33 | 1.05 | 3.47 | 1.69 | 4.4 | S | 193 |
| Cu₂(FTF) | Requ | 6.341 | 6.331 | -3.68 | 3.34 | -3.94 | 4.97 | 0 | W | 216 |
| Cu₂(FTF) | Agg | 8.756 | 8.760 | 7.54 | 0.28 | 4.44 | 7.55 | 6.3 | W | 216 |
| Fe(Proto IX)Cl | Dimer | 6.075 | 6.665 | 4.80 | 0.86 | 3.63 | 4.87 | 0 | W | 217 |
| Fe(Meso IX DME)(OMe) | Dimer | 4.843 | 5.549 | 2.33 | 2.30 | 3.57 | 3.27 | 0 | I | 218 |
| Fe(OEC) | Agg | 4.879 | 4.886 | 2.41 | 2.35 | 3.54 | 3.36 | 0 | I | 62 |
| Fe(OEP) | Agg | 4.812 | 4.812 | 3.35 | 0.53 | 3.42 | 3.39 | 0 | I | 62 |
| Fe(OEP)(2-MeHIm)⁺ | Dimer | 3.635 | 4.280 | 1.51 | 0.07 | 3.31 | 1.51 | 0 | S | 16 |
| Fe(OEP)(2-MeHIm)⁺ C2/c | Dimer | 4.039 | 4.602 | 2.04 | 0.66 | 3.43 | 2.14 | 0.9 | S | 16 |
| Fe(OEP)(3-ClPy)⁺ | Dimer | 4.822 | 5.137 | 3.04 | 1.25 | 3.53 | 3.28 | 4.6 | I | 17 |
| Fe(OEP)(SPh) | Dimer | 8.221 | 8.670 | 7.43 | 0.28 | 3.51 | 7.43 | 0 | W | 219 |
| Fe(OEP)(CS) | Dimer | 4.961 | 5.286 | 3.32 | 1.03 | 3.54 | 3.48 | 0 | I | 220 |
| Fe(OEP)(NCS) | Dimer | 4.813 | 5.475 | 2.61 | 1.86 | 3.59 | 3.20 | 0 | I | 307 |
| Fe(OEP)(NO)⁺ | Dimer | 3.652 | 4.238 | 1.59 | 0.04 | 3.29 | 1.59 | 0 | S | 30 |
| Fe(OEP)(OClO₃) | Dimer | 4.897 | 5.259 | 3.37 | 0.25 | 3.55 | 3.38 | 0 | I | 221 |
| Fe(ProtoIXDME)(SPhNO₂) | Dimer | 4.688 | 5.354 | 2.82 | 1.51 | 3.43 | 3.20 | 0 | I | 37 |
| Mg(EtChl a) | Agg | 8.859 | 8.859 | 8.05 | 0.94 | 3.58 | 8.11 | 0 | W | 139 |
| Mg(EtChl b) | Agg | 8.760 | 8.760 | 8.00 | 1.01 | 3.42 | 8.07 | 0 | W | 140 |
| Mg(MeChl a) | Agg | 8.470 | 8.470 | 7.70 | 0.51 | 3.50 | 7.71 | 0 | W | 138 |
| Mg(MePyrChl a) | Agg | 8.420 | 8.420 | 7.74 | 0.82 | 3.20 | 7.79 | 0 | W | 137 |
| Mn(OEP)(2-MeHIm)⁺ | Dimer | 3.756 | 4.144 | 1.25 | 0.79 | 3.45 | 1.48 | 0 | S | 194 |
| Nb(OEP)(O)(F) | Dimer | 8.204 | 9.208 | 6.92 | 2.09 | 3.88 | 7.23 | 0 | W | 222 |
| Ni(Deut IX DME) | Dimer | 3.814 | 3.770 | 1.66 | 0.23 | 3.43 | 1.67 | 0 | S | 223 |
| Ni(DPEP) | Dimer | 4.877 | 4.864 | 3.13 | 1.27 | 3.52 | 3.38 | 0 | I | 224 |
| Ni(OEP) (tricl) | Agg | 7.617 | 7.617 | 6.32 | 2.38 | 3.51 | 6.76 | 0 | W | 225 |
| Ni(OEP) (tricl II) | Agg | 4.814 | 4.814 | 3.36 | 0.47 | 3.41 | 3.39 | 0 | I | 198 |
| Ni(OEP) (tetrag) | Agg | 8.228 | 8.228 | 7.45 | 0.46 | 3.46 | 7.47 | 0 | W | 146 |
| Ni(OxoOEP) | Dimer | 6.473 | 6.467 | 5.44 | 1.26 | 3.27 | 5.59 | 0 | W | 148 |
| Ni(TMC) | Agg | 5.093 | 5.091 | 3.36 | 0.63 | 3.77 | 3.42 | 16.1 | I | 143 |

**Table XV** (Continued)

| Compound | Type | Ct–Ct[a] | M–M[a] | x[a] | y[a] | z[a] | L.S.[a] | Dihed. angle[b] | Group | Ref. |
|---|---|---|---|---|---|---|---|---|---|---|
| Ni(TMP) | Agg | 5.664 | 5.664 | 4.55 | 0.10 | 3.38 | 4.55 | 0 | W | 143 |
| Ni(TMiBC)A | Agg | 4.673 | 4.675 | 2.60 | 1.59 | 3.54 | 3.05 | 15.0 | I | 144 |
| Ni(TMiBC)B | Agg | 5.109 | 5.115 | 2.86 | 2.80 | 3.12 | 4.00 | 21.4 | I | 144 |
| Ni$_2$(DP-A) | Requ | 4.593 | 4.566 | 2.14 | 1.58 | 3.74 | 2.66 | 3.9 | I | 193 |
| Pb(TPrP) | Dimer | 5.468 | 7.291 | 3.51 | 1.78 | 3.80 | 3.93 | 0 | I | 226 |
| Rh(Etio I)(PhCO) | Dimer | 4.690 | 4.843 | 3.15 | 0.91 | 3.36 | 3.28 | 0 | I | 227 |
| Rh(OEP)(CH$_3$) | Dimer | 7.166 | 7.224 | 5.89 | 2.09 | 3.51 | 6.25 | 0 | W | 228 |
| [Ru(OEP)]$_2$ | Requ | 3.005 | 2.408 | 0.05 | 0.00 | 3.01 | 0.05 | 0.2 | | 150 |
| Ti(OEP)(O$_2$) | Dimer | 7.243 | 7.959 | 6.14 | 0.66 | 3.78 | 6.18 | 0 | W | 69 |
| Ti(OEP)(O) | Dimer | 7.091 | 7.738 | 5.98 | 0.67 | 3.75 | 6.02 | 0 | W | 69 |
| V(O)(DPEP) | Dimer | 5.089 | 5.820 | 3.34 | 1.27 | 3.62 | 3.58 | 0 | I | 229 |
| V(O)(Etio I) | Dimer | 4.975 | 5.727 | 3.39 | 0.76 | 3.56 | 3.47 | 0 | I | 190 |
| V(O)(Etio I) H$_2$Quin | Dimer | 4.909 | 5.726 | 3.11 | 1.09 | 3.64 | 3.29 | 0 | I | 190 |
| V(O)(OEP) | Dimer | 7.045 | 7.678 | 5.93 | 0.67 | 3.75 | 5.97 | 0 | W | 230 |
| Zn(5-OxoProto IX DME)Cl | Dimer | 5.484 | 6.316 | 4.07 | 0.38 | 3.66 | 4.09 | 0 | I | 231 |
| Zn(5-NO$_2$OEP) | Agg-A | 4.78 | 4.75 | 3.26 | 0.79 | 3.39 | 3.36 | 0 | I | 232 |
| Zn(5-NO$_2$OEP) | Agg-B | 4.95 | 4.97 | 3.40 | -0.64 | -3.54 | 3.46 | 0 | I | 232 |

[a] Value in Å;   [b] Value in degrees

With this definition, x and y are approximately in the plane of the macrocycle and z is approximately perpendicular to the mean plane of the ring. (Deviations from the idealized relationship can result from nonplanarity of the macrocycle and nonorthogonality of the vectors joining opposite pairs of nitrogen atoms.) This new orthogonal coordinate system is used to define the position of the center (Ct) of the adjacent molecule(s) in the π-π interaction. As before, Ct is the center of the four nitrogen atoms. The results of this description are given in Table XV. In this table, the z displacement roughly corresponds to M.P.S. of Table XIV and $x^2 + y^2$ to (L.S.)$^2$ of Table XIV. This coordinate system is always defined so that x > y. For solid-state structures involving linear chains, the same coordinate system was used for both interactions (Agg-A and Agg-B). Differences between the analogous quantities of Table XIV and XV principally result from nonideal geometry of the macrocycle and the noncoincidence of the center of the 24-atom core and the center of the four nitrogen atoms (Ct). In part, this is a result of the frequent small "saucering" of the cores in the π-π interaction.

This detailed information for Groups S and I is collected in Tables XVI and XVII, respectively. Several points need to be mentioned. The average lateral shifts between the $N_4$ ring center (Ct) is 1.63(26) Å in Group S[197]. The largest lateral shift of any species assigned to this group is 2.14 Å. The smallest lateral shifts (vide infra) in the compounds assigned to Group I is 3.1–3.2 Å. The greater than 1 Å gap would thus appear to confirm the appropriateness of the separation into two groups. In Group I, the free base derivatives appear to be present more frequently relative to their number in the full set of π-π interactions (Tables XIV, XV). Conversely, the metallo derivatives appear relatively less frequently. The relative orientations of the two π-π complexed rings are, with three exceptions discussed below, laterally slipped along one opposite pair of nitrogen atoms as was illustrated in Fig. 6. Thus in most cases of Group S x ≫ y. The distinct prevalence of this orientation of the interacting rings suggests that the effect may be so based because of electronic factors. The matter would appear worthy of theoretical investigation.

The three violations of this apparent pattern require comment. The $Cu_2$(DP–B) complex has the particular π-π interaction geometry observed because of the constraints of the diporphyrin ligand. The observed structure was somewhat unexpected. The rigid biphenylene pillar was designed to hold the two halves of the biporphyrin at a fixed spacing and with eclipsed geometry. However, the observed conformation has the two halves of the diporphyrin slipped with respect to each other as shown in Fig. 11. The rigid

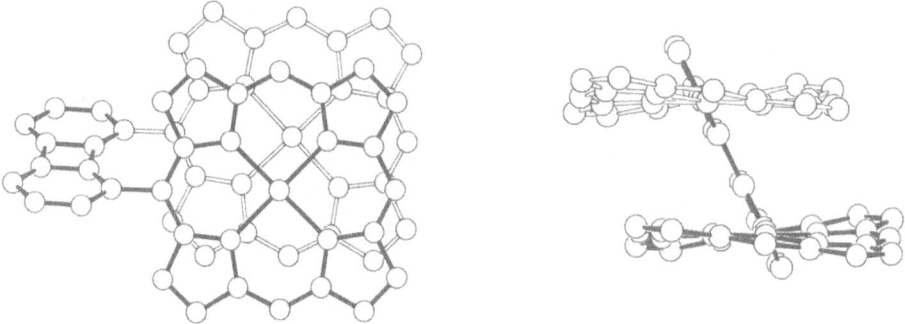

**Fig. 11.** Edge-on and overlap views of the $Cu_2$(DP–B) molecule illustrating how the rigid biphenylene pillar controls aspects of the π-π overlap. Drawn from the coordinates reported in Ref. 193

**Table XVI.** The group S parameters

| Compound | Type | Ct–Ct[a] | M–M[a] | x[a] | y[a] | z[a] | L.S.[a] | Dihed. angle[b] | Group | Ref. |
|---|---|---|---|---|---|---|---|---|---|---|
| 21-EtO–HOEP | Dimer | 3.738 | – | 1.89 | 0.23 | 3.22 | 1.91 | 0 | S | 105 |
| 25-EtO–H₂OEP | Dimer | 3.592 | – | 1.15 | 0.35 | 3.39 | 1.20 | 0 | S | 115 |
| 4-Bu-5-Et MeBPheo d | Agg-A | 3.859 | – | 1.12 | 0.78 | -3.61 | 1.37 | 1.4 | S | 133 |
| 4,5DiEt MeBPheo d | Agg-A | 3.852 | – | 1.84 | -0.20 | 3.38 | 1.85 | 1.7 | S | 133 |
| H₃OEP⁺ I₃⁻ (tricl) | Dimer | 3.488 | – | 1.41 | 0.13 | 3.19 | 1.42 | 0 | S | 233 |
| H₃OEP⁺ I₃⁻ (ortho) | Dimer | 3.563 | – | 1.37 | 0.00 | 3.29 | 1.37 | 0 | S | 233 |
| NO₂Proto IX DME | Dimer | 3.711 | – | 1.63 | 0.12 | 3.33 | 1.64 | 0 | S | 206 |
| Porphine | Dimer | 3.776 | | 1.52 | 0.49 | 3.42 | 1.59 | 0 | S | 209 |
| Cu₂(DP–B) | Requ | 3.861 | 3.807 | 1.33 | 1.05 | 3.47 | 1.69 | 4.4 | S | 193 |
| Fe(OEP)(2-MeHIm)⁺ | Dimer | 3.635 | 4.280 | 1.51 | 0.07 | 3.31 | 1.51 | 0 | S | 16 |
| Fe(OEP)(2-MeHIm)⁺ C2/c | Dimer | 4.039 | 4.602 | 2.04 | 0.66 | 3.43 | 2.14 | 0.9 | S | 16 |
| Fe(OEP)(NO)⁺ | Dimer | 3.652 | 4.238 | 1.59 | 0.04 | 3.29 | 1.59 | 0 | S | 30 |
| Mn(OEP)(2-MeHIm)⁺ | Dimer | 3.756 | 4.144 | 1.25 | 0.79 | 3.45 | 1.48 | 0 | S | 194 |
| Ni(Deut IX DME) | Dimer | 3.814 | 3.770 | 1.66 | 0.23 | 3.43 | 1.67 | 0 | S | 223 |

[a] Value in Å;  [b] Value in degrees

**Table XVII.** The group I parameters

| Compound | Type | Ct-Ct[a] | M-M[a] | x[a] | y[a] | z[a] | L.S.[a] | Dihed. angle[b] | Group | Ref. |
|---|---|---|---|---|---|---|---|---|---|---|
| 4,5DiEt MeBPheo d | Agg-B | 5.358 | — | 0.34 | 3.82 | -3.74 | 3.84 | 1.7 | I | 133 |
| Siro-iBC | Agg-A | 5.229 | — | -3.80 | -0.28 | -3.58 | 3.81 | 0 | I | 132 |
| H2TPrP | Agg | 5.078 | — | 3.52 | 1.40 | 3.38 | 3.79 | 0 | I | 211 |
| H3OEP+ [Re2(CO)6Cl3]^- | Dimer | 4.673 | — | 3.11 | 0.13 | 3.49 | 3.11 | 0 | I | 234 |
| Cu(5-F-10-BBV-OEP) | Dimer | 5.323 | 5.299 | 3.63 | 1.00 | 3.76 | 3.77 | 0 | I | 213 |
| Cu(TPrP) | Agg | 5.010 | 5.010 | 3.42 | 1.36 | 3.40 | 3.68 | 0 | I | 214 |
| Cu2(DP-7) | Agg | 4.67 | 4.69 | -3.02 | 1.15 | -3.36 | 3.24 | 0 | I | 215 |
| Cu2(DP-7) | Requ | 5.23 | 5.21 | 3.18 | 2.16 | 3.55 | 3.84 | 0 | I | 215 |
| Fe(Meso IX DME)(OMe) | Dimer | 4.843 | 5.549 | 2.33 | 2.30 | 3.57 | 3.27 | 0 | I | 218 |
| Fe(OEC) | Agg | 4.879 | 4.886 | 2.41 | 2.35 | 3.54 | 3.36 | 0 | I | 62 |
| Fe(OEP) | Agg | 4.812 | 4.812 | 3.35 | 0.53 | 3.42 | 3.39 | 0 | I | 62 |
| Fe(OEP)(3-ClPy)^+ | Dimer | 4.822 | 5.137 | 3.04 | 1.25 | 3.53 | 3.28 | 4.6 | I | 17 |
| Fe(OEP)(CS) | Dimer | 4.961 | 5.286 | 3.32 | 1.03 | 3.54 | 3.48 | 0 | I | 220 |
| Fe(OEP)(NCS) | Dimer | 4.813 | 5.475 | 2.61 | 1.86 | 3.59 | 3.20 | 0 | I | 307 |
| Fe(OEP)(OClO3) | Dimer | 4.897 | 5.259 | 3.37 | 0.25 | 3.55 | 3.38 | 0 | I | 221 |
| Fe(ProtoIXDME)(SPhNO2) Dimer | 4.688 | 5.354 | 2.82 | 1.51 | 3.43 | 3.20 | 0 | I | 37 |
| Ni(DPEP) | Dimer | 4.877 | 4.864 | 3.13 | 1.27 | 3.52 | 3.38 | 0 | I | 224 |
| Ni(OEP) (tricl II) | Agg | 4.814 | 4.814 | 3.36 | 0.47 | 3.41 | 3.39 | 0 | I | 225 |
| Ni(TMC) | Agg | 5.093 | 5.091 | 3.36 | 0.63 | 3.77 | 3.42 | 16.1 | I | 143 |
| Ni(TMiBC)A | Agg | 4.673 | 4.675 | 2.60 | 1.59 | 3.54 | 3.05 | 15.0 | I | 144 |
| Ni(TMiBC)B | Agg | 5.109 | 5.115 | 2.86 | 2.80 | 3.12 | 4.00 | 21.4 | I | 144 |
| Ni2(DP-A) | Requ | 4.593 | 4.566 | 2.14 | 1.58 | 3.74 | 2.66 | 3.9 | I | 193 |
| Pb(TPrP) | Dimer | 5.468 | 7.291 | 3.51 | 1.78 | 3.80 | 3.93 | 0 | I | 226 |
| Rh(Etio I)(PhCO) | Dimer | 4.690 | 4.843 | 3.15 | 0.91 | 3.36 | 3.28 | 0 | I | 227 |
| V(O)(DPEP) | Dimer | 5.089 | 5.820 | 3.34 | 1.27 | 3.62 | 3.58 | 0 | I | 229 |
| V(O)(Etio I) | Dimer | 4.975 | 5.727 | 3.39 | 0.76 | 3.56 | 3.47 | 0 | I | 190 |
| V(O)(Etio I) H2Quin | Dimer | 4.909 | 5.726 | 3.11 | 1.09 | 3.64 | 3.29 | 0 | I | 190 |
| Zn(5-OxoProto IX DME)Cl Dimer | 5.484 | 6.316 | 4.07 | 0.38 | 3.66 | 4.09 | 0 | I | 231 |
| Zn(5-NO2OEP) | Agg-A | 4.78 | 4.75 | 3.26 | 0.79 | 3.39 | 3.36 | 0 | I | 232 |
| Zn(5-NO2OEP) | Agg-B | 4.95 | 4.97 | 3.40 | -0.64 | -3.54 | 3.46 | 0 | I | 232 |

[a] Value in Å;  [b] Value in degrees

meso biphenylene connection requires a lateral slip if the separation between the two cores is to be less than the 3.80 Å required with the biphenylene plane perpendicular to both cores. This lateral slip must occur with x ≈ y. The nature of the linkage in the other diporphyrin species listed in Tables XIV and XV requires larger lateral shifts to achieve mean plane separations equal to or greater than that observed in $Cu_2$(DP–B). However, the necessity of x ≈ y remains in all of these diporphyrin complexes. The unexpected, strongly apparent, π-π interaction provides strong evidence for the reality of the interactions. The structures of these species show the difficulties of controlling the interplanar spacing.

No such geometric constraints apply to the two remaining violations: 4-Bu-5-Et MeB-Pheo d and [Mn(OEP)(2-MeHIm)]$^+$. However, this diagonal lateral shift does lead to a geometry that appears to be more eclipsed than is typical. The nature of the overlap for the bacterial pheophorbide derivative is illustrated in Fig. 12.

Similar lateral shift patterns apply to the members of Group I as detailed in Table XVII. As can be seen, there are few members of the group for which x is not greater than 3 y. Again a number of these violations result from the steric constraints of diporphyrin ligands. A nickel(III) derivative of anthracene pillared diporphyrin[193] has a lateral slipping between centers of the two rings of 2.66 Å. All other derivatives that have been classified as belonging to Group I have lateral separations > 3.1 Å. The average lateral slip of the compounds listed in Table XVII is 3.50(32) Å. The perpendicular separation between ring centers is 3.54 Å and is slightly larger than the 3.39 Å value found for the Group S class. This is consistent with a slightly weaker interaction in Group I. Typical views of the overlap patterns are shown in Figs. 8 and 9.

A typical interaction between the porphinato cores of a member of Group W is given in Figs. 13 and 14 which illustrates the relatively small overlap areas. Nonetheless, the orientation of the peripheral groups is such to allow the close approach of the partial structure.

An interesting question, as yet incompletely answered, is whether π-π interactions will have any effect of the geometrical parameters of a tetrapyrrole derivative. A recently obtained crystalline form of Ni(OEP) (triclinic II)[198], currently under investigation by X-ray diffraction and resonance Raman studies, suggests that π complexation phenomena could effect the Ni–N bond lengths. Values for this complex reported in Tables XIV, XV,

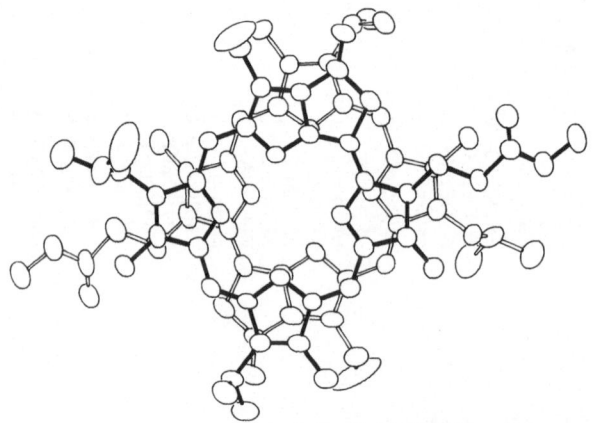

4-Bu-5-Et MeBPheo d Agg-A

**Fig. 12.** An overlap diagram for 4-Bu-5-Et MeBPheo d drawn from the coordinates reported in Ref. 133

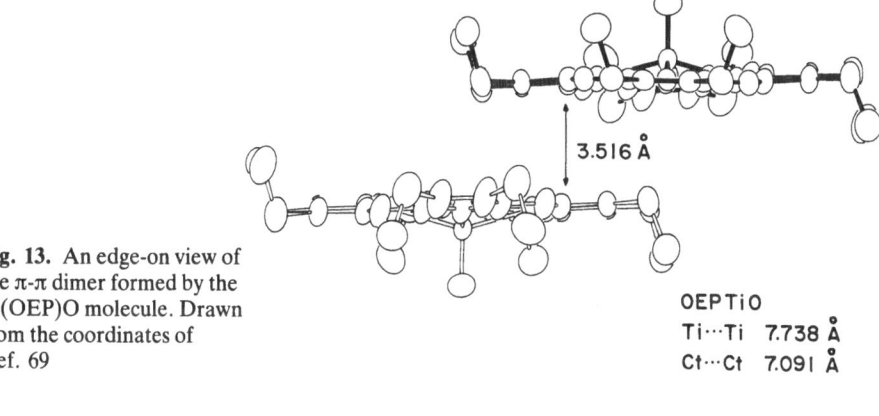

**Fig. 13.** An edge-on view of the π-π dimer formed by the Ti(OEP)O molecule. Drawn from the coordinates of Ref. 69

3.516 Å

OEPTiO
Ti⋯Ti 7.738 Å
Ct⋯Ct 7.091 Å

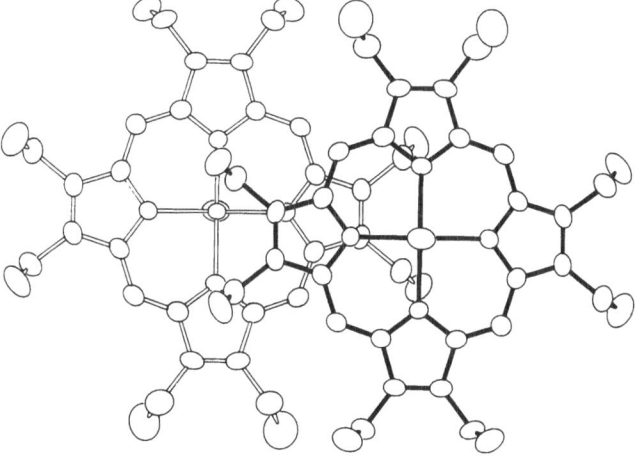

**Fig. 14.** An overlap view of the π-π dimer formed by the Ti(OEP)O molecule. Drawn from the coordinates of Ref. 69

OEPTiO

and XVII are based on a not quite final set of x-ray coordinates. A combined X-ray diffraction and theoretical study[199] of Ni(TMP) (Group W, Tables XIV and XV) has been carried out. In this slipped stack aggregate species, the nickel(II) atom lies almost directly above and below pairs of $C_b$–$C_b$ bonds from adjacent Ni(TMP) molecules. The distance between the midpoint of the $C_b$–$C_b$ bond and nickel atom is 3.30 Å. The calculations in this combined experiment were undertaken to see if this slipped stacked complex could explain the 0.02 Å variation in the Ni–N bond distances. The conclusion of the investigators is that the induced charge density shifts are too small to account for the variation in Ni–N bonds.

There are in addition a number of π-complexes formed between porphyrins and smaller aromatic molecules. There is a series of bis(toluene) solvates of Zn(TPP)[200], Mn(TPP)[172] and Cr(TPP)[201]. Two other species that show π-complex geometry are Fe(TPP)(FSbF$_5$) · $C_6H_5F$[21] and Fe(TPP)(OClO$_3$) · 1/2 $C_8H_{10}$[19]. This last complex forms a "triple decker" sandwich with the m-xylene molecule between the two five-coordinate porphyrin molecules. In all five cases, the plane of the aromatic solvate is close to parallel to the porphyrin plane. The metal atom is always an apparent point of the π interaction.

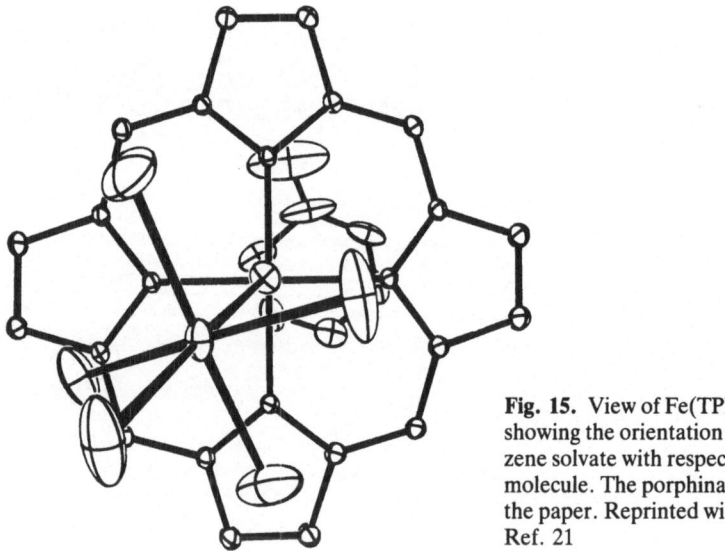

**Fig. 15.** View of Fe(TPP)(FSbF$_5$) · C$_6$H$_5$F showing the orientation of the fluoroben-zene solvate with respect to the porphyrin molecule. The porphinato core is in plane of the paper. Reprinted with permission from Ref. 21

An "electron rich" carbon atom of the aromatic solvate is always close to the metal atom with M–C distances ranging from 3.05 Å in Mn(TPP) to ~3.30 Å. The relative orienta-tion of the fluorobenzene molecule to [Fe(TPP)(FSbF$_5$)] is shown in Fig. 15; the geomet-ries are similar in all of the other complexes. In the aromatic complexes, the metallopor-phyrin acts as an acceptor and the solvate molecule acts as a π-donor.

There are also a limited number of charge transfer complexes that have been structur-ally characterized. The structure[202] of Ni(TMP)(TCNQ) consists of stacks of alternating parallel molecules of Ni(TMP) and TCNQ with interplanar separations of ~3.30 Å. Spectroscopic parameters for the complex suggest that the amount of charge transfer between Ni(TMP) and TCNQ is small. (Bond distances changes in TCNQ suggest that the charge transfer is less than 0.2 e.) There has also been a brief report[203] of a 2 : 1 complex: Ni(Etio I) · 2tetranitrofluorenone. There are two fluorenone rings that sand-wich the Ni(Etio I) molecule with interplanar angles of 11.4°. The nickel atom is 3.33 Å from the fluorenone plane.

Finally it can be noted that there are porphyrin-porphyrin interactions between tetra-aryl porphyrin derivatives that probably result from some kind of π-π interaction. The discussion of these species will be deferred to Sect. D.II. Other references in the tables are[204–234].

# D. Conformational Aspects

## I. Axial Ligand Orientations

A number of recent studies have demonstrated the importance of axial ligand orienta-tions in "fine-tuning" the physical properties of iron porphyrinate derivatives. Properties affected include structure, spin state, EPR and redox potentials. A convenient measure of the orientation is the dihedral angle formed by the coordinate plane containing oppo-

**Table XVIII.** Relative orientation of imidazole ligands in tetrapyrrole derivatives

| Complex | M–N$_p^a$ | M–N(Im)$^a$ | $\phi^b$ | Relative orientation$^{b,c}$ | Spin$^d$ state, d$^n$ | Ref. |
|---|---|---|---|---|---|---|
| [Mg(TPP)(1-MeIm)$_2$] | 2.078(6) | 2.297(8) | 13.4 | 0 | L.S., d$^0$ | 173 |
| [Mn(TPP)(1-MeIm)$_2$]$^+$ | 2.014(3) | 2.308(3) | 18.9 | 0 | H.S., d$^4$ | 240 |
| [Mn(TPP)(Im$^-$)]$_n^e$ | 2.019(4) | 2.186(5) | 26 | 0 | L.S., d$^4$ | 176 |
| | | 2.280(4) | 28 | 0 | H.S., d$^4$ | |
| [Mn(TPP)(1-MeIm)] | 2.128(7) | 2.192(2) | 15.4 | | H.S., d$^5$ | 241 |
| [Fe(TPP)(HIm)$_2$]Cl | 1.989(8) | 1.991(5) | 18 | 57 | L.S., d$^5$ | 235 |
| | | 1.957(4) | 39 | | | |
| [Fe(TPP)(HIm)$_2$]Cl · H$_2$O$^g$ | 1.993(7) | 1.977(3) | 6 | 0 | L.S., d$^5$ | 237 |
| | | 1.964(3) | 41 | 0 | L.S., d$^5$ | |
| [Fe(TPP)(c-MU)]$^+$ | 1.996(10) | 1.979(7) | 16 | 0 | L.S., d$^5$ | 308 |
| | | 1.967(7) | 29 | 0 | | |
| [Fe(TPP)(t-MU)$_2$]$^+$ | 1.992(5) | 1.983(4) | 22 | 0 | L.S., d$^5$ | 308 |
| [Fe(Proto IX)(1-MeIm)$_2$] | 1.990(16) | 1.988(5) | 3 | 13 | L.S., d$^5$ | 242 |
| | | 1.966(5) | 16 | | | |
| [Fe(TPP)(2-MeHIm)$_2$]$^+$ | 1.971(5) | 2.013(4) | 32 | 89.3 | L.S., d$^5$ | 238 |
| | | | 32 | | | |
| [Fe(OEP)(2-MeHIm)$_2$]$^+$ | 2.041(11) | 2.275(1) | 22 | 0 | H.S., d$^5$ | 41 |
| [Fe(TPP)(BzHIm)$_2$]$^+$ | 2.049(14) | 2.216(5) | 10 | 0 | H.S., d$^5$ | 42 |
| [Fe(TPP)(4-MeIm$^-$)$_2$]$^-$ | 1.998(25) | 1.958(12) | 1 | 18 | L.S., d$^5$ | 13 |
| | | 1.928(12) | 17 | | | |
| [Fe(OEP)(2-MeHIm)]$^{+f}$ | 2.038(6) | 2.068(4) | 3.9, 3, 7 | | H.S., d$^5$ | 16 |
| [Fe(TPP)(1-MeIm)$_2$] | 1.997(4) | 2.014(5) | 10 | 0 | L.S., d$^6$ | 240 |
| [Fe(TPP)(NO)(1-MeIm) | 2.008(12) | 2.180(4) | 25.2 | | L.S., d$^6$ | 243 |
| [Fe(TPP–C$_5$Im)(THT)] | 1.993(8) | 2.002(5) | 3.5 | | L.S., d$^6$ | 261 |
| [Fe(TPP)(2-MeHIm)] | 2.086(4) | 2.161(5) | 7.4 | | H.S., d$^6$ | 2 |
| [Fe(TpivPP)(2-MeHIm)] | 2.072(5) | 2.095(6) | 22.8 | | H.S., d$^6$ | 67 |
| [Fe(TpivPP)(2-MeHIm)(O$_2$)] | 1.996(1) | 2.107(4) | 22.2 | | L.S., d$^6$ | 67 |
| [Fe(TpivPP)(1-MeIm)(O$_2$)] | 1.98(1) | 2.07(2) | 20 | | L.S., d$^6$ | 66 |
| [Co(TPP)(HIm)$_2$](OAc)$^g$ | 1.982(11) | 1.93(2) | 43 | 0 | L.S., d$^6$ | 244 |
| | | | 42 | 0 | L.S., d$^6$ | |
| [Co(TPP)(1-MeIm)] | 1.977(6) | 2.157(3) | 3.8 | | L.S., d$^7$ | 245 |
| [Co(OEP)(1-MeIm)] | 1.96(1) | 2.15(1) | 10 | | L.S., d$^7$ | 246 |
| [Co(TPP)(1,2-Me$_2$Im)] | 1.985(3) | 2.216(2) | 20 | | L.S., d$^7$ | 247 |
| [Ni(TMPyP)(HIm)$_2$]$^{4+}$ | 2.038(20) | 2.160(4) | 23 | 0 | H.S., d$^8$ | 248 |
| [Zn(OEP)(1-MeIm)] | 2.068(7) | 2.106(4) | 10.5 | | L.S., d$^{10}$ | 309 |

$^a$ Value in Å; $^b$ Value in degrees; $^c$ Relative orientation of two axial planes (when present). A value of "0" indicates planes required to be parallel by crystallographic symmetry; $^d$ L.S. indicates minimum number of unpaired electrons. I.S. indicates an intermediate number of unpaired electrons. H.S. indicates maximum number of unpaired electrons; $^e$ Two crystallographically independent molecules in polymer; $^f$ Three independent molecules in two crystalline forms; $^g$ Two independent molecules

**Table XIX.** Relative orientation of pyridine ligands in tetrapyrrole derivatives

| Complex | $M–N_p{}^a$ | $M–N(Py)^a$ | $\phi^b$ | Relative orienta- tion$^{b,c}$ | Spin$^d$ state, $d^n$ | Ref. |
|---|---|---|---|---|---|---|
| [Mg(OEP)(Py)$_2$] | 2.070(8) | 2.389(2) | 13.2 | 0 | L.S., $d^0$ | 174 |
| [Mg(TPP)(4-MePy)$_2$] | 2.071(2) | 2.386(2) | 41.5 | 0 | L.S., $d^0$ | 173 |
| [Mn(TPP)(Py)Cl] | 2.010(9) | 2.444(4) | 37.1 | | H.S., $d^4$ | 249 |
| [Cr(TPP)(Py)$_2$]$^g$ | 2.019(10) | 2.141(8) | 21 | 0 | L.S., $d^4$ | 250 |
| | 2.035(13) | 2.121(8) | 32 | 0 | | |
| [Mo(TTP)(Py)$_2$] | 2.071(6) | 2.215(3) | 34 | 0 | L.S., $d^4$ | 83 |
| [Fe(TPP)(N$_3$)(Py)] | 1.989(6) | 2.089(6) | 40 | | L.S., $d^5$ | 29 |
| [Fe(TPP)(CN)(Py)] | 1.970(14) | 2.075(3) | 39.8 | | L.S., $d^5$ | 28 |
| [Fe(TPP)(NCS)(Py)] | 1.988(9) | 2.082(3) | 39 | | L.S., $d^5$ | 27 |
| [Fe(OEP)(NCS)(Py)] | 2.048(4) | 2.442(2) | 4.2 | | H.S., $d^5$ | 27 |
| [FeTPP(NCS)(Py)]$^e$ | | | | | | |
| 293 K HS ↔ L.S., | 2.03 | 2.22 | 44 | | L.S. ↔ H.S., $d^5$ | 43 |
| 96 K HS ↔ L.S., | 1.99 | 2.03 | 44 | | L.S., $d^5$ | |
| 96 K        H.S. | 2.06 | 2.33 | 30 | | H.S., $d^5$ | |
| tri-[Fe(OEP)(3-ClPy)$_2$]$^+$ | | | | | | 39 |
| 100 K | 1.995(6) | 2.031(2) | 41 | 0 | L.S., $d^5$ | |
| RT$^f$ | 2.014(4) | 2.194 | 41 | 0 | L.S. ↔ H.S., $d^5$ | |
| mono[Fe(OEP)(3-ClPy)$_2$]$^{+\,g}$ | | | | | | 24 |
| molecule 1 | 2.003(1) | 2.314(8) | 5.7, 7.0 | 1 | I.S., $d^5$ | |
| molecule 2 | 2.008(8) | 2.303(27) | 11.5, 13.9 | 2.5 | I.S., $d^5$ | |
| [Fe(OEP)(3-ClPy)$_2$]$^+$ · CHCl$_3$ | 2.006(8) | 2.304(27) | 8.7, 6.1 | 2.6 | I.S., $d^5$ | 40 |
| [Fe(OEP)(3-ClPy)]$^+$ | 1.979(6) | 2.126(5) | 41.1 | | I.S., $d^5$ | 17 |
| [Fe(TPP)(CO)(Py)] | 2.02(3) | 2.10(1) | 45 | | L.S., $d^6$ | 251 |
| [Fe(TPP)(Py)$_2$] | 1.993(6) | 2.039(1) | 34.4 | 0 | L.S., $d^6$ | 56 |
| [Ru(TPP)(Py)(CO)] | 2.052(9) | 2.193(4) | 26.1 | | L.S., $d^6$ | 263 |
| [Ru(OEP)(Py)$_2$]$^g$ | 2.047(2) | 2.100(6) | 44.1 | 0 | L.S., $d^6$ | 252 |
| | 2.036(6) | 2.089(6) | 30.0 | 0 | L.S., $d^6$ | |
| [Os(OEPMe$_2$)(Py)(CO)] | 2.067(3) | 2.223(40) | 45 | | L.S., $d^6$ | 253 |
| [Co(TPP)(3,5-Lut)(NO$_2$)] | 1.954(6) | 2.036(4) | 36.4 | | L.S., $d^6$ | 254 |
| [Co(TPP)(Py)Cl] | 1.976(12) | 1.978(8) | 42 | | L.S., $d^6$ | 255 |
| [Co(TPP)(Py)(OCH$_3$)] | 1.96(1) | 1.99 | 32 | | L.S., $d^6$ | 256 |
| [Co(TPP)(3,5-Lut)] | 2.000(4) | 2.161(5) | 41 | | L.S., $d^7$ | 257 |
| [Co(OEP)(3-MePy)$_2$] | 1.992(1) | 2.386(2) | 10 | 0 | L.S., $d^7$ | 258 |
| [Zn(TPyP)(Py)] | 2.073(9) | 2.143(4) | 23 | | L.S., $d^{10}$ | 259 |
| [Zn(OEP)(Py)] | 2.067(6) | 2.200(3) | 4 | | L.S., $d^{10}$ | 260 |
| [Zn(TPC)(Py)] | 2.081(33)$^n$ | 2.171(2) | 18.6 | | L.S., $d^{10}$ | 141 |
| [Zn(TPiBC)(Py)] | 2.086(23) | 2.155 | 20.5 | | L.S., $d^{10}$ | 142 |
| [Zn(TPP–C$_3$Py)] | 2.059(10) | 2.147(7) | 5.8 | | L.S., $d^{10}$ | 262 |

site nitrogen atoms and the axial ligand plane as shown in Fig. 16. A value of 0° for $\phi$ corresponds to an eclipsed conformation and a value of 45° to a staggered conformation. We have summarized the $\phi$ values of all imidazole (Table XVIII) and pyridine (Table XIX) complexes recorded in the literature. The $\phi$ value is not always reported and we have calculated the value when required if the atomic coordinates were available. We would encourage the routine reporting of $\phi$ in all structural reports. We have also recorded, for each species, the average $M-N_p$ value, the axial $M-N$ bond distance(s), the relative orientation of the two axial ligand planes (when two are present) and the spin state and $d^n$ configuration of the metal ion. The relative orientation is the unsigned dihedral angle between the two ligand planes. In all such calculation of dihedral angles, we have assumed idealized geometry. This assumption should have relatively small effects on the values reported in the two tables.

There are two important generalizations that should be recognized. An examination of the entries in Tables XVIII and XIX reveals that most bis-ligated species have crystallographically required parallel planes; almost all of the remaining derivatives have relative orientations that deviate little from that idealized geometry. Secondly, the average $\phi$ value, especially for the imidazole derivatives, is smaller than would be expected on simple steric grounds. The average $\phi$ value for all of the five-coordinate imidazole complexes is 10.4 (7.4)°, where the value given in parentheses is the estimated standard deviation. The esd is quoted to give a measure of the dispersion in the values. The value for the six-coordinate imidazole species is 21.0 (12.1)°. The values for the five- and six-coordinate pyridine complexes are 22.0 (14.9) and 28.5 (14.3)°, respectively.

The use of $\phi$ as a measure of ligand orientation was originally suggested by Hoard[235] in his classic structural analysis of $[Fe(TPP)(HIm)_2]Cl$. It was pointed out that the smaller value of $\phi$ was associated with, and responsible for, the longer axial $Fe-N$ bond. Of

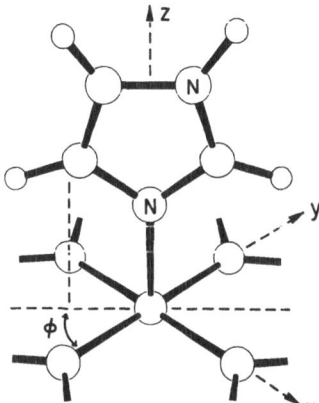

**Fig. 16.** Diagram illustrating the orientation of an axial ligand with respect to the porphinato nitrogen atoms and the definition of the angle $\phi$

---

[a] Value in Å; [b] Value in degrees; [c] Relative orientation of two axial planes (when present). A value of "0" indicates planes required to be parallel by crystallographic symmetry; [d] L.S. indicates minimum number of unpaired electrons. I.S. indicates an intermediate number of unpaired electrons. H.S. indicates maximum number of unpaired electrons; [e] Two independent molecules in crystal with different spin states; [f] Spin-equilibrium, average structure; [g] Two independent molecules

course, the shorter axial Fe–N bond was associated with the larger value of $\phi$. An explanation was based on the effect of nonbonded interactions that exist between the $\alpha$-hydrogen atoms of the ligand and the porphinato core that are at a maximum at $\phi = 0$. As can be seen from an inspection of the bis(imidazole-ligated) complexes listed in Table XVIII, this pattern is always observed. It should also be noted that the value of $\phi$ is *not* a quantitative predictor of the axial bond length, even in such a closely related series as that of the low-spin iron(III) complexes.

In a recent report, we demonstrated[236] that the appearance of an unusual "strong $g_{max}$" EPR signal in low-spin ferric porphyrinates is related to the attainment of axial electronic symmetry. The strong $g_{max}$ signal is a signal in which $g_{max}$ is greater than 3. For bis-ligated imidazole complexes, this is achieved by a perpendicular, rather than the usual parallel relative orientation of the two ligands. A second crystalline form of $[Fe(TPP)(HIm)_2]Cl$[237], with two independent units, provides evidence that EPR spectra with "normal" g values will show small, but significant, variations in the g tensor as a function of different geometries, i.e., different $\phi$ values. We have also estimated[236] that the effect of a perpendicular rather than a parallel orientation of the axial ligands could lead to a 50 mv positive shift in the redox potential, all other structural and environmental factors being equal.

Another physical property that can be controlled (or is at least modulated) by the axial ligand orientation is the spin state of the central metal ion. It would appear that the compounds must be near the spin-crossover point and thus far only iron(III) compounds are known to be affected. As mentioned earlier, we have published the structural and magnetic characterization of $[Fe(OEP)(3-ClPy)_2]ClO_4$ in different crystalline forms. One form[39] *(tri-)* is a high-spin low-spin thermal equilibrium. A second crystalline form[24] *(mono-)* is an intermediate-spin complex. We suggested[24] that the spin-state is controlled by the axial ligand orientation. As is readily apparent from Table XIX, the orientation of the axial 3-chloropyridine ligands are quite different in the two complexes. The small value of $\phi$ in *mono-* $[Fe(OEP)(3-ClPy)_2]ClO_4$ is incompatible with the short axial Fe–N distance ($\sim 2.00$ Å) required for a low-spin state because of severe nonbonded interactions between pyridine hydrogen atoms and core atoms. The observed value (41°) in *tri-* does allow for the requisite axial distance of a low-spin complex. This type of analysis demands that the thermodynamically stable electronic state be the low-spin state. Confirmation of the analysis has been given by the isolation[40] and characterization of still a third form of $[Fe(OEP)(3-ClPy)_2]ClO_4$. This complex also has small values of $\phi$ and an intermediate-spin state.

Additional confirmation and emphasis of the effect is given by other examples that are summarized in Tables XVIII and XIX. These are: two[27, 43] crystalline forms of $[Fe(TPP)(Py)(NCS)]$, the differences in pyridine orientation in $[Fe(TPP)(Py)(NCS)]$ and $[Fe(OEP)(Py)(NCS)]$[43] and the high- and low-spin forms of bis(2-MeHIm)iron(III)[41, 238] complexes. This last pair demonstrates the subtleties of nonbonded interactions and ligand orientation: a coupling of a ligand orientation difference of only 10 degrees and core conformation changes yield essentially equivalent nonbonded interactions in these species even though they have radically different axial bond lengths and spin states. Finally, it should be noted that there are no pyridine-ligated species (Table XIX) or bulky imidazole ligated species (Table XVIII) that have, simultaneously, small $\phi$ values and short axial M–N bonds.

A final topic concerns the axial ligand orientations and an attempt to understand the reasons for the small $\phi$ values found for imidazole complexes. The data in Table XVIII is taken to clearly indicate a preferred orientation of imidazole ligand(s) that are sterically unfavorable. A series of charge iterative Huckel theory calculations[239] for a broadly representative series of complexes reveals a $\pi$ bonding effect that favors eclipsed orientations of the ligand(s). The $\pi$ bond is dominated by the metal $p\pi$-imidazole $p\pi$ interaction. The results provide an explanation of why the orientation effect of imidazole complexes seems insensitive to metal $d^n$ configuration, spin state, oxidation state, and the presence or absence of a sixth axial ligand.

Additional references in the tables are[240–263].


## II. Core Conformations and Other Conformational Aspects

Our detailed examination of core conformations, especially those of the *meso*-tetraaryl-substituted species was a consequence of attempting to understand why an unusual conformation is found for the cores in several porphyrin $\pi$-cation radicals. The porphinato cores of these complexes are found to have an unusual $D_{2d}$-ruffled geometry. The $D_{2d}$-ruffled core in these saddle-shaped species have the pyrrole rings displaced, alternately, above and below the mean plane of the core. The meso carbon atoms are found to be in, or nearly in, the plane or the core. This conformation differs from the more usual $D_{2d}$-ruffled cores by a 45° rotation of the point group symmetry operators around the major twofold axis (perpendicular to the mean plane). The displacements of the atoms of the 24-atom cores in the two nearly idealized $D_{2d}$-ruffled forms are illustrated in Fig. 17 with actual values from two complexes. We will denote these two forms by the labels "Ruf" and "Sad". Other idealized deformations from planarity include $C_{4v}$ doming, a slight "stepping" of the core appropriate to inversion symmetry and a "roof"[264] folding along a line joining opposite meso carbon atoms.

We first offer a few qualitative observations concerning core conformations in tetra-pyrrole derivatives. As is generally known, the 24-atom core is readily deformed in a direction perpendicular to the mean plane, but less so in the radial (in plane) direction. Further, the pyrrole ring subunits are always planar. The first observation is a very

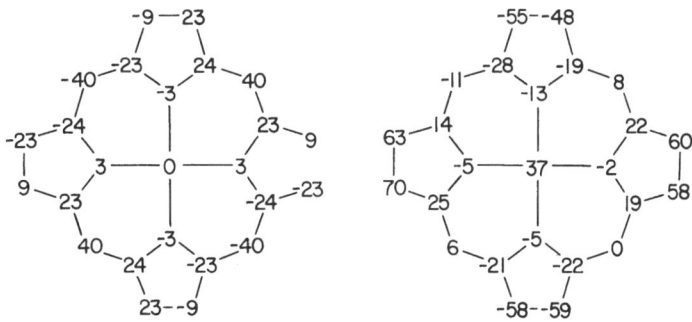

**Fig. 17.** Formal diagrams of the porphinato cores of Fe(TPP) *(left)* and [Fe(TPP)(Cl)]$^+$ *(right)*. The perpendicular displacement of each atom, in units of 0.01 Å, from the mean plane of 24-atom core is given

qualitative one: *meso*-tetraaryl substituted species (i.e., $H_2TPP$ derivatives) tend to deviate more from near planarity than β-substituted species (i.e., $H_2OEP$ derivatives). It is to be emphasized that both types of species are *capable* of substantial and equivalent distortions from planarity. Second, steric crowding at the periphery always appears to lead to substantial nonplanarity of the core. Thus a series of 5- or 5,15-substituted octaalkylporphyrinates (for example, a substituted octaethylporphyrin) are observed[4b, 204, 208, 213] to result in $S_4$ ruffling. This ruffling minimizes the steric crowding at the periphery. Third, deformations can be induced by "short-strap" bridging[265], a covalent linking of opposite pyrrole rings.

In trying to understand the unusual core conformations of the π-cation radicals, we considered a number of possible factors. None appeared reasonable. We did note that

**Table XX.** Summary of "planar" tetraarylporphyrin complexes

| Compound | Symmetry[a] | Ref. | Compound | Symmetry[a] | Ref. |
|---|---|---|---|---|---|
| [Mg(TPP)(1-MeIm)₂] | $C_i$ | 173 | [Fe(O₂)(TpivPP)(2-MeIm)] | $C_2$ | 67 |
| [Mg(TPP)(Pip)₂] | $C_i$ | 173 | [Fe(TPP)(Py)₂][h] | $C_i$ | 56 |
| [Mg(TPP)(4-MePy)₂] | $C_i$ | 173 | [Fe(TPP)(Pip)₂] | $C_i$ | 274 |
| [Mg(TPP)(H₂O)][b] | $C_{4h}$ | 266 | [Fe(TPP)(NO)][b] | $C_{4h}$ | 275 |
| [Ti(TPP)Br₂][b] | $C_{4h}$ | 267 | [Fe(TPP)(CO)(Py)] | – | 251 |
| [Ti(TPP)(OCH₃)] | S.D.[c] | 76 | [Fe(TPP)(4-MePip)(NO)][i] | – | 276 |
| [Mo(TTP)Cl₂] | – | 78 | [Fe(TpivPP)(THT)] | – | 277 |
| [Cr(TPP)O][b] | $C_{4h}$ | 96b | [Fe(TpivPP)(1-MeIm)(O₂)] | $C_2$ | 66 |
| [Cr(TPP)] | $C_i$ | 201 | [Fe(TPP)(t-BuNC)₂] | Sad | 278 |
| [Mn(TPP)(OCH₃)₂] | $C_i$ | 91 | [Fe(TPP)(THF)₂] | $C_i$ | 15, 125 |
| [Mn(TPP)(DMF)₂]ClO₄[d] | $C_i$ | 178 | [Fe(TPP)]⁻ | $C_i$ | 63 |
| [Mn(TPP)(CH₃OH)₂]ClO₄[e] | $C_i$ | 177 | [Fe(TPP)]²⁻ | $C_i$ | 63 |
| [Mn(TPP)(Im)]ₙ[e] | $C_i$ | 176 | [Ru(TPP)(CO)(Py)] | | 286 |
| [Mn(TPP)] | $C_i$ | 172 | [Ru(TPP)(Ph₂PCH₂PPh₂)₂] | $C_i$ | 182 |
| [Fe(TPP)(FSbF₅)][f] | – | 21 | [Co(TPP)(SC₆HF₄)₂]⁻ [e] | $C_i$ | 188 |
| [Fe(TPP)(SPh)₂]⁻ | $C_i$ | 34 | [Co(TPP)(CH₂CHO)] | Ruf | 184 |
| [Fe(TPP)Cl][b] | $C_{4h}$ | 268 | [Co(TPP)(Pip)₂] | $C_i$ | 279 |
| [Fe(TPP)(TMSO)₂]ClO₄[e] | $C_i$ | 269 | [Co(TPP)Cl] | $C_4$ | 280 |
| [Fe(TPP)Br] | Sad | 270 | [Co(TPP)(1,2-Me₂Im)] | – | 247 |
| [Fe(TPP)(C(CN)₃)]ₙ[b] | $C_s$ | 271 | [Co(TPP)(HIm)₂][OAc][e] | $C_i$ | 244 |
| [Fe(TPP)(H₂O)₂]ClO₄ | $C_i$ | 272 | [Co(TPP)(Py)Cl] | – | 255 |
| [Fe(TPP)I] | Sad | 22 | [Co(TPP)(3,5-Lut)] | – | 257 |
| [Fe(TPP)(CN)₂]⁻ | $C_i$ | 12 | [Co(TPP)(1-MeIm)] | – | 245 |
| [Fe(TPP)F][b] | $C_{4h}$ | 31 | [CoTPP]⁻ | $C_i$ | 64 |
| [Fe(TPP)(NO₃)] | Sad | 273 | [In(TPP)(SO₃CH₃)]ₙ[e, i, j] | $C_i$ | 169 |
| [FeCl(C₂-Cap)] | – | 52 | [Zn(TPP)] | $C_i$ | 200 |
| [Fe(TPP)(F)₂]⁻ [g] | $C_i$ | 14 | [Zn(TPP)(THF)₂] | $C_i$ | 157 |
| [Fe(TPP)(OCH₃)] | S.D.[c] | 124 | [Zn(TPP)(OH₂)][b] | $C_{4h}$ | 284 |
| [Fe(TPP)(SO₃Ph)] | – | 50 | [Cd(TPP)(Pip)] | S.D.[c] | 170 |
| [Fe(P–N₄)Cl] | Ruf | 44 | [Sn(TPP)Cl₂][b] | $C_{4h}$ | 281 |
| [Fe(TPP)(NO)(H₂O)]ClO₄ | $C_i$ | 30 | [Sn(TPP)(OH)₂] | – | 282 |
| [Fe(TpivPP)(2-MeIm)] | $C_2$ | 67 | [Zn(TPP)] | $C_i$ | 283 |
| | | | [Ag(TPP)] | $C_i$ | 283 |

[a] Crystallographic symmetry required except for Ruf, Sad, S.D. notations;   [b] Exact planarity and phenyl dihedral angles of 90° required;   [c] S.D. = slightly domed with idealized $C_{4v}$ symmetry;   [d] Dihedral angle (D.A.) 59.8°;   [e] Two independent half units in asymmetric unit;   [f] D.A., 59.4°;   [g] D.A., 55.3°;   [h] D.A., 58.3°;   [i] unsolvated form;   [j] D.A., 59.2°

**Table XXI.** Summary of core conformations (not $D_2d$)

| Complex[a] | Average absolute displacement | | Max $C_b$[b] | Symmetry[c] | Ref. |
|---|---|---|---|---|---|
| | $C_m$[b] | $C_b$[b] | | | |
| $[\{Nb(TPP)\}_2O_3]$ | 5(6) | 9(4) | 17 | $C_{4v}$ Domed | 155 |
| $[Tl(TPP)Cl]$ | 2(1) | 10(7) | 19 | $C_{2v}$ Domed | 165 |
| $[Tl(TPP)(CH_3)]$ | 1(1) | 11(4) | 16 | $C_{4v}$ Domed | 165 |
| $[In(TPP)Cl]$ | 2(1) | 11(7) | 21 | $C_{2v}$ Domed | 167 |
| $[In(TPP)(CH_3)]$ | 3(2) | 12(7) | 28 | $C_{2v}$ Domed | 168 |
| $[Mo(TPP)(N_2C_6H_5)_2]$[d] | 11(6) | 12(9) | 23 | $C_i$ | 79 |
| $TPP[Tc(CO)_3]_2$ | 8(5) | 14(9) | 25 | $C_i$ | 285 |
| $TPP[Re(CO)_3]_2$ | 8(5) | 14(8) | 23 | $C_i$ | 285 |
| $[Tl(TPP)(NOR)]$ | 18(12) | 18(15) | 48 | Domed | 166 |
| | 13(14) | 23(6) | 29 | Domed | |
| $[Mo(TPP)(CO)_2]$[d] | 14(14) | 26(18) | 57 | $C_{2v}$ Domed | 83 |

[a] Complexes ordered on averaged absolute displacements of $C_b$; [b] Value in $\text{Å} \times 10^2$; [c] Except for $C_i$, all are idealized descriptions with no required symmetry; [d] Dihedral angles not reported, displacements based on 4 N plane

the π-cation radical species, all tetraarylporphyrin derivatives, had in addition to the unusual (and accentuated) Sad core conformation unusually small values for the dihedral angles formed by the peripheral phenyl rings and the core. These have generally been recognized to range from 60–90° but no systematic quantitative information concerning the distribution of small values was available. This led us to completely survey the literature for core conformations and dihedral angles of tetraarylporphyrin derivatives. We have recorded, or calculated, as required, the deviation of the atoms from the mean plane of the 24-atom core and the values of the dihedral angles formed by the peripheral phenyl groups and the core. A large number of such complexes can be described as "planar". These species also have unremarkable values for the dihedral angles. We have recorded these species in Table XX. No derivative listed in Table XX has any atom displaced by more than 0.15 Å from the mean plane of the core. In addition, except for those few noted instances, no phenyl group dihedral angle is less than 60°. We have further listed in Table XX the crystallographically required symmetry of the core. A blank indicates no required symmetry. In a few instances where there is no required symmetry, a description of the idealized geometry of the core is given.

Table XXI summarizes core conformational data for all tetraaryl substituted complexes that have atomic deviations from the mean plane greater than 0.15 Å and are *not* appropriately described as either Ruf or Sad. In this table we report the average *absolute* value of displacements from the 24-atom planes for two types of atoms: $C_m$ and $C_b$. Estimated standard deviations are reported to provide a measure of the dispersion in the values. Also reported is the absolute value of the maximum displacement of a $C_b$ atom from the 24-atom mean plane. We hope that these values, along with a report of the idealized core conformation, will provide adequate information to permit a "visualization" somewhat equivalent to that provided by formal diagrams like those of Fig. 17.

None of the complexes listed in Table XXI have phenyl group dihedral angles less than 60°. Furthermore, very few observations of phenyl group dihedral angles less than 60° are found for the several complexes with a Ruf core conformation. These species are listed in Table XXII which gives the same information as Table XXI. An examination of the values listed in Table XXII also reveals that the average $C_m$ displacement is generally larger than the average $C_b$ displacement. The complexes listed in this table are all judged to have ruffled core conformations approximating that of idealized Ruf. Average displacements of $C_m$ are seen to increase up to 0.83 Å in the phosphorous(V) complex. The ruffling found in many of the complexes listed in Table XXII is the probable result of achieving shorter M–N bonds than would be possible with planar cores. Additional references for Tables XX–XXII are[266–290].

**Table XXII.** Summary complexes with "Ruf" symmetry

| Complex[a] | Average absolute displacement | | Max $C_b$[b] | Ref. |
|---|---|---|---|---|
| | $C_m$[b] | $C_b$[b] | | |
| [Nb(TPP)]₂O₃[c] | 7(5) | 13(8) | 26 | 154 |
| | 7(2) | 9(4) | 14 | |
| [Fe(TTOP)]₂ | 14(3) | 11(7) | 20 | 54 |
| [Fe(TPP)(OClO₃)] | 18(2) | 18(8) | 29 | 19 |
| [Mo(TPP)(σ-C₆H₅)Cl] | 19(5) | 18(9) | 31 | 80 |
| [Zn(TPP–C₃Py)] | 20(3) | 9(7) | 23 | 262 |
| [Fe(TPP)(N₃)(Py)] | 29(5) | 21(12) | 35 | 29 |
| [Mn(TPP)(Py)Cl][d] | 30(5) | 26(13) | 42 | 249 |
| [Fe(TPP)(HIm)₂]Cl | 31(3) | 12(9) | 24 | 235 |
| [Fe(TPP)(NCS)(Py)] | 35(6) | 22(15) | 40 | 27 |
| [Co(TPP)(Py)(OCH₃)] | 36(4) | 17(9) | 27 | 256 |
| Pd(TPP)[e] | 38(0) | 14(10) | 23 | 287 |
| [Fe(TPP)(CN)(Py)] | 38(7) | 23(16) | 41 | 28 |
| Pt(TPP)[e] | 39(0) | 16(8) | 23 | 288 |
| Fe(TPP)[c] | 40(0) | 16(8) | 23 | 289 |
| Cu(TPP)[e] | 42(0) | 16(8) | 23 | 287 |
| Co(TPP)[e] | 42(0) | 17(9) | 24 | 290 |
| [Mn(TPP)(2,6-LutNO)₂]ClO₄[f] | 44(3) | 17(6) | 25 | 179 |
| [Co(TPP)(H₂O)₂]ClO₄[g] | 47(6) | 26(20) | 49 | 187 |
| [Mn(TPP)(NCO)₂][h] | 59(2) | 23(14) | 38 | 92 |
| [Co(TPP)(3,5-Lut)(NO₂)][i] | 60(5) | 24(2) | 26 | 254 |
| [P(TPP)(OH)₂]OH[j] | 83(4) | 32(2) | 35 | 163 |

[a] Complexes are ordered in increasing value of the $C_m$ displacement;  [b] Value in Å × 10²;  [c] Two independent rings; one Dihedral angle (D.A.) of 59.8°;  [d] D.A. of 59.8°;  [e] Crystallographically required S₄ symmetry;  [f] D.A., 58.2;  [g] D.A., 59°;  [h] Values reported on 4N plane;  [i] Required two fold axis;  [j] D.A., 58.9°

**Table XXIII.** Summary of core conformations with "Sad" symmetry

| Complex[a] | Average absolute displacement | | Max $C_b$[b] | Ph Dihedral Angles[c] | Ref. |
|---|---|---|---|---|---|
| | $C_m$[b] | $C_b$[b] | | | |
| [Fe(TPP-C$_5$Im)(THT)] | 6(3) | 11(4) | 17 | 67.5, 84.4, 65.2, 87.9 | 261 |
| [Fe(TPP)]$_2$SO$_4$ | 4(3) | 14(3) | 17 | 65.3, 64.3, 81.7, 63.7 | 47 |
| [Zn(TPyP)(Py)][d] | 1(0) | 15(3) | 17 | 65, 59 | 259 |
| [Mn(TPP)(4-MePip)(NO)] | 2(1) | 16(4) | 23 | 64.1, 81.0, 82.5, 67.7 | 175 |
| [Fe(TPP)(4-MePip)(NO)][e] | 2(1) | 16(3) | 20 | 65.5, 81.6, 81.5, 69.2 | 276 |
| [Mn(TPP)(1-MeIm)] | 5(3) | 18(7) | 25 | 85.1, 81.0, 56.4, 78.9 | 241 |
| [Fe(TPP)(PMS)$_2$]ClO$_4$ | 2(1) | 18(2) | 20 | 63.3, 74.4, 82.5, 63.9 | 261 |
| [Fe(TPP)(4-MeIm$^-$)$_2$]$^-$ | 5(3) | 19(3) | 23 | 64.0, 59.3, 69.8, 70.6 | 13 |
| [Fe(TPP)(NO)(1-MeIm)] | 1(2) | 22(2) | 24 | 60.7, 79.9, 80.5, 75.1 | 243 |
| [Fe(TAP)(SH)] | 18(5) | 25(9) | 35 | 82.8. 55.0, 67.8. 57.4 | 33 |
| [Co(TPP)(SC$_6$HF$_4$)$_2$]$^-$ [f] | 16(2) | 28(8) | 37 | 57.4, 82.4, 68.9, 66.6 | 188 |
| [Mo(TTP)O] | 13(9) | 32(23) | 57 | 70.5, 56.1, 81.9, 61.8 | 78 |
| [Cr(TTP)N] | 15(6) | 35(10) | 48 | 66.3, 51.2, 89.7, 53.7 | 97 |
| [Mn(TTP)(NO)] | 16(5) | 35(10) | 48 | 67.5, 55.2, 85.8, 54.6 | 175 |
| [Cr(TTP)O] | 13(6) | 39(9) | 50 | 68.7, 50.9, 86.9, 52.3 | 96a |
| H$_4$TPyP$^{2+}$ [d] | 6(2) | 86(4) | 90 | 35.6, 33.0 | 291 |
| H$_4$TPP$^{2+}$ | 6(1) | 90(4) | 96 | 30.6, 30.8, 35.7, 34.9 | 292 |
| H$_4$TPP$^{2+}$ [g] | 28(0) | 107(14) | 117 | 21.1 | 291 |
| [Fe(TPP)]$_2$O[h] | 6(4) | 15(8) | 30 | 82.8, 82.7, 77.0, 53.3 | 293 |
| [Fe$_2$(μ-O)FF · H$_2$O][i] | 23(3) | 24(14) | 45 | 70.5, 59.1, 60.5, 62.1 | 45 |
| | 8(4) | 8(5) | 16 | 70.8, 77.2, 58.8, 71.0 | |
| [Mn(TPP)(N$_3$)]$_2$O[j] | 11(0) | 37(9) | 50 | 51.3, 68.8 | 88 |
| | 22(1) | 50(11) | 62 | 44.4, 56.3 | |
| [Mo(TTP)(O)]$_2$O[k] | 3(1) | 29(6) | 37 | 53.6, 64.0, 53.8, 54.2 | 154 |
| [Fe(TPP)]$_2$N[l] | 26(2) | 45(14) | 64 | 54.0, 65.4 | 89 |

[a] Complexes ordered on average absolute value of the displacements of $C_b$ except for the final five entries (see text); [b] Values in Å × 10$^2$; [c] Value in degrees; [d] $C_2$ symmetry; [e] CHCl$_3$ solvate; [f] Second molecule; [g] $S_4$ symmetry; [h] M.P.S. = 4.58 Å; [i] Two independent rings at angle of 15.8° cf. Fig. 4; [j] Two independent half rings each with $C_2$ symmetry, M.P.S. = 3.89 Å; [k] M.P.S. = 3.85 Å; [l] M.P.S. = 4.15 Å

Selected species with the Sad core conformation are listed in Table XXIII. The complexes are listed in order of increasing average displacement of the $C_b$ atoms except for the final five entries in the table, which we exclude for subsequent discussion. It can be seen that as the $C_b$ displacement increases in these Sad species there is a strong tendency for at least some of the dihedral angles to have small values. This trend culminates in the extremely small values found[291, 292] for the porphyrin diacids. The vertical tilting of the pyrrole rings in the porphyrin diacids was suggested by Fleischer[291] to be the result of the steric hindrance engendered by the positioning of the four hydrogen atoms at the center of the porphyrin ring. Possibly the electrostatic repulsion of the positively charged nitrogen atoms also contributes. The extreme tilting of the pyrrole rings allows for the phenyl or pyridyl rings to become much more nearly coplanar with the porphine nucleus. We would suggest that the converse relationship is also true. If for some reason the aryl groups need to become more nearly coplanar with the porphine nucleus, the most reasonable stereochemical path is through a Sad deformation of the core. The data presented thus far, plus results to be described subsequently, are in accord with this idea. We thus suggest that a major driving force for the Sad core conformation is structural. The importance of Sad conformation in allowing near coplanarity of the peripheral aryl groups is shown in Fig. 18 which shows the C···C contact distances between ortho phenyl

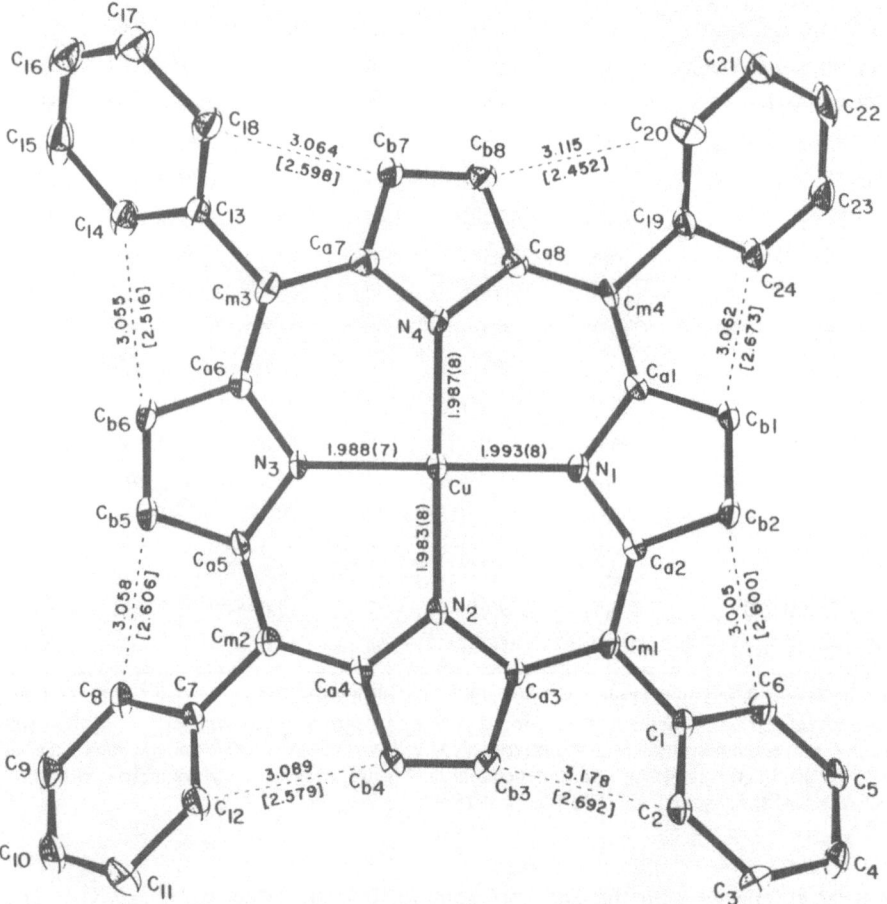

**Fig. 18.** Drawing illustrating the importance of the Sad core conformation in minimizing intramolecular contacts between nearly coplanar phenyl groups and core atoms. The values displayed are taken from the structural data for [Cu(TPP)]⁺. The numbers over the *dashed lines* are the distances between the $C_b$ and ortho phenyl carbon atoms. The numbers in the *square brackets* are the analogous hydrogen hydrogen separations

**Table XXIV.** Summary of core conformations with "Sad" symmetry and solid-state interactions

| Complex[a] | Average absolute displacement | | Max $C_b$[b] | Ph. dihedral angles[c] | M.P.S. | Ct–Ct | M–M | S.A. (Ct) | S.A. (M) | Ref. |
|---|---|---|---|---|---|---|---|---|---|---|
| | $C_m$[b] | $C_b$[b] | | | | | | | | |
| [Zn(TPC)(Py)] | 4(3) | 9(7) | 20 | 86.5, 66.6, 87.5, 86.3 | 3.99 | 7.010 | 7.433 | 55.3 | 57.5 | 141 |
| [Zn(TPiBC)(Py)] | 5(3) | 10(6) | 18 | 89.0, 87.5, 69.2, 87.5 | 4.02 | 7.000 | 7.419 | 55.0 | 57.2 | 142 |
| [Sn(TPP)]MnHgMn | 8(3) | 27(7) | 39 | 54.6, 56.9, 77.7, 87.1 | 3.81 | 5.077 | 6.474 | 41.4 | 53.9 | 151 |
| [Mg(TPP)(OClO₃)] | 3(3) | 35(9) | 45 | 45.0, 63.4, 53.3, 50.9 | 3.77 | 6.672 | 7.193 | 55.6 | 58.4 | 126 |
| [Mn(TPP)(CN)] | 14(3) | 36(6) | 46 | 50.8, 57.1, 54.4, 80.4 | 3.82 | 5.409 | 5.780 | 45.1 | 48.6 | 28 |
| [Zn(TPP)(OClO₃)] | 2(1) | 38(7) | 45 | 44.3, 63.0, 52.2, 50.6 | 3.70 | 6.539 | 6.932 | 55.5 | 57.7 | 127 |
| [Au(TPP)]Cl | 16(5) | 39(8) | 54 | 51.3, 52.6, 50.6, 70.4 | 3.76 | 5.482 | 5.502 | 46.7 | 46.9 | 294 |
| [Nb(TPP)(O)(OAc)] | 7(6) | 39(13) | 58 | 70.9, 70.7, 65.0, 60.2 | 3.92 | 6.136 | 7.606 | 50.3 | 59.0 | 155 |
| [Fe(TPP)(Ph)] | 7(3) | 40(4) | 43 | 60.2, 66.1, 75.7, 62.8 | 3.87 | 5.252 | 5.516 | 42.5 | 45.4 | 32 |
| [Mn(TPP)Cl] | 14(2) | 41(6) | 49 | 54.2, 56.5, 49.7, 77.4 | 3.80 | 5.258 | 5.655 | 43.7 | 47.8 | 295 |
| [Fe(TPP)(THF)]⁺ | 26(1) | 47(11) | 59 | 66.6, 47.1, 56.5, 49.1 | 3.85 | 5.494 | 5.763 | 45.5 | 48.1 | 18 |
| [Mo(TTP)(O)₂] | 5(3) | 49(10) | 62 | 56.7, 65.6, 80.6, 50.0 | 3.89 | 5.617 | 7.166 | 46.2 | 57.1 | 93 |
| [Fe(TTP)Cl]⁺ | 6(5) | 59(6) | 70 | 43.7, 48.8, 39.5, 46.2 | 3.68 | 4.703 | 5.393 | 38.5 | 47.0 | 128 |
| [Fe(TPP)(B₁₁CH₁₂)] | 12(3) | 60(6) | 69 | 58.2, 44.5, 66.8, 45.4 | 3.83 | 5.305 | 5.489 | 43.8 | 45.8 | 20 |
| [Cu(TPP)]⁺ | 16(4) | 65(8) | 80 | 41.7, 42.6, 43.0, 38.6 | 3.84 | 5.396 | 5.433 | 44.7 | 45.0 | 129 |
| Co(TPP) | 4(4) | 67(4) | 72 | 85.0, 86.7, 55.7, 87.6 | 3.84 | 5.420 | 5.418 | 45.2 | 45.2 | 296 |

a Complexes ordered on average absolute value of the $C_b$ displacement;  b Values in Å $\times$ 10²;  c Value in degrees

carbons and $C_b$ core carbon atoms found in the π-cation species, $[Cu(TPP)]^+$. The figure makes clear the importance of the vertical tilt of the pyrrole rings in maximizing phenyl core atom contacts for groups with small dihedral angles. It is to be emphasized that the constraints of the macrocyclic ring probably requires the propagation of the ruffling around the entire core perimeter once it commences between a pair of pyrrole rings. It should also be noted that a Sad core conformation does not require that all or even any phenyl dihedral angles be small. We do not believe the converse relationship to be true, rather, one or more small dihedral angles does require the Sad core conformation. We have found no exceptions to this in our survey. Every compound for which we note a dihedral angle less than 55 degrees has the Sad conformation.

The final five compounds listed Table XXIII are binuclear species with interplanar spacings that can lead to steric problems with the bulky phenyl groups of the two porphyrin rings. These steric contacts can be ameliorated by the core conformation and smaller dihedral angles that roughly correlate with decreasing interplanar spacings. Further, the four TTP derivatives listed in Table XXIII also seem to have intermolecular contacts that lead to smaller phenyl group dihedral angles and a Sad core conformation. The similar core conformations and dihedral angles are the probable consequence of nearly equivalent solid state environments in this isomorphous series of complexes.

A most interesting set of compounds that have the Sad core conformation are the species listed in Table XXIV that includes all of the π-cation species. The species in this table are collected together because they all display solid-state interactions that resemble those described in the π-π section. All of these species save Co(TPP) form dimers in the solid state. Co(TPP) forms a linear aggregate. All species have planes that are related to each other by inversion centers and hence have exactly parallel arrangements. Table XXIV lists these interacting species ordered on increasing displacements of the $C_b$ atoms. Also given is the mean plane separation and the Ct···Ct distance. It can be seen that

**Table XXV.** Summary of x, y, z displacements between Ct's for tetraarylporphyrins

| Compound | Type | Ct–Ct[a] | M–M[a] | x[a] | y[a] | z[a] | L.S.[a] | Group | Ref. |
|---|---|---|---|---|---|---|---|---|---|
| [Zn(TPC)(Py)] | Dimer | 7.010 | 7.433 | 4.98 | 2.98 | 3.93 | 5.80 | W | 141 |
| [Zn(TPiBC)(Py)] | Dimer | 7.000 | 7.419 | 4.99 | 2.90 | 3.96 | 5.77 | W | 142 |
| [Sn(TPP)]MnHgMn | Dimer | 5.077 | 6.474 | 2.92 | 1.33 | 3.94 | 3.21 | I | 151 |
| [Mg(TPP)(OClO₃)] | Dimer | 6.672 | 7.193 | 4.23 | 3.40 | 3.87 | 5.43 | W | 126 |
| [Mn(TPP)(CN)] | Dimer | 5.409 | 5.780 | 3.49 | 1.54 | 3.84 | 3.81 | I | 28 |
| [Zn(TPP)(OClO₃)] | Dimer | 6.539 | 6.932 | 4.13 | 3.34 | 3.81 | 5.32 | W | 127 |
| [Au(TPP)]Cl | Dimer | 5.482 | 5.502 | 3.47 | 1.73 | 3.88 | 3.87 | I | 294 |
| [Nb(TPP)(O)(OAc)] | Dimer | 6.136 | 7.606 | 4.16 | 1.75 | 4.16 | 4.51 | W | 155 |
| [Fe(TPP)(Ph)] | Dimer | 5.252 | 5.516 | 3.23 | 1.54 | 3.85 | 3.58 | I | 32 |
| [Mn(TPP)Cl] | Dimer | 5.258 | 5.655 | 3.31 | 1.42 | 3.83 | 3.61 | I | 295 |
| [Fe(TPP)(THF)]⁺ | Dimer | 5.494 | 5.763 | 3.71 | 1.47 | 3.78 | 3.99 | I | 18 |
| [Mo(TTP)O₂] | Dimer | 5.617 | 7.166 | 3.67 | 1.07 | 4.11 | 3.82 | I | 93 |
| [Fe(TTP)Cl]⁺ | Dimer | 4.703 | 5.393 | 2.72 | 1.56 | 3.51 | 3.13 | I | 128 |
| [Fe(TPP)(B₁₁CH₁₂)] | Dimer | 5.305 | 5.489 | 3.69 | 0.39 | 3.79 | 3.71 | I | 20 |
| [Cu(TPP)]⁺ | Dimer | 5.396 | 5.433 | 3.39 | 1.77 | 3.81 | 3.82 | I | 129 |
| Co(TPP) | Agg-A | 5.420 | 5.418 | 3.85 | −0.22 | −3.81 | 3.86 | I | 296 |
| Co(TPP) | Agg-B | 5.457 | 5.452 | 0.17 | −3.88 | 3.84 | 3.88 | I | 296 |

[a] Value in Å

these values lead to placing the compounds into the previously defined groups I and W of the $\pi$-$\pi$ classification. It would appear that the bulky aryl groups preclude the possibility of ever achieving the geometry of Group S. Table XXV presents a detailed geometrical analysis for these species and is similar to that reported in Table XV of Sect. C. The importance of the small dihedral angles for dimer formation can be clearly seen in Fig. 19. The saddle-shaped conformation also generally seems to lead to slightly tighter interring interactions than implied by the mean plane separation. As can be seen in Fig. 19, the overall interaction between the two porphyrin rings has individual pyrrole rings nesting into each other. Indeed, owing to the rather nonplanar conformations of the species of Table XXV, there are interatomic contacts (between dimer rings) that are significantly smaller than the mean plane spacing. This is not typical of the species described earlier in Sect. C. We take this geometric feature as strong evidence for the driving force towards dimerization or aggregation. It should be noted forming extended structures appears to be rather more difficult stereochemically for the *meso*-substituted aryls than for the $\beta$-substituted alkyl derivatives. Only one example is known to us, the Co(TPP) entry of Tables XXIV and XXV. In this structure[296], the cobalt atom is directly above and below the midpoint of the $C_b$–$C_b$ bonds of two adjacent molecules and the resulting infinite chain has 90° steps. Finally, it should be noted that all species with the Sad conformation and substantial $C_b$ displacements (Tables XXIII and XXIV) are binuclear, or four- or five-coordinate species. This fact clearly allows for strong interporphyrin interactions in the solid state and is a feature not likely to be present in species with axial ligands on both sides of the porphyrin plane.

Our detailed survey of core conformations leads us to conclude that the reason for the unusual Sad conformation of the $\pi$-cation radical complexes is not electronic but steric in origin. The conformation is a reaction to effects that lead to more nearly coplanar *meso*-aryl groups required by porphyrin-porphyrin interactions. These interactions presumably result from the same conditions that led to the interactions of the compounds described in Sect. C. The driving force towards porphyrin-porphyrin interactions appears to be especially pronounced for $\pi$-cation species. We thus regard any electronic factors leading to Sad conformations to be secondary to those that favor the formation of interporphyrin interactions. As noted earlier in the $\pi$-cation section the core conformations may have profound influences on the observed magnetic properties. Finally, this analysis leads to

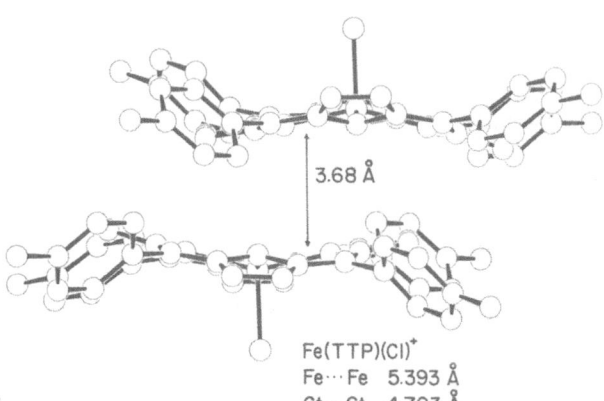

**Fig. 19.** Plot showing the dimeric interaction in the structure of the $\pi$-cation radical [Fe(TPP)Cl]$^+$. Drawn from the coordinates reported in Ref. 128

3.68 Å

Fe(TTP)(Cl)$^+$
Fe$\cdots$Fe  5.393 Å
Ct$\cdots$Ct  4.703 Å

predictions concerning the structures of other π-cation radical species. The Sad conformation will not be observed for *meso*-aryl species with sufficiently bulky ortho substituents that do not allow near coplanarity of the peripheral aryl. β-Alkyl substituted
species are unlikely to show such conformations. References found in Tables
XXIII–XXV include[293–296].

## III. Crystal Packing and Isomorphous Series

Recently, we have published structure determinations for unsolvated, four-coordinate
Ag(TPP) and Zn(TPP)[283]. We noted that the probable cause of the ~ 0.02 Å variation in
the $M–N_p$ bond distances within each complex was crystal packing effects. Details of the
argument can be found in the original publication. During the course of our investigations of metalloporphyrin structure, we have often noted the possible effects of crystal
packing on the structural parameters of the target molecules. We have already mentioned[39, 43] the two crystalline forms of [Fe(OEP)(3-ClPy)$_2$]ClO$_4$. The orientation of the
axial ligand must be controlled in part as a result of crystal packing. Two different
crystalline forms of [Fe(TPP)(NO)(4-MePip)] have also been noted[276]: the two forms
have radically different Fe–N(4-MePip) bond distances. Again the probable explanation
was cited as crystal packing effects. Intuitively, it seems likely that a large tetrapyrrole
core could dominate the packing interactions and the axial ligand geometry, counterion,
or other "minor" structural portions of the molecule would adapt to these constraints. It
is even possible that the packing unit that dominates is the macrocyclic ligand plus axial
ligand(s). The adaptation to the required packing could include such features as a
required disorder, solvate molecule incorporation or crystallization of a species that
forms only a minor component of an equilibrium solution.

It seems clear that all cases of structure "control" by crystal packing effects are
unlikely to be noted much less proven. However, we think that the probable importance
of the concept can be buttressed by noting effects that can be catalogued under this
general rubric. In this section we will note the occurrence of a number of complexes
having, after appropriate cell reduction, similar cell constants and form isomorphous
series. We will comment on members of the series that have features that are reasonably
ascribed to crystal packing effects. We start with the highest symmetry crystal systems
and work towards the triclinic systems.

Table XXVI summarizes two different series of *meso*-tetraphenylporphyrin derivatives that crystallize in the tetragonal crystal system. The compound, space group, *a, c,*
and unit cell volume are listed for each member. Series a, the four-coordinate species,
has H$_2$TPP as the parent compound. The existence of the series is expected and solid
solutions of H$_2$TPP and the metallo derivatives listed are readily formed[297]. All members
of this series have Ruf core conformations with substantial displacements of the C$_m$
carbon atoms. The second series is a substantial series of five- or six-coordinate complexes of which [Sn(TPP)Cl$_2$][281] is the prototypical ordered example with $C_{4h}$ symmetry.
We first concern ourselves with the subset that have been catalogued as having I4/m as
the space group. The phenyl groups of adjacent I-centered molecules make equivalent
ligand pockets above and below the porphyrin plane. The pocket size is adequate for
small (mono- or diatomic) axial ligands. Most of the species listed in Table XXVI(b) are
seen to be five-coordinate; hence the complexes display disorder in the crystal and

**Table XXVI.** The isomorphous tetraphenylporphyrins in the tetragonal crystal system

| Complex | Sp.Gr. | $a^a$ | $c^a$ | Vol.$^b$ | Ref. |
|---|---|---|---|---|---|
| a. The I$\bar{4}$2d Series (Z = 2) | | | | | |
| Ni(TPP) | I$\bar{4}$2d | 15.065 | 13.887 | 3151.7 | 147 |
| Cu(TPP) | I$\bar{4}$2d | 15.040 | 13.993 | 3165.2 | 287 |
| Co(TPP) | I$\bar{4}$2d | 15.062 | 13.954 | 3165.7 | 290 |
| Pt(TPP) | I$\bar{4}$2d | 15.073 | 13.988 | 3178.0 | 288 |
| Pd(TPP) | I$\bar{4}$2d | 15.088 | 13.987 | 3184.1 | 287 |
| H$_2$TPP | I$\bar{4}$2d | 15.125 | 13.940 | 3189.0 | 291 |
| Fe(TPP) | I$\bar{4}$2d | 15.080 | 14.043 | 3193.5 | 289 |
| b. The I4, I$\bar{4}$, I4/m Series (Z = 4) | | | | | |
| [V(TPP)O] | I4/m | 13.345 | 9.745 | 1735.5 | 190 |
| [Cr(TPP)O] | I4/m | 13.351 | 9.749 | 1737.8 | 96b |
| [Ag(TPP)$_{0.54}$(TPP)$_{0.46}$] | I4/m | 13.384 | 9.717 | 1740.6 | 298 |
| [Fe(TPP)F] | I4/m | 13.381 | 9.767 | 1748.8 | 31 |
| [Mg(TPP)(H$_2$O)] | I4/m | 13.460 | 9.680 | 1753.7 | 266 |
| [Zn(TPP)(H$_2$O)] | I4/m | 13.440 | 9.715 | 1754.9 | 284 |
| [Co(TPP)(NO)] | I4/m | 13.434 | 9.754 | 1760.3 | 299 |
| [Fe(TPP)(NO)] | I4/m | 13.468 | 9.755 | 1769.4 | 275 |
| [Co(TPP)Cl] | I4/m | 13.489 | 9.779 | 1779.3 | 300 |
| [Co(TPP)Br] | I4/m | 13.460 | 9.840 | 1782.7 | 300 |
| [Mo(TPP)(O)Cl] | I4/m | 13.469 | 9.852 | 1787.3 | 301 |
| [Fe(TPP)Cl] | I4/m | 13.534 | 9.820 | 1798.7 | 268 |
| [Co(TPP)Cl] | I4 | 13.693 | 9.701 | 1818.9 | 280 |
| [Sn(TPP)Cl$_2$] | I4/m | 13.673 | 9.961 | 1862.2 | 281 |
| [Ti(TPP)Br$_2$] | I4/m | 13.757 | 9.880 | 1869.8 | 267 |
| [Mo(TPP)(O$_2$)$_2$] | I$\bar{4}$ | 14.660 | 9.571 | 2057.0 | 70 |

$^a$ Value in Å; $^b$ Cell volume in Å$^3$

possess the crystallographically required $C_{4h}$ symmetry only in a statistical sense. Indeed, obtaining ordered examples of the complex with these axial ligands can usually only be effected by changing to a different porphyrin derivative. The two ordered species, [Ti(TPP)Br$_2$] and [Sn(TPP)Cl$_2$] illustrate that the cell size and packing can be modulated, in a minor way, by the species themselves. The increase in the $a$ cell length in the titanium derivative relative to the tin(IV) complex leads to a small increase in pocket size in the $ab$ plane that is just sufficient to maintain van der Waals contact with the axial ligand and the adjacent molecules. The [Co(TPP)Cl] entries appear to truly reflect two crystalline phases of the substance that differ trivially in orientation (relative to the cell coordinate system) except in phenyl group dihedral angle. It is presumed that this is also true for [Mo(TPP)(O$_2$)$_2$][70], this has not been checked owing to the unavailability of atomic coordinates.

Table XXVII reports six series we have found in the monoclinic crystal system. Four of these are tetraphenylporphyrin derivatives; there is in addition one series each of octaethyl- and tetratolylporphyrin derivatives. Table XXVIII presents seven isomorphous series found in the triclinic crystal system. With one exception, all triclinic space groups are reported to be P$\bar{1}$. Tabulated are the unique cell constants and the cell volumes. In these two tables, an asterisk following the compound name indicates that the literature cell constants had to be transformed to obtain the cell constants reported. In each series of Tables XXVII and XXVIII, the porphinato core has been found to have a similar orientation with respect to the unit cell axes. Furthermore, similar core conformations and (when appropriate) similar aryl dihedral angles are observed. Certainly the fundamental packing is similar within each series.

**Table XXVII.** The isomorphous porphyrins in the monoclinic crystal system

| Complex | $a^a$ | $b^a$ | $c^a$ | beta$^b$ | Vol.$^c$ | Ref. |
|---|---|---|---|---|---|---|
| a. P2$_1$/n Series (Z = 4) | | | | | | |
| [Fe(TPP)(CN)(Py)] · H$_2$O | 13.227 | 23.575 | 14.065 | 102.64 | 4279.5 | 28 |
| [Fe(TPP)(N$_3$)(Py)] · 1/2 Py* | 13.030 | 23.651 | 14.519 | 103.94 | 4342.6 | 29 |
| [Co(TPP)(Py)Cl] · 1/2 C$_6$H$_6$* | 13.120 | 23.420 | 14.500 | 102.10 | 4356.4 | 255 |
| [Fe(TPP)(Py)(NCS)] · 1/2 Py | 13.238 | 23.917 | 14.269 | 104.74 | 4369.1 | 27 |
| [Co(TPP)(H$_2$O)$_2$]ClO$_4$ · Actn. 3/2 H$_2$O* | 13.720 | 23.460 | 13.930 | 102.28 | 4381.1 | 187 |
| [Mn(TPP)Cl(Py)] · C$_6$H$_6$ | 13.149 | 23.380 | 14.786 | 100.50 | 4469.4 | 249 |
| [Mo(TPP)(σ-Ph)Cl] · 1/2 C$_6$H$_6$ | 14.778 | 23.513 | 13.163 | 101.70 | 4478.8 | 80 |
| b. The P2$_1$/n Series (Z = 2) | | | | | | |
| [Fe(TPP)(OEt)$_2$]BF$_4$ (99 K) | 10.447 | 16.611 | 11.798 | 108.65 | 1939.9 | 302 |
| [Fe(TPP)(OEt)$_2$]BF$_4$* | 10.561 | 16.826 | 11.969 | 109.53 | 2004.5 | 303 |
| [Fe(TPP)(OEt)$_2$]ClO$_4$ | 10.600 | 16.792 | 11.963 | 108.79 | 2015.9 | 304 |
| [Fe(TPP)(C$_6$H$_5$)(THF)] | 11.182 | 16.367 | 12.079 | 111.38 | 2058.5 | 305 |
| c. The P2$_1$/c Series (Z = 4) | | | | | | |
| [V(OEP)O] | 14.334 | 23.061 | 9.815 | 104.18 | 3145.6 | 230 |
| [Ti(OEP)O] | 14.407 | 23.126 | 9.793 | 104.20 | 3163.1 | 69 |
| [Ti(OEP)O$_2$] | 14.603 | 23.266 | 9.793 | 105.00 | 3213.8 | 69 |
| d. The P2$_1$/n Series (Z = 4) | | | | | | |
| [Fe(TPP)(OCH$_3$)]* | 10.219 | 15.927 | 20.811 | 95.16 | 3373.4 | 124 |
| [Tl(TPP)Cl]* | 10.046 | 16.177 | 21.114 | 90.18 | 3431.3 | 165 |
| [In(TPP)Cl] | 10.099 | 16.117 | 21.090 | 90.70 | 3432.5 | 167 |
| [In(TPP)(CH$_3$)]* | 10.064 | 16.221 | 21.085 | 90.06 | 3442.1 | 168 |
| [Tl(TPP)(CH$_3$)]* | 10.046 | 16.244 | 21.096 | 90.04 | 3442.6 | 165 |
| [Fe(TPP)Br]* | 10.191 | 16.121 | 20.990 | 90.69 | 3448.2 | 270 |
| [Ti(TPP)(OCH$_3$)]* | 10.150 | 16.290 | 20.930 | 92.27 | 3457.9 | 76 |
| [Fe(TPP)(NO$_3$)] | 10.279 | 16.232 | 20.951 | 90.50 | 3495.5 | 273 |
| [Fe(TPP)I] | 10.118 | 16.352 | 21.211 | 90.44 | 3509.2 | 22 |
| e. P2$_1$/n Series (Z = 4) | | | | | | |
| [Mn(TPP)Cl] · (CH$_3$)$_2$CO* | 12.150 | 21.765 | 14.588 | 101.50 | 3780.3 | 295 |
| [Au(TPP)Cl] · CHCl$_3$* | 12.060 | 22.350 | 14.500 | 102.91 | 3809.6 | 294 |
| [Mn(TPP)(CN)] · CHCl$_3$ | 12.433 | 22.023 | 14.460 | 102.01 | 3872.7 | 28 |
| f. P2$_1$/c Series (Z = 4) | | | | | | |
| [Mo(TTP)O] · C$_6$H$_6$ | 17.420 | 16.851 | 15.734 | 112.23 | 4275.3 | 78 |
| [CN(TTP)N] · C$_6$H$_6$ | 17.365 | 16.938 | 15.736 | 112.35 | 4280.7 | 97 |
| [CO(TTP)O] · C$_6$H$_6$ | 17.342 | 16.964 | 15.804 | 112.52 | 4294.8 | 96a |
| [Mn(TTP)(NO)] · C$_6$H$_6$* | 17.389 | 16.979 | 15.755 | 112.43 | 4299.7 | 175 |

* Transformed cell constants
$^a$ Value in Å;   $^b$ Value in degrees;   $^c$ Cell volume in Å$^3$

**Table XXVIII.** The isomorphous porphyrins in the triclinic crystal system

| Complex | $a^a$ | $b^a$ | $c^a$ | alpha$^b$ | beta$^b$ | gamma$^b$ | Vol.$^c$ | Ref. |
|---|---|---|---|---|---|---|---|---|
| OEP N-Oxide* | 9.740 | 10.566 | 7.612 | 98.49 | 108.43 | 91.61 | 732.7 | 113 |
| Ni(OEP) | 9.924 | 10.564 | 7.617 | 97.66 | 109.47 | 92.35 | 743.0 | 225 |
| H₂OEP | 9.791 | 10.771 | 7.483 | 97.43 | 106.85 | 93.25 | 745.2 | 207 |
| Fe(OEP)* | 13.252 | 13.369 | 4.812 | 92.20 | 93.20 | 113.13 | 781.1 | 62 |
| Ni(OEP) | 13.302 | 13.344 | 4.814 | 92.26 | 93.63 | 113.63 | 781.5 | 198 |
| [Zn(TPP)] | 10.382 | 12.421 | 6.443 | 98.30 | 101.15 | 96.47 | 798.1 | 283 |
| H₂TPP* | 10.420 | 12.410 | 6.440 | 99.14 | 101.12 | 96.06 | 798.7 | 306 |
| [Ag(TPP)] | 10.503 | 12.485 | 6.351 | 97.72 | 100.68 | 97.15 | 801.4 | 283 |
| [Zn(TPP)(THF)₂]* | 11.115 | 11.720 | 9.572 | 103.78 | 115.01 | 102.71 | 1022.0 | 157 |
| [Fe(TPP)(THF)₂] | 11.354 | 11.804 | 9.688 | 103.92 | 115.91 | 102.38 | 1055.0 | 15 |
| [Cd(TPP)] · 2 Dioxan* | 11.573 | 11.614 | 9.845 | 102.59 | 116.75 | 102.74 | 1073.5 | 171 |
| [Fe(TPP)(Pip)₂] | 11.113 | 12.071 | 9.797 | 105.67 | 113.70 | 101.02 | 1089.6 | 274 |
| [Co(TPP)(Pip)₂] | 11.503 | 11.830 | 9.934 | 104.99 | 115.64 | 101.49 | 1099.5 | 279 |
| [Mg(TPP)(Pip)₂]* | 11.463 | 11.914 | 9.944 | 104.59 | 115.60 | 101.78 | 1106.4 | 173 |
| [Mg(TPP)(4-MePy)₂]* | 11.885 | 11.643 | 10.146 | 103.68 | 119.44 | 100.79 | 1109.0 | 173 |
| [ZnTPP] · 2 C₇H₈ | 11.349 | 11.404 | 10.502 | 110.48 | 103.87 | 107.65 | 1118.6 | 200 |
| [MnTPP] · 2 C₇H₈ | 11.320 | 11.465 | 10.487 | 110.63 | 103.34 | 107.80 | 1121.9 | 172 |
| [Co(N-CH₃TPP)Cl]* | 14.972 | 17.398 | 7.484 | 97.03 | 94.13 | 102.93 | 1875.8 | 98 |
| [Fe(N-CH₃TPP)Cl]* | 14.961 | 17.434 | 7.562 | 97.12 | 93.84 | 103.15 | 1896.6 | 101 |
| [Mn(N-CH₃TPP)Cl]* | 14.993 | 17.476 | 7.558 | 97.09 | 94.00 | 103.13 | 1904.0 | 99 |
| [Zn(TPiBC)(Py)] · C₆H₆* | 13.332 | 14.718 | 11.452 | 94.37 | 104.88 | 99.05 | 2128.7 | 142 |
| [Zn(TPC)(Py)] · C₆H₆* | 13.334 | 14.809 | 11.414 | 94.14 | 105.08 | 99.18 | 2133.0 | 141 |

* Transformed cell constants;  $^a$ Value in Å;  $^b$ Value in degrees;  $^c$ Cell volume in Å$^3$

The most interesting of the monoclinic series is that of the group of molecules, [M(TPP)(Py)X] · solvate, (series a) where X is a small anion and a variety of solvate molecules are found to be incorporated. The wide variety, both in type and stoichiometry, of solvate molecules are strongly suggestive of the crystal packing being driven by the [M(TPP)(Py)X] unit. The appearance of the diaquocobalt(III) complex is perhaps surprising but the compound does indeed show the expected orientation of the porphyrin plane. The appearance of this compound in the series suggests the importance of the ruffled macrocycle in defining the lattice packing for the series. It is to be noted that almost all of the mixed axial ligand iron(III) complexes, all difficult to prepare, are members of this series. In one case, that of [Fe(TPP)(Py)(NCS)][27], NMR measurements suggest that the mixed ligand species that crystallized is a very minor component of the solution under the crystallizing conditions. We regard this as evidence that solid state effects play an important role in the isolation of the iron(III) species and perhaps the $Mn^{III}$ compound as well. There is some subtlety involved as a second crystalline form of [Fe(TPP)(Py)(NCS)] has also been prepared[43]. It is also probable that the partially "bent" Fe–NCS group noted[27] is caused by the packing. No other species in this series has such a bulky ligand as a linear triatomic. Finally it can be noted that the orientation of all the planar axial ligands in this series are found to be similar.

The isomorphous series (b) that has members that differ only in the nature of the anion ($BF_4^-$ or $ClO_4^-$) would be reasonably expected. However, the occurence of the σ-bonded phenyl complex in this series is unexpected. All remaining members of the series in the monoclinic systems do not appear unusual or unexpected in any sense.

The first two isomorphous groups listed in Table XXVIII may be noted to have one complex common to both: Ni(OEP). Members of the two series differ in the extent of their π-π interactions: members of the first series were catalogued into Group W and members of the second series into Group I. The obvious question of why Ni(OEP) crystallizes in both series is not clear, however, preliminary evidence suggests that the occurence of the compound in the first series results from a template effect. In our hands[198], the first crystalline form of Ni(OEP) is found to always contain small amounts of $H_2OEP$; the second contains less or none. Further work is needed to completely clarify the matter. The crystal packing effects of the third series were mentioned earlier and are detailed elsewhere[283].

Members of the fourth triclinic series have the general formula [M(TPP)(L)$_2$]. The first two members of the series appear to have limited stability in solution; five-coordinate complexes represent the majority species for both. Accordingly, we attribute their isolation and characterization to be a direct result of the extra stability gained as a result of crystal packing. The third substance in the group was reported to have the cadmium(II) ion centered in the porphinato plane with very long axial "bonds" of 2.65 and 2.80 Å, i.e., a quasi four-coordinate complex. We think that the solid-state system is better described in space group $P\bar{1}$, as for all other members of the series, with a disordered cadmium ion similar to the disorder described[172] for Mn(TPP). In our view, the solvate dioxane molecules are held close to the $Cd^{II}$ ion simply through packing effects. The remaining members of this series and the subsequent series of Table XXVIII are unremarkable[298–306].

# E. Abbreviations

## I. Abbreviations for Porphyrins and Other Tetrapyrroles

| | |
|---|---|
| (C$_2$-Cap) | dianion of 5,10,15,20[pyrromellitoyltetrakis(o-(oxyethoxy)-phenyl)]porphyrin |
| Deut IX DME | dianion of Deuteroporphyrin IX dimethyl ester |
| DioxoOEP | dianion of 3,8-dioxo-2,2,7,7,12,13,17,18-octaethylporphyrin |
| DP-A | tetraanion of anthracene pillared diporphyrin |
| DP-B | tetraanion of biphenylene pillared diporphyrin |
| DPEP | dianion of Deoxophylloerythroetioporphyrin |
| DP-7 | tetraanion of hexyldiporphyrin-7 |
| Etio I | dianion of Etioporphyrin I |
| Formylporphyrin | 3,8-diformyl-13,17-di-n-pentyl-2,7,12,18-tetramethylporphyrin |
| FF | tetraanion of N,N'-Bis(5-(o-phenylen)-10,15,20-triphenylporphy-rin)urea |
| FTF | tetraanion of a bisporphyrin, face-to-face (6-3,2-NH-diamid)[216] |
| HHOEP | dianion of 2,3,7,8,12,13,17,18-octaethyl-1,2,3,7,8,20,-hexahydro-porphyrin |
| MeBPheo a | Methyl Bacteriopheophorbide a |
| 4-Bu-5-Et MeBPheo d | 4-isobutyl-5-ethyl methylbacteriopheophorbide d |
| 4,5-DiEt MeBPheo d | 4,5-diethyl methylbacteriopheophorbide d |
| H$_2$Meso IX DME | mesoporphyrin IX dimethyl ester |
| MePheo a | Methyl Pheophorbide a |
| Mg(EtChl a, b) | Ethyl Chlorophyllide a or b |
| Mg(MeChl a) | Methyl Chlorophyllide a |
| Mg(MePyrChl a) | Methyl Pyrochlorophyllide a |
| NO$_2$Proto IX DME | 3-vinyl-8-(E-2-nitrovinyl-1)deutero |
| N–CH$_3$TPP | monoanion of 21-Methyl-5,10,15,20-tetraphenylporphyrin |
| N–PhTPP | monoanion of 21-Phenyl-5,10,15,20-tetraphenylporphyrin |
| ODM | dianion of 5,15-dimethyl-2,3,7,8,12,13,17,18-octaethylporphyrin |
| (P–N$_4$) | dianion of 5,10,15,20-α,α,α,α-tetrakis(o-nicotinamidophenyl)-porphyrin |
| 5-NO$_2$OEP | dianion of 5-nitro-2,3,7,8,12,13,17,18-octaethylporphyrin |
| OEC | dianion of trans-7,8-Dihydro-2,3,7,8,12,13,17,18-octaethylpor-phyrin |
| OEP | dianion of 2,3,7,8,12,13,17,18-octaethylporphyrin |

OEPiBC — dianion of 2,3,7,8,12,13,17,18-octaethyl-2,3,12,13-tetrahydroporphyrin

OEPBC — dianion of 2,3,7,8,12,13,17,18-octaethyl-2,3,7,8-tetrahydroporphyrin

5-F-10-BBV-OEP — dianion of 2,3,7,8,12,13,17,18-octaethyl-5-formyl-10-[2,2-bis-(benzyloxycarbonyl)vinyl]-porphyrin

5-Bis(EtO)H₂OEP — 2,3,7,8,12,13,17,18-octaethyl-5-[2,2-bis(ethoxycarbonyl)vinyl]-porphyrin

21-EtO-HOEP — 21-Ethoxycarbonylmethyl-2,3,7,8,12,13,17,18-octaethylporphyrin

25-EtO-H₂OEP — 25-Ethoxycarbonyl-2,3,7,8,12,13,17,18-octaethyl-21,22-methano-21H, 22H-porphyrin

OEPMe₂ — dianion of 5,15-dimethyl,5,15-dihydro-2,3,7,8,12,13,17,18-octaethylporphyrin

OEPBu₂ — dianion of 5,15-dibutyl,5,15-dihydro-2,3,7,8,12,13,17,18-octaethylporphyrin

OxoOEP — dianion of 2-oxo-3,3,7,8,12,13,17,18-octaethylporphyrin

5-OxoProto IX DME — monoanion of 5-Oxoniaprotoporphyrin IX dimethyl ester

Pc — phthalocyanine dianion

PCOEP — dianion of 2,3,7,8,12,13,17,18-octaethyl-2,3,7,8,12,13-hexahydroporphyrin

Porph — dianion of a generalized porphyrin

Proto IX DME — dianion of protoporphyrin IX dimethyl ester

Siro-iBC — 3,7-Dimethyl-3,7-dihydro-2,2,8,8,12,13,17,18-octaethylporphyrin

TAP — dianion of 5,10,15,20-tetrakis(p-methoxyphenyl)porphyrin

TMiBC — dianion of 5,10,15,20-tetramethylisobacteriochlorin

TMC — dianion of 5,10,15,20-tetramethylchlorin

TMP — dianion of 5,10,15,20-tetramethylporphyrin

TMPyP — dianion of 5,10,15,20-tetrakis(N-methyl-4-pyridyl)porphyrin

TTOP — trianion of 5-(2-hydroxyphenyl)-10,15,20-tri-p-tolylporphyrin

Tp-ClPP — dianion of 5,10,15,20-tetrakis(p-chlorophenyl)porphyrin

TpivPP — dianion of 5,10,15,20-tetrakis(α,α,α,α-o-pivalamidophenyl)porphyrin

TPrP — dianion of 5,10,15,20-tetra-n-propylporphyrin

TPiBC — dianion of 5,10,15,20-tetraphenylisobacteriochlorin

TPC — dianion of 5,10,15,20-tetraphenylchlorin

TPP — dianion of 5,10,15,20-tetraphenylporphyrin

| TPyP | dianion of 5,10,15,20-tetra(4-pyridyl)porphyrin |
|------|------|
| TTP | dianion of 5,10,15,20-tetra-p-tolylporphyrin |
| TPP-C$_3$Py | dianion of 5-[2-[2-(3-pyridyl)ethylcarbonylamino]-1-phenyl]-10,15,20-triphenylporphyrin |
| TPP–C$_5$Im | dianion of 5-mono[2-(5-(N-imidazolyl)valeramido)-1-phenyl]-10,15,20-triphenylporphyrin |

## II. Abbreviations for Axial Ligands and Other Species

| Bcage | the $B_{11}CH_{12}^-$ anion |
|------|------|
| BzHIm | benzimidazole |
| 3-ClPy | 3-chloropyridine |
| Ct | center of the four nitrogen atoms of a tetrapyrrole |
| Ct′ | center of the 24-atom core of a tetrapyrrole |
| DMF | dimethylformamide |
| EPR | electron paramagnetic resonance |
| HIm | imidazole |
| Im$^-$ | imidazolate anion |
| L | generalized neutral axial ligand |
| 3,5-Lut | 3,5-dimethylpyridine |
| 2,6-LutNO, | 2,6-dimethylpyridine-N-oxide |
| 2-MeHIm | 2-methylimidazole |
| 1-MeIm | 1-methylimidazole |
| 1,2-Me$_2$Im | 1,2-dimethylimidazole |
| 4-MeIm$^-$ | 4-methylimidazolate anion |
| 4-MePip | 4-methylpiperidine |
| 3-MePy | 3-methylpyridine |
| 4-MePy | 4-methylpyridine |
| MNT | cis-1,2-dicyano-1,2-ethylenedithiolate |
| MU | methyl urocanate (cis or trans) |
| NMR | nuclear magnetic resonance |
| NOR | norbornenyl anion |
| OAc | acetate ion |
| Ph | phenyl |
| PhCO | benzoyl |
| Pip | piperidine |

| Py | pyridine |
|---|---|
| H$_2$Quin | para-hydroquinone |
| Ruf | Ruffled Core Conformation |
| Sad | Saddle-shaped Core Conformation |
| SC$_6$HF$_4$ | 2,3,5,6-tetrafluorobenzene thiolate |
| Sn(TPP)MnHgMn | the (TPP)Sn–Mn(CO)$_4$–Hg–Mn(CO)$_5$ molecule |
| SPh | phenyl thiolate |
| SPhNO$_2$ | p-nitrophenyl thiolate |
| TCNQ | 7,7,8,8-tetracyano-p-quinodimethane |
| THF | tetrahydrofuran |
| TMSO | tetramethylene sulfoxide |
| X | generalized anionic axial ligand |

*Acknowledgement.* We gratefully acknowledge the National Institutes of Health (Grant HL-15627) for support of porphyrin research at the University of Notre Dame.

# F. Notes Added in Proof

Elliot et al. (*Inorg. Chem. 25*, 1891 (1986)) have now provided complete physical details of the structure reported in Ref. 55. Complete details of the unusual iron structure reported in Ref. 118 are now available (*J. Am. Chem. Soc. 108*, 5598 (1986)). An unusual tungsten(VI) complex containing both an oxo and a peroxo ligand (cis) has been reported (Yang, C.-H., Dzugan, S. J., Goedken, V. L.: *J. Chem. Soc., Chem. Commun.* 1425 (1985)). The structure of a cobalt(0) complex ([Co(TPP)]$^{2-}$) has been reported (Ciurli, S., Gambarotta, S., Floriani, C., Chiesi-Villa, A., Guastini, C.: *Angew. Chem. Int. Ed. Eng. 25*, 553 (1986)) and is similar in all respects to the analogous iron(0) species. A tripledecker dilanthanide, containing two phthalocyanine rings and one porphyrin ring (Nd$_2$Pc$_2$TAP) has been reported (Moussavi, M., De Cian, A., Fischer, J., Weiss, R.: *Inorg. Chem. 25*, 2107 (1986)). A thorium porphyrin structure (Dormon, A., Belkalem, B., Charpin, P., Lance, D., Vigner, D., Folcher, G., Guilard, R.: *Inorg. Chem. 25*, 4785 (1986)) has also been reported. The copper(II) derivative of a potentially binucleating porphyrin has been reported recently (Larsen, N. G., Boyd, P. D. W., Rodgers, S. J., Wuenschell, G. E., Koch, C. A., Rasmussen, S., Tate, J. R., Erler, B. S., Reed, C. A.: *J. Am. Chem. Soc. 108*, 6950 (1986)). The recent report by Strouse et al.[308] of bis(substituted imidazole)iron(III) derivatives confirms results reported in the body of this review concerning axial ligand orientation effects. The crystal structure of a 5-azaporphyrin derivative (Abeysekera, A. M., Grigg, R., Malone, J. F., King, T. J., Morley, J. O.: *J. Chem. Soc., Perkin Trans. 2*, 395 (1985)) displays π-π interaction classified in this review as type S. The solid state structure of Cd(TPP) (Hazell, A.: *Acta Crystallogr., Sect. C C 42*, 296 (1986)) shows interesting core conformations and intermolecular interactions.

The dimer interaction in Cd(TPP) is quite unusual. The cadmium(II) ion is displaced 0.74 Å from the 24-atom plane and 0.58 Å from the four nitrogen plane. The core conformation shows the essential saddle conformation (Sad) expected for a dimer although it is somewhat less regular than most of the other species described earlier in this review. A most unexpected feature of the structure is the unusually close approach of the cadmium ion to an α-pyrrole carbon atom of the adjacent molecule with a Cd⋯C distance of 2.84 Å, a distance considerably shorter than the average mean plane separation of 3.64 Å. As expected, the dihedral angles between peripheral phenyls (three) and the mean plane of the core are less than 60°.

# G. References and Notes

1. Crute, M. B.: Acta Crystallogr. *12*, 24 (1959)
2. Hoard, J. L.: in "Porphyrins and Metalloporphyrins", (Smith, K. M., Ed.), Elsevier: Amsterdam, 1975, pp. 317–380
3. Scheidt, W. R.: in "The Porphyrins", (Dolphin, D., Ed.), Academic Press, Vol. III, 1978, pp. 463–511
4. Glusker, J. P.: in Vitamin $B_{12}$, (Dolphin, D., Ed.), John Wiley, Vol. 1, 1982, pp. 23–106
5. Hoard, J. L.: Science (Washington, D.C.) *174*, 1295 (1971)
6. Scheidt, W. R.: Acc. Chem. Res. *10*, 339 (1977)
7. Scheidt, W. R., Gouterman, M.: Article in "Iron Porphyrins, Part One", (Lever, A. B. P., Gray, H. B., Eds.), Addison-Wesley, Reading MA, 1983, pp. 89–139
8. Scheidt, W. R., Reed, C. A.: Chem. Rev. *81*, 543 (1981)
9. Hoffman, B. M., Ibers, J. A.: Acc. Chem. Res. *16*, 15 (1983)
10. Ibers, J. A., Pace, L. J., Martinsen, J., Hoffman, B. M.: Struct. Bond. *50*, 1 (1982)
11. Hoard, J. L.: Ann. N.Y. Acad. Sci. *206*, 18 (1973)
12. Scheidt, W. R., Haller, K. J., Hatano, K.: J. Am. Chem. Soc. *102*, 3017 (1980)
13. Quinn, R., Strouse, C. E., Valentine, J. S.: Inorg. Chem. *22*, 3934 (1983)
14. Scheidt, W. R., Lee, Y. J., Tamai, S., Hatano, K.: J. Am. Chem. Soc. *105*, 778 (1983)
15. Reed, C. A., Mashiko, T., Scheidt, W. R., Spartalian, K., Lang, G.: ibid. *102*, 2302 (1980)
16. Scheidt, W. R., Geiger, D. K., Lee, Y. J., Reed, C. A., Lang, G.: ibid. *107*, 5693 (1985)
17. Scheidt, W. R., Geiger, D. K., Lee, Y. J., Reed, C. A., Lang, G.: Inorg. Chem. *26*, 1039 (1987)
18. Shelly, K., Reed, C. A., Lee, Y. J., Scheidt, W. R.: to be submitted for publication
19. Reed, C. A., Mashiko, T., Bentley, S. P., Kastner, M. E., Scheidt, W. R., Spartalian, K., Lang, G.: J. Am. Chem. Soc. *101*, 2948 (1979)
20. Shelly, K., Reed, C. A., Lee, Y. J., Scheidt, W. R.: ibid. *108*, 3117 (1986)
21. Shelly, K., Bartzcak, T., Scheidt, W. R., Reed, C. A.: Inorg. Chem. *24*, 4325 (1985)
22. Hatano, K., Scheidt, W. R.: ibid. *18*, 877 (1979)
23. Masuda, H., Taga, T., Osaki, K., Sugimoto, H., Yoshida, Z.-I., Ogoshi, H.: Bull. Chem. Soc. Jpn. *55*, 3891 (1982)
24. Scheidt, W. R., Geiger, D. K., Hayes, R. G., Lang, G.: J. Am. Chem. Soc. *105*, 2625 (1983)
25. Any possible, subtle changes in the nature of the admixed intermediate-spin state are ignored here
26. Bloom, A., Hoard, J. L.: unpublished results
27. Scheidt, W. R., Lee, Y. J., Geiger, D. K., Taylor, K., Hatano, K.: J. Am. Chem. Soc. *104*, 3367 (1982)
28. Scheidt, W. R., Lee, Y. J., Luangdilok, W., Haller, K. J., Anzai, K., Hatano, K.: Inorg. Chem. *22*, 1516 (1983)
29. Adams, K. M., Rasmussen, P. G., Scheidt, W. R., Hatano, K.: ibid. *18*, 1892 (1979)
30. Scheidt, W. R., Lee, Y. J., Hatano, K.: J. Am. Chem. Soc. *106*, 3191 (1984)
31. Anzai, K., Hatano, K., Lee, Y. J., Scheidt, W. R.: Inorg. Chem. *20*, 2337 (1981)

32. Doppelt, P.: ibid. *23*, 4009 (1984)
33. English, D. R., Hendrickson, D. N., Suslick, K. S., Eigenbrot, C. W., Scheidt, W. R.: J. Am. Chem. Soc. *106*, 7258 (1984)
34. Byrn, M. P., Strouse, C. E.: ibid. *103*, 2633 (1981)
35. Miller, K. M., Strouse, C. E.: Acta Crystallogr., Sect. C *C40*, 1324 (1984)
36. Miller, K. M., Strouse, C. E.: Inorg. Chem. *23*, 2395 (1984)
37. Tang, S. C., Koch, S., Papaefthymiou, G. C., Foner, S., Frankel, R. B., Ibers, J. A., Holm, R. H.: J. Am. Chem. Soc. *98*, 2414 (1974)
38. This form appears to be the only phase obtained in other laboratories. See for example: Hill, H. A. O., Skyte, P. D., Buchler, J. W., Lueken, H., Tonn, M., Gregson, A. K., Pellizer, G.: J. Chem. Soc., Chem. Commun. 151 (1979)
    Gregson, A. K.: Inorg. Chem. *20*, 81 (1981)
39. Scheidt, W. R., Geiger, D. K., Haller, K. J.: J. Am. Chem. Soc. *104*, 495 (1982)
40. Scheidt, W. R., Geiger, D. K., Lee, Y. J., Reed, C. A., Lang, G.: Inorg. Chem. *26*, 1039 (1987)
41. Geiger, D. K., Lee, Y. J., Scheidt, W. R.: J. Am. Chem. Soc. *106*, 6339 (1984)
42. Levan, K. R., Strouse, C. E.: "Abstracts of Papers", American Crystallographic Association Summer Meeting, Snowmass, CO, Aug 1–5, 1983, Abstract H1
    Levan, K. R.: Ph.D. Thesis, UCLA, 1984
43. Geiger, D. K., Chunplang, V., Scheidt, W. R.: Inorg. Chem. *24*, 4736 (1985)
44. Gunter, M. J., McLaughlin, G. M., Berry, K. J., Murray, K. S., Irving, M., Clark, P. E.: ibid. *23*, 283 (1984)
45. Landrum, J. T., Grimmett, D., Haller, K. J., Scheidt, W. R., Reed, C. A.: J. Am. Chem. Soc. *103*, 2640 (1981)
46. Scheidt, W. R., Kenny, J. E., Lay, K.-L., Buchler, J. W.: Inorg. Chim. Acta *123*, 91 (1986)
47. Scheidt, W. R., Lee, Y. J., Bartzcak, T., Hatano, K.: Inorg. Chem. *23*, 2252 (1984)
48. Scheidt, W. R., Lee, Y. J., Finnegan, M. F.: to be submitted for publication
49. Oumous, H., Lecomte, C., Protas, J., Cocolios, P., Guilard, R.: Polyhedron *3*, 651 (1984)
50. Cocolios, P., Lagrange, G., Guilard, R., Oumous, H., Lecomte, C.: J. Chem. Soc., Dalton Trans. 567 (1984)
51. Ricard, L., Fischer, J., Weiss, R., Momenteau, M.: Nouv. J. Chim. *8*, 639 (1984)
52. Sabat, M., Ibers, J. A.: J. Am. Chem. Soc. *104*, 3715 (1982)
53. Buchler, J. W., Lay, K. L., Lee, Y. J., Scheidt, W. R.: Angew. Chem. *96*, 456 (1982), Angew. Chem., Int. Ed. Eng. *21*, 432 (1982)
54. Goff, H. M., Shimomura, E. T., Lee, Y. J., Scheidt, W. R.: Inorg. Chem. *23*, 315 (1984)
55. Schauer, C. K., Akabori, K., Elliott, C. M., Anderson, O. P.: J. Am. Chem. Soc. *106*, 1127 (1984)
56. Li, N., Petricek, V., Coppens, P., Landrum, J.: Acta Crystallogr., Sect. C *C41*, 902 (1985)
57. Mansuy, D., Battioni, P., Chottard, J.-C., Riche, C., Charioni, A.: J. Am. Chem. Soc. *105*, 455 (1983)
58. Mahy, J.-P., Battioni, P., Mansuy, D., Fischer, J., Weiss, R., Mispelter, J., Morgenstern-Badarau, I., Gans, P.: ibid. *106*, 1699 (1984)
59. Caron, C., Mitschler, A., Riviere, G., Ricard, L., Schappacher, M., Weiss, R.: ibid. *101*, 7401 (1979)
60. Jameson, G. B., Molinaro, F. S., Ibers, J. A., Collman, J. P., Brauman, J. I., Rose, E., Suslick, K. S.: ibid. *100*, 6769 (1978)
61. Schappacher, M., Ricard, L., Weiss, R., Montiel-Montoya, R., Gonser, U., Bill, E., Trautwein, A.: Inorg. Chim. Acta *78*, L9 (1983)
62. Strauss, S. H., Silver, M. E., Long, K. M., Thompson, R. G., Hudgens, R. A., Spartalian, K., Ibers, J. A.: J. Am. Chem. Soc. *107*, 4207 (1985)
63. Mashiko, T., Reed, C. A., Haller, K. J., Scheidt, W. R.: Inorg. Chem. *23*, 3192 (1984)
64. Doppelt, P., Fischer, J., Weiss, R.: ibid. *23*, 2958 (1984)
65. Collman, J. P., Gagne, R. R., Reed, C. A., Robinson, W., Rodley, G. A.: Proc. Natl. Acad. Sci. U.S.A. *71*, 1326 (1974)
66. Jameson, G. B., Rodley, G. A., Robinson, W. T., Gagne, R. R., Reed, C. A., Collman, J. P.: Inorg. Chem. *17*, 850 (1978)
67. Jameson, G. B., Molinaro, F. S., Ibers, J. A., Collman, J. P., Brauman, J. I., Rose, E., Suslick, K.: J. Am. Chem. Soc. *102*, 3224 (1980)

68. Ricard, L., Schappacher, M., Weiss, R., Montiel-Montoya, R., Bill, E., Gonser, U., Trautwein, A.: Nouv. J. Chim. *7*, 405 (1983)
69. Guilard, R., Latour, J.-M., Lecomte, C., Marchon, J. C., Protas, J., Ripoll, D.: Inorg. Chem. *17*, 1228 (1978)
70. Chevrier, B., Diebold, Th., Weiss, R.: Inorg. Chim. Acta *19*, L57 (1976)
71. Friant, P., Goulon, J., Fischer, J., Ricard, L., Schappacher, M., Weiss, R., Momenteau, M.: Nouv. J. Chim. *9*, 33 (1985)
72. (a) McCandlish, E., Valentine, J. S.: in Frontiers of Biological Energetics, Vol. 2, (Dutton, P. L., Leigh, J. S., Scarpa, A., Eds.), Academic Press, New York 1978, pp. 933–940
    (b) McCandlish, E., Mikostai, A. R., Nappa, M., Sprenger, A. Q., Valentine, J. S., Strong, J. D., Spiro, T. G.: J. Am. Chem. Soc. *102*, 4268 (1980)
73. Reed, C. A.: Adv. Chem. Ser. *201*, 333 (1982)
74. Welborn, H. C., Dolphin, D., James, B. R.: J. Am. Chem. Soc. *103*, 2869 (1981)
75. VanAtta, R. B., Strouse, C. E., Hanson, L. K., Valentine, J. S.: to be submitted for publication
76. Boreham, C. J., Buisson, G., Duee, E., Jordanov, J., Latour, J.-M., Marchon, J.-C.: Inorg. Chim. Acta *70*, 77 (1983)
77. Oumous, H., Lecomte, C., Protas, J., Poncet, J. L., Barbe, J. M., Guilard, R.: J. Chem. Soc., Dalton Trans. 2677 (1984)
78. Diebold, T., Chevrier, B., Weiss, R.: Inorg. Chem. *18*, 1193 (1979)
79. Colin, J., Butler, G., Weiss, R.: ibid. *19*, 3828 (1980)
80. Colin, J., Weiss, R.: Organometallics *4*, 1090 (1985)
81. De Cian, A., Colin, J., Schappacher, M., Richard, L., Weiss, R.: J. Am. Chem. Soc. *103*, 1850 (1980)
82. Diebold, T., Schappacher, M., Chevrier, B., Weiss, R.: J. Chem. Soc., Chem. Commun., 693 (1979)
83. Colin, J., Strich, A., Schappacher, M., Chevrier, B., Veillard, A., Weiss, R.: Nouv. J. Chim. *8*, 55 (1984)
84. Masuda, H., Taga, T., Osaki, K., Sugimoto, H., Mori, M., Ogoshi, H.: J. Am. Chem. Soc. *103*, 2199 (1981)
85. Collman, J. P., Barnes, C. E., Brothers, P. J., Collins, T. J., Ozawa, T., Gallucci, J. C.; Ibers, J. A.: ibid *106*, 5151 (1984)
86. Masuda, H., Taga, T., Osaki, K., Sugimoto, H., Mori, M., Ogoshi, H.: Bull. Chem. Soc. Jpn. *55*, 3887 (1982)
87. Masuda, H., Taga, T., Osaki, K., Sugimoto, H., Mori, M.: ibid. *57*, 2345 (1984)
88. Schardt, B. C., Hollander, F. J., Hill, C. L.: J. Am. Chem. Soc. *104*, 3964 (1982)
89. Scheidt, W. R., Summerville, D. A., Cohen, I. A.: ibid. *98*, 6623 (1976)
90. Goedken, V. L., Deakin, M. R., Bottomely, L. A.: J. Chem. Soc., Chem. Commun. 607 (1982)
91. Camenzind, M. J., Hollander, F. J., Hill, C. L.: Inorg. Chem. *21*, 4301 (1982)
92. Camenzind, M. J., Hollander, F. J., Hill, C. L.: ibid. *22*, 3776 (1983)
93. Mentzen, B. F., Bonnet, M. C., Ledon, H. J.: ibid. *19*, 2061 (1980)
94. Hill, C. L., Hollander, F. J.: J. Am. Chem. Soc. *104*, 7318 (1982)
95. Buchler, J. W., Dreher, C., Lay, K.-L., Lee, Y. J., Scheidt, W. R.: Inorg. Chem. *22*, 888 (1983)
96. (a) Groves, J. T., Kruper, W. J., Jr., Haushalter, R. C., Butler, W. M.: ibid. *21*, 1363 (1982)
    (b) Budge, J. R., Gatehouse, B. M. K., Nesbit, M. C., West, B. O.: J. Chem. Soc., Chem. Commun. 370 (1981)
97. Groves, J. T., Takahashi, T., Butler, W. M.: Inorg. Chem. *22*, 884 (1983)
98. Anderson, O. P., Lavallee, D. K.: J. Am. Chem. Soc. *99*, 1404 (1977)
99. Anderson, O. P., Lavallee, D. K.: Inorg. Chem. *16*, 1634 (1977)
100. Lavallee, D. K., Kopelove, A. B., Anderson, O. P.: J. Am. Chem. Soc. *100*, 3025 (1978)
101. Anderson, O. P., Kopelove, A. B., Lavallee, D. K.: Inorg. Chem. *19*, 2101 (1977)
102. Lavallee, D. K., Anderson, O. P.: J. Am. Chem. Soc. *104*, 4707 (1982)
103. Kuila, D., Lavallee, D. K., Schauer, C. K., Anderson, O. P.: ibid. *106*, 448 (1984)
104. Goldberg, D. E., Thomas, K. M.: ibid. *98*, 913 (1976)
105. McLaughlin, G. M.: J. Chem. Soc., Perkin Trans. II, 136 (1974)

106. Abeysekera, A. M., Grigg, R., Trocha-Grimshaw, J., Henrick, K.: Tetrahedron *36*, 1857 (1980)
107. Chevrier, B., Weiss, R.: J. Am. Chem. Soc. *98*, 2985 (1976)
108. Batten, P., Hamilton, A. L., Johnson, A. W., Mahendran, M., Ward, D., King, T. J.: J. Chem. Soc., Perkin Trans. I, 1623 (1977)
109. Chevrier, B., Weiss, R., Lange, M., Chottard, J.-C., Mansuy, D.: J. Am. Chem. Soc. *103*, 2899 (1981)
110. Olmstead, M. M., Cheng, R., Balch, A. L.: Inorg. Chem. *21*, 4143 (1982)
111. Callot, H. J., Chevrier, B., Weiss, R.: J. Am. Chem. Soc. *100*, 4733 (1978)
112. Balch, A. L., Chan, Y.-W., Olmstead, M. M.: ibid. *107*, 6510 (1985)
113. Balch, A. L., Chan, Y.-W., Olmstead, M. M., Renner, M. W.: ibid. *107*, 2393 (1985)
114. Callot, H. J., Fischer, J., Weiss, R.: ibid. *104*, 1272 (1982)
115. Bartczak, T. J.: Acta Crystallogr., Sect. C *C41*, 865 (1985)
116. Bartczak, T. J.: Acta Crystallogr., Sect. C *C41*, 604 (1985)
117. Callot, H. J., Chevrier, B., Weiss, R.: J. Am. Chem. Soc. *101*, 7729 (1979)
118. Mansuy, D., Battioni, J.-P., Akhrem, I., Dupre, D., Fischer, J., Weiss, R., Morgenstern-Badarau, I.: ibid. *106*, 6112 (1984)
119. Chan, Y.-W., Wood, F. E., Renner, M. W., Hape, H., Balch, A.: ibid. *106*, 3380 (1984)
120. Hirayama, N., Takenaka, A., Sasada, Y., Watanabe, E., Ogoshi, H., Yoshida, Z.: Bull. Chem. Soc. Jpn. *54*, 998 (1981)
121. Stevens, E. D.: J. Am. Chem. Soc. *103*, 5087 (1981)
122. Coppens, P., Li, L.: J. Chem. Phys. *81*, 1983 (1984)
123. Tanaka, K., Elkaim, E., Li, L., Jue, Z. N., Coppens, P., Landrum, J.: ibid., *84*, 6969 (1986)
124. Lecomte, C., Chadwick, D. L., Coppens, P., Stevens, E. D.: Inorg. Chem. *22*, 2982 (1983)
125. Lecomte, C., Blessing, R. H., Coppens, P., Tabard, A.: J. Am. Chem. Soc. *108*, 6942 (1986)
126. Barkigia, K. M., Spaulding, L. D., Fajer, J.: Inorg. Chem. *22*, 349 (1983)
127. Spaulding, L. D., Eller, P. G., Bertrand, J. A., Felton, R. H.: J. Am. Chem. Soc. *96*, 982 (1974)
128. Buisson, G., Deronzier, A., Duee, E., Gans, P., Marchon, J.-C., Regnard, J.-R.: ibid. *104*, 6793 (1982)
     Gans, P., Buisson, G., Duee, E., Marchon, J.-C., Erler, B. S., Scholz, W. F., Reed, C. A.: ibid. *108*, 1223 (1986)
129. Scholz, W. F., Reed, C. A., Lee, Y. J., Scheidt, W. R., Lang, G.: ibid. *104*, 6791 (1982)
130. Hoppe, W., Will, G., Gassmann, J., Weichselgartner, H.: Z. Kristallogr. *128*, 18 (1969)
131. Fisher, M. S., Templeton, D. H., Zalkin, A., Calvin, M.: J. Am. Chem. Soc. *94*, 3613 (1972)
132. Barkigia, K. M., Fajer, J., Chang, C. K., Williams, G. J. B.: ibid. *104*, 315 (1982)
133. Smith, K. M., Goff, D. A., Fajer, J., Barkigia, K. M.: ibid. *104*, 3747 (1982)
134. Barkigia, K. M., Fajer, J., Smith, K. M., Williams, G. J. B.: ibid. *103*, 5890 (1981)
135. Barkigia, K. M., Fajer, J., Chang, C. K., Young, R.: ibid. *106*, 6457 (1984)
136. Cruse, W. B. T., Harrison, P. J., Kennard, O.: ibid. *104*, 2376 (1982)
137. Kratky, H. P., Isenring, H. P., Dunitz, J. D.: Acta Crystallogr., Sect. B *B33*, 547 (1977)
138. Kratky, C., Dunitz, J. D.: Acta Crystallogr., Sect. B *B33*, 545 (1977)
139. Chow, H.-C., Serilin, R., Strouse, C. E.: J. Am. Chem. Soc. *97*, 7230 (1975)
140. Serlin, R., Chow, H.-C., Strouse, C. E.: ibid. *97*, 7237 (1975)
141. Spaulding, L. D., Andrews, L. C., Williams, G. J. B.: ibid. *99*, 6918 (1977)
142. Barkigia, K. M., Fajer, J., Spaulding, L. D., Williams, G. J. B.: ibid. *103*, 983 (1981)
143. Gallucci, J. C., Swepston, P. N., Ibers, J. A.: Acta Crystallogr., Sect. B *B38*, 2134 (1982)
144. Suh, M. P., Swepston, P. N., Ibers, J. A.: J. Am. Chem. Soc. *106*, 5164 (1984)
145. Kratky, C., Waditschatka, R., Angst, C., Johansen, J. E., Plaquevent, J. C., Schreiber, J., Eschenmoser, A.: Helv. Chim. Acta *68*, 1312 (1985)
146. Meyer, Jr., E. F.: Acta Crystallogr., Sect. B *B28*, 2162 (1972)
147. Hoard, J. L., Sayler, A.: unpublished results
148. Stolzenberg, A. M., Glazer, P. A., Foxman, B. M.: Inorg. Chem. *25*, 983 (1986)
149. Chang, C. K., Barkigia, K. M., Hanson, L. K., Fajer, J.: J. Am. Chem. Soc. *108*, 1352 (1986)
150. Collman, J. P., Barnes, C. E., Swepston, P. N., Ibers, J. A.: ibid. *106*, 3500 (1984)
151. Onaka, S., Kondo, Y., Toriumi, K., Ito, T.: Chem. Letts. 1605 (1980)
     Onaka, S., Kondo, Y., Yamashita, M., Tatematsu, Y., Kato, Y., Goto, M., Ito, T.: Inorg. Chem. *24*, 1070 (1985)

152. Barbe, J.-M., Guilard, R., Lecomte, C., Gerardin, R.: Polyhedron 3, 889 (1984)
153. Kato, S., Noda, I., Mizuta, M., Itoh, Y.: Angew. Chem. Int. Ed. Engl. 18, 82 (1979)
154. Johnson, J. F., Scheidt, W. R.: Inorg. Chem. 17, 1280 (1978)
155. Lecomte, C., Protas, J., Guilard, R., Fliniaux, R., Fournari, P.: J. Chem. Soc., Dalton Trans. 1306 (1979)
156. Lecomte, C., Protas, J., Richard, P., Barbe, J. M., Guilard, R.: ibid. 247 (1982)
157. Schauer, C. K., Anderson, O. P., Eaton, S. S., Eaton, G. R.: Inorg. Chem. 24, 4082 (1985)
158. Scheidt, W. R., Eigenbrot, C. W., Hatano, K.: to be submitted for publication
159. Buchler, J. W., De Cian, A., Fischer, J., Kihn-Botulinski, M., Paulus, H., Weiss, R.: J. Am. Chem. Soc. 108, 3652 (1986)
160. Buchler, J. W., Fischer, J., Weiss, R.: to be submitted for publication
161. Hitchcock, P. B.: J. Chem. Soc., Dalton Trans. 2127 (1983)
162. Sayer, P., Gouterman, M., Connell, C. R.: J. Am. Chem. Soc. 99, 1082 (1977)
163. Mangani, S., Meyer, E. F., Jr., Cullen, D. L., Tsutsui, M., Carrano, C. J.: Inorg. Chem. 22, 400 (1983)
164. Cullen, D. L., Meyer, E. F., Jr., Smith, K. M.: ibid. 16, 1179 (1977)
165. Henrick, K., Matthews, R. W., Tasker, P. A.: ibid. 16, 3293 (1977)
166. Brady, F., Henrick, K., Matthews, R. W.: J. Organomet. Chem. 210, 281 (1981)
167. Ball, R. G., Lee, K. M., Marshall, A. G., Trotter, J.: Inorg. Chem. 19, 1463 (1980)
168. Lecomte, C., Protas, J., Cocolios, P., Guilard, R.: Acta Crystallogr., Sect. B B36, 2769 (1980)
169. Cocolios, P., Fournari, P., Guilard, R., Lecomte, C., Protas, J., Boubel, J. C.: J. Chem. Soc., Dalton Trans. 2081 (1980)
170. Rodesiler, P. F., Griffith, E. A. H., Charles, N. G., Lebioda, L., Amma, E. L.: Inorg. Chem. 24, 4595 (1985)
171. Rodesiler, P. F., Griffith, E. A. H., Ellis, P. D., Amma, E. L.: J. Chem. Soc., Chem. Commun. 492 (1980)
172. Kirner, J. F., Reed, C. A., Scheidt, W. R.: J. Am. Chem. Soc. 99, 1093 (1977)
173. McKee, V., Ong, C. C., Rodley, G. A.: Inorg. Chem. 23, 4242 (1984)
174. Bonnett, R., Hursthouse, M. B., Malik, K. M. A., Mateen, B.: J. Chem. Soc., Perkin Trans. 2, 2072 (1977)
175. Scheidt, W. R., Hatano, K., Rupprecht, G. A., Piciulo, P. L.: Inorg. Chem. 18, 292 (1979)
176. Landrum, J. T., Hatano, K., Scheidt, W. R., Reed, C. A.: J. Am. Chem. Soc. 102, 6729 (1980)
177. Hatano, K., Anzai, K., Iitaka, Y.: Bull. Chem. Soc. Jpn. 56, 422 (1983)
178. Hill, C. L., Williamson, M. M.: Inorg. Chem. 24, 2836 (1985)
179. Hill, C. L., Williamson, M. M.: ibid. 24, 3024 (1985)
180. Wayland, B. B., Woods, B. A., Pierce, R.: J. Am. Chem. Soc. 104, 302 (1982)
181. Grigg, R., Trocha-Grimshaw, J., Henrick, K.: Acta Crystallogr., Sect. B B38, 2455 (1982)
182. Ball, R. G., Domazetis, G., Dolphin, D., James, B. R., Trotter, J.: Inorg. Chem. 20, 1556 (1981)
183. James, B. R., Dolphin, D., Leung, T. W., Einstein, F. W. B., Willis, A. C.: Can. J. Chem. 62, 1238 (1984)
184. Masuda, H., Taga, T., Sugimoto, H., Mori, M.: J. Organomet. Chem. 273, 385 (1984)
185. Michalowicz, A., Huet, J., Gaudemer, A.: Nouv. J. Chim. 6, 79 (1982)
186. Frisse, M., Scheidt, W. R.: unpublished
187. Masuda, H., Taga, T., Osaki, K., Sugimoto, H., Mori, M.: Bull. Chem. Soc. Jpn. 55, 4 (1982)
188. Doppelt, P., Fischer, J., Weiss, R.: J. Am. Chem. Soc. 106, 5188 (1984)
189. Gupta, G. P., Lang, G., Scheidt, W. R., Geiger, D. K., Reed, C. A.: J. Chem. Phys. 83, 5945 (1985)
190. Drew, M. G. B., Mitchell, P. C. H., Scott, C. E.: Inorg. Chim. Acta 82, 63 (1984)
191. Abraham, R. J., Burbridge, P. A., Jackson, A. H., MacDonald, D. B.: J. Chem. Soc. (B) 620 (1966)
     Abraham, R. J., Evans, B., Smith, K. M.: Tetrahedron 34, 1213 (1978)
     Snyder, R. V., LaMar, G. N.: J. Am. Chem. Soc. 99, 7178 (1977)
192. Blumberg, W. E., Peisach, J.: J. Biol. Chem. 240, 870 (1965)
     Chikira, M., Kon, H., Hawley, R. A., Smith, K. M.: J. Chem. Soc., Dalton Trans. 245 (1979)
     MacCragh, A., Storm, C. B., Koski, W. S.: J. Am. Chem. Soc. 87, 1470 (1965)

Boas, J. F., Pilbrow, J. R., Smith, T. D.: J. Chem. Soc., (A) 721 (1969)
Boyd, P. D. W., Smith, T. D., Price, J. H., Pilbrow, J. R.: J. Chem. Phys. *56,* 1253 (1972)
193. Fillers, J. P., Ravichandran, K. G., Abdalmuhdi, I., Tulinsky, A., Chang, C. K.: J. Am. Chem. Soc. *108,* 417 (1986)
194. Scheidt, W. R., Mondal, J. U., Eigenbrot, C. W.: to be submitted for publication
195. Deisenhofer, J., Epp, O., Miki, K., Huber, R., Michel, H.: J. Mol. Biol. *180,* 385 (1984)
Zinth, W., Knapp, E. W., Fischer, S. F., Kaiser, W., Deisenhofer, J., Michel, W.: Chem. Phys. Letters *119,* 1 (1985)
196. A complete geometric description of the two rings would require that the lateral shift be broken into two translational and one rotational component
197. The esd gives a measure of the dispersion in the values of the sample
198. Brennan, T., Shelnutt, J. A., Scheidt, W. R.: work in progress. This is the third crystalline modification of Ni(OEP) to be obtained and studied
199. Kutzler, F. W., Swepston, P. N., Berkovitch-Yellin, Z., Ellis, D. E., Ibers, J. A.: J. Am. Chem. Soc. *105,* 2996 (1983)
200. Scheidt, W. R., Kastner, M. E., Hatano, K.: Inorg. Chem. *17,* 706 (1978)
201. Scheidt, W. R., Reed, C. A.: ibid. *17,* 710 (1978)
202. Pace, L. J., Ulman, A., Ibers, J. A.: ibid. *21,* 199 (1982)
203. Grigg, R., Trocha-Grimshaw, J., King, T. J.: J. Chem. Soc., Chem. Commun. 571 (1978)
204. Chang, C. K., Hatada, M. H., Tulinsky, A.: J. Chem. Soc., Perkin Trans. II, 371 (1983)
205. Little, R. G., Ibers, J. A.: J. Am. Chem. Soc. *97,* 5363 (1975)
206. Bonnett, R., Hursthouse, M. B., Scourides, P. A., Trotter, J.: J. Chem. Soc., Perkin Trans. I, 490 (1980)
207. Lauher, J. W., Ibers, J. A.: J. Am. Chem. Soc. *95,* 5148 (1973)
208. Sheldrick, W. S.: Acta Crystallogr., Sect. B *B33,* 3967 (1977)
209. Webb, L. E., Fleischer, E. B.: J. Chem. Phys. *43,* 3100 (1965)
Chen, B. M. L., Tulinsky, A.: J. Am. Chem. Soc. *94,* 4144 (1972)
210. Caughey, W. S., Ibers, J. A.: ibid. *99,* 6639 (1977)
211. Codding, P. W., Tulinsky, A.: ibid. *94,* 4151 (1972)
212. Little, R. G., Ibers, J. A.: ibid. *96,* 4452 (1974)
213. Sheldrick, W. S.: Acta Crystallogr., Sect. B *B34,* 663 (1978)
214. Moustakali, I., Tulinsky, A.: J. Am. Chem. Soc. *95,* 6811 (1973)
215. Hatada, M. H., Tulinsky, A., Chang, C. K.: ibid. *102,* 7115 (1980)
216. Collman, J. P., Chong, A. O., Jameson, G. B., Oakley, R. T., Rose, E., Schmittou, E. R., Ibers, J. A.: ibid. *103,* 516 (1981)
217. Koenig, D. F.: Acta Crystallogr. *18,* 663 (1965)
218. Hoard, J. L., Hamor, M. J., Hamor, T. A., Caughey, W. S.: J. Am. Chem. Soc. *87,* 2312 (1965)
219. Miller, K. M., Strouse, C. E.: Acta Crystallogr., Sect. C *40,* 1324 (1984)
220. Scheidt, W. R., Geiger, D. K.: Inorg. Chem. *21,* 1208 (1982)
221. Masuda, H., Taga, T., Osaki, K., Sugimoto, H., Yoshida, Z.-I.: ibid. *19,* 950 (1980)
222. Lecomte, C., Protas, J., Richard, P., Barbe, J. M., Guilard, R.: J. Chem. Soc., Dalton Trans. 247 (1982)
223. Hamor, T. A., Caughey, W. S., Hoard, J. L.: J. Am. Chem. Soc. *87,* 2305 (1965)
224. Pettersen, R. C.: ibid. *93,* 5629 (1971)
225. Cullen, D. L., Meyer, Jr., E. F.: ibid. *96,* 2095 (1974)
226. Barkigia, K. M., Fajer, J., Adler, A. D., Williams, G. J. B.: Inorg. Chem. *19,* 2057 (1980)
227. Grigg, R., Trocha-Grimshaw, J.: Acta Crystallogr., Sect. B *B38,* 2455 (1982)
228. Takenaka, A., Syal, S. K., Sasada, Y., Omura, T., Ogoshi, H., Yoshida Z.-I.: Acta Crystallogr., Sect. B *B32,* 62 (1976)
229. Pettersen, R. C.: Acta Crystallogr., Sect. B *B25,* 2527 (1969)
230. Molinaro, F. S., Ibers, J. A.: Inorg. Chem. *15,* 2278 (1976)
231. Fuhrhop, J.-H., Krüger, P., Sheldrick, W. S.: Liebigs Ann. Chem., 339 (1977)
232. Scheidt, W. R., Eigenbrot, C. W., Smith, K. M.: to be submitted for publication
233. Hirayama, N., Takenaka, A., Sasada, Y., Watanabe, E., Ogoshi, H., Yoshida, Z. I.: Bull. Chem. Soc. Jpn. *54,* 998 (1981)
234. Hrung, C. P., Tsutsui, M., Cullen, D. L., Meyer, E. F., Jr., Morimoto, C. N.: J. Am. Chem. Soc. *100,* 6068 (1978)

235. Collins, D. M., Countryman, R., Hoard, J. L.: ibid. *94*, 2066 (1972)
236. Walker, F. A., Huynh, B. H., Scheidt, W. R., Osvath, S. R.: ibid. *108*, 5288 (1986)
237. Scheidt, W. R., Osvath, S. R., Lee, Y. J.: ibid. *109*, 1958 (1987)
238. Kirner, J. F., Hoard, J. L., Reed, C. A.: "Abstracts of Papers", 175th National Meeting of the American Chemical Society, Anaheim, CA March 13–17, 1978, American Chemical Society, Washington, D.C. 1978, INOR 14.
   Scheidt, W. R., Kirner, J. F., Hoard, J. L., Reed, C. A.: J. Am. Chem. Soc. *109*, 1963 (1987)
239. Scheidt, W. R., Chipman, D. M.: J. Am. Chem. Soc. *108*, 1163 (1986)
240. Steffen, W. L., Chun, H. K., Hoard, J. L., Reed, C. A.: "Abstracts of Papers", 175th National Meeting of the American Chemical Society, Anaheim, CA March 13–17, 1978, American Chemical Society, Washington, D.C. 1978, INOR 15
241. Kirner, J. F., Reed, C. A., Scheidt, W. R.: J. Am. Chem. Soc. *99*, 2557 (1977)
242. Little, R. G., Dymock, K. R., Ibers, J. A.: ibid. *97*, 4532 (1975)
243. Piciulo, P. L., Scheidt, W. R.: ibid. *98*, 1913 (1976)
244. Lauher, J. W., Ibers, J. A.: ibid. *96*, 4447 (1974)
245. Scheidt, W. R.: ibid. *96*, 90 (1974)
246. Little, R. G., Ibers, J. A.: ibid. *96*, 4452 (1974)
247. Dwyer, P. N., Madura, P., Scheidt, W. R.: ibid. *96*, 4815 (1974)
248. Kirner, J. F., Garofalo, J., Jr., Scheidt, W. R.: Inorg. and Nucl. Chem. Letts. *11*, 107 (1975)
249. Kirner, J. F., Scheidt, W. R.: Inorg. Chem. *14*, 2081 (1975)
250. Scheidt, W. R., Brinegar, A. C., Kirner, J. F., Reed, C. A.: ibid. *18*, 3610 (1979)
251. Peng, S.-M., Ibers, J. A.: J. Am. Chem. Soc. *98*, 8032 (1976)
252. Hopf, F. R., O'Brien, T. P., Scheidt, W. R., Whitten, D. G.: ibid. *97*, 277 (1975)
253. Buchler, J. W., Lay, K. L., Smith, P. D., Scheidt, W. R., Rupprecht, G. A., Kenny, J. A.: J. Organomet. Chem. *110*, 109 (1976)
254. Kaduk, J. A., Scheidt, W. R.: Inorg. Chem. *13*, 1875 (1974)
255. Sakurai, T., Yamamoto, K., Seino, N., Katsuta, M.: Acta Crystallogr., Sect. B *B31*, 2514 (1975)
256. Riche, C., Chiarnoi, A., Fauvet-Perree, M., Gaudemer, A.: Acta Crystallogr., Sect. B *B34*, 1868 (1978)
257. Scheidt, W. R., Ramanuja, J. A.: Inorg. Chem. *14*, 2643 (1975)
258. Little, R. G., Ibers, J. A.: J. Am. Chem. Soc. *96*, 4440 (1974)
259. Collins, D. M., Hoard, J. L.: ibid. *92*, 3761 (1970)
260. Cullen, D. L., Meyer, E. F., Jr.: Acta Crystallogr., Sect. B *B32*, 2259 (1976)
261. Mashiko, T., Reed, C. A., Kastner, M. E., Haller, K. J., Scheidt, W. R.: J. Am. Chem. Soc. *103*, 5758 (1981)
262. Bobrik, M. A., Walker, F. A.: Inorg. Chem. *19*, 3383 (1980)
263. Little, R. G., Ibers, J. A.: J. Am. Chem. Soc. *95*, 8573 (1973)
264. Barkigia, K. M., Fajer, J., Adler, A. D., Williams, G. J. B.: Inorg. Chem. *19*, 2057 (1980)
265. Wijesekera, T. P., Paine, J. B., III, Dolphin, D., Einstein, F. W. B., Jones, T.: J. Am. Chem. Soc. *105*, 6747 (1983)
266. Timkovich, R., Tulinsky, A.: ibid. *91*, 4430 (1969)
267. Lecomte, C., Protas, J., Marchon, J. C., Nakajima, M.: Acta Crystallogr., Sect. B *B34*, 2856 (1978)
268. Hoard, J. L., Cohen, G. H., Glick, M. D.: J. Am. Chem. Soc. *89*, 1992 (1967)
269. Mashiko, T., Kastner, M. E., Spartalian, K., Scheidt, W. R., Reed, C. A.: ibid. *100*, 6354 (1978)
270. Skelton, B. W., White, A. H.: Aust. J. Chem. *30*, 2655 (1977)
271. Summerville, D. A., Cohen, I. A., Hatano, K., Scheidt, W. R.: Inorg. Chem. *17*, 2906 (1978)
272. Scheidt, W. R., Cohen, I. A., Kastner, M. E.: Biochemistry *18*, 3546 (1979)
273. Phillippi, M. A., Baenziger, N., Goff, H. M.: Inorg. Chem. *20*, 3904 (1981)
274. Radonovich, L. J., Bloom, A., Hoard, J. L.: J. Am. Chem. Soc. *94*, 2073 (1972)
275. Scheidt, W. R., Frisse, M. E.: ibid. *97*, 17 (1975)
276. Scheidt, W. R., Brinegar, A. C., Ferro, E. B., Kirner, J. F.: ibid. *99*, 7315 (1977)
277. Jameson, G. B., Robinson, W. R., Collman, J. P., Sorrel, T. N.: Inorg. Chem. *17*, 858 (1978)
278. Jameson, G. B., Ibers, J. A.: ibid. *18*, 1200 (1979)
279. Scheidt, W. R.: J. Am. Chem. Soc. *96*, 84 (1974)
280. Sakurai, T., Yamamoto, K., Naito, H., Nakamoto, N.: Bull. Chem. Soc. Jpn. *49*, 3042 (1976)

281. Collins, D. M., Scheidt, W. R., Hoard, J. L.: J. Am. Chem. Soc. *94*, 6689 (1972)
282. Harrison, P. G., Molloy, K., Thornton, E. W.: Inorg. Chim. Acta *33*, 137 (1979)
283. Scheidt, W. R., Mondal, J. U., Eigenbrot, C. W., Adler, A., Radonovich, L. J., Hoard, J. L.:
     Inorg. Chem. *25*, 795 (1986)
284. Glick, M. D., Cohen, G. H., Hoard, J. L.: J. Am. Chem. Soc. *89*, 1996 (1967)
285. Tsutsui, M., Hrung, C. P., Ostfeld, D., Srivastava, T. S., Cullen, D. L., Meyer, E. F., Jr.:
     ibid. *97*, 3952 (1975)
286. Little, R. G., Ibers, J. A.: ibid. *95*, 8583 (1973)
287. Fleischer, E. B., Miller, C. K., Webb, L. E.: ibid. *86*, 2342 (1964)
288. Hazell, A.: Acta Crystallogr., Sect. C *C40*, 751 (1984)
289. Collman, J. P., Hoard, J. L., Kim, N., Lang, G., Reed, C. A.: J. Am. Chem. Soc. *97*, 2676
     (1975)
290. Madura, P., Scheidt, W. R.: Inorg. Chem. *15*, 3182 (1976)
291. Stone, A. B., Fleischer, E. B.: J. Am. Chem. Soc. *90*, 2735 (1968)
292. Navaza, A., de Rango, C., Charpin, P.: Acta Crystallogr., Sect. C *C39*, 1625 (1983)
293. Hoffman, A. B., Collins, D. M., Day, V. W., Fleischer, E. B., Srivastava, T. S., Hoard, J. L.:
     J. Am. Chem. Soc. *94*, 3620 (1972)
294. Timkovich, R., Tulinsky, A.: Inorg. Chem. *16*, 962 (1977)
295. Tulinsky, A., Chen, B. M. L.: J. Am. Chem. Soc. *99*, 3647 (1977)
296. Hanson, L. K., Kabuto, C., Silverton, J. V., Kon, H.: to be submitted for publication
297. Madura, P., Scheidt, W. R.: unpublished results
298. Schneider, M. L.: J. Chem. Soc., Dalton Trans. 1093 (1972)
299. Scheidt, W. R., Hoard, J. L.: J. Am. Chem. Soc. *95*, 8281 (1973)
300. Sakurai, T., Yamamoto, K.: Sci. Pap. Inst. Phys. Chem. Res. *70*, 31 (1976)
301. Ledon, H., Mentzen, B.: Inorg. Chim. Acta *31*, L393 (1978)
302. Lee, Y. J., Scheidt, W. R., Marchon, J. C.: unpublished
303. Gans, P., Buisson, G., Duee, E., Regnard, J.-R., Marchon, J. C.: J. Chem. Soc., Chem.
     Commun. 393 (1979)
304. Lee, Y. J., Geiger, D. K., Scheidt, W. R., Marchon, J. C., Strouse, C. E.: unpublished results
305. Haller, K. J., Luangdilok, W., Scheidt, W. R., Mashiko, T., Scholz, W. F., Reed, C. A.:
     unpublished results
306. Silver, S. J., Tulinsky, A.: J. Am. Chem. Soc. *89*, 3331 (1967)
307. Strauss, S. H.: personal communication
308. Quinn, R., Valentine, J. S., Byrn, M. P., Strouse, C. E.: J. Am. Chem. Soc., in press
309. Brennan, T. D., Scheidt, W. R.: to be submitted for publication

# Infrared and Raman Spectra of Metalloporphyrins

**Teizo Kitagawa[1] and Yukihiro Ozaki[2]**

[1] Institute for Molecular Science, Okazaki National Research Institutes, Myodaiji, Okazaki 444, Japan
[2] The Jikei University School of Medicine, Minato-ku, Tokyo 105, Japan

Infrared and Raman spectra of metalloporphyrins, metallochlorins, and metallophthalocyanines are comprehensively reviewed. The review starts from general explanations of basic properties of resonance Raman (RR) spectra of these compounds. Vibrational assignments of IR and RR bands of symmetric metalloporphyrins and their application to asymmetric protoporphyrin IX are interpreted on the basis of the isotopic shift data observed for specifically isotope-labeled porphyrins and of the normal coordinate calculations. For iron porphyrins, empirical rules with regard to the relation between the RR spectra and the spin-, oxidation-, and coordination states are explained. Fe$^I$- and Fe$^{IV}$-porphyrins and porphyrin cation-radical states are also included. The internal modes of an axial ligand (L) coordinated to the metal ion (M) as well as the M–L stretching modes and their difference between organic solution and protein matrix are described. Based on these data, differences and similarities in the M–L bonds among the three types of metal complexes of the conjugated macrocycles are discussed.

Structure and Bonding 64
© Springer-Verlag Berlin Heidelberg 1987

# 1 Introduction

The biological and chemical importance of metalloporphyrins leads our interest to the nature of the metal ligand linkages in the complexes as well as to physicochemical properties of the macrocycle. The macrocycle may function as a reservoir of electrons and control a reactivity at the axial position of metal, which usually serves as a catalytic site in heme enzymes. Since vibrational spectroscopy is one of the most powerful tools to gain insight into the nature of chemical bonds and geometrical structure of molecules, infrared (IR) and Raman spectra of metalloporphyrins have been extensively investigated. Particularly after the pioneering work on the resonance Raman (RR) spectra of hemoglobin (Hb) by Spiro and Strekas[1, 2], and Brunner et al.[3] and of chlorophyll by Lutz[4], this technique has held such general attention that hitherto a number of papers have been published regarding heme proteins and related compounds. Since RR spectra of the biological systems have conveyed unique and important information about the structure-function relationship, they have been occasionally reviewed[5-12]. Therefore, we exclude the biological systems from this paper except for the Fe-axial ligand mode. Here we try to review the vibrational spectra of metalloporphyrins, metallochlorins, and metallophthalocyanines. These spectra provide a basis for the analysis of RR spectra of biological systems. Previously infrared spectra of metalloporphyrins were reviewed by Bürger[13] and comprehensive reviews on vibrational spectra of metalloporphyrins up to 1983 were published by Spiro[14], and Solovyov et al.[15] but in this article the topics up to 1986 besides the fundamental observations will be covered.

# 2 General Considerations on Vibrational Spectroscopy

## 2.1 IR and Raman Spectra

The energy of vibrational transitions is roughly one tenth of the energy of electronic transitions. Supposed that $|m\rangle$ and $|n\rangle$ in Fig. 1 are two quantum states of a given normal vibration in the ground electronic state and $|e\rangle$ is an electronically excited state, IR

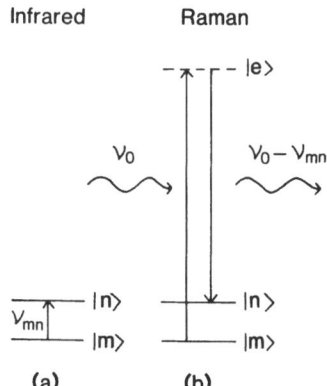

**Fig. 1a, b.** Molecular processes of infrared absorption (a) and Raman scattering (b). $|m\rangle$ and $|n\rangle$ are two quantum states of a given vibration in the electronic ground state, and $|e\rangle$ is an electronically excited state, which can be a virtual state

absorption arises from a direct transition from $|m\rangle$ to $|n\rangle$, but Raman scattering involves two simultaneous transitions, from $|m\rangle$ to $|e\rangle$ and from $|e\rangle$ to $|n\rangle$. If the two transitions were independent, each would correspond to light absorption and emission. Both IR and Raman techniques reveal the energy separation between $|m\rangle$ and $|n\rangle$ and thus the vibrational frequency in the electronic ground state. However, the selection rule is different between the two methods. When the transition from $|m\rangle$ to $|n\rangle$ is accompanied by a change of a molecular dipole moment, the vibration is infrared active, and when it is accompanied by a change of molecular polarizability, the vibration is Raman active. For molecules having inversion symmetry the infrared and Raman active modes should be antisymmetric and symmetric to a center of symmetry, respectively. Accordingly, the two techniques are complementary to each other. Even if there is no inversion symmetry on the molecule, vibrations which give rise to strong IR absorption or Raman scattering are generally different. Hence measurements of both kinds of spectra are desirable for the vibrational analysis.

## 2.2 Choice of IR and Raman Techniques

It may be helpful to point out here some advantages and disadvantages of RR and IR spectroscopies for studying metalloporphyrins. Advantages of RR spectroscopy are as follows; 1) It can be applied not only to metalloporphyrins but also to heme proteins. 2) Dilute solutions with concentrations between $10^{-4}$–$10^{-6}$ M suffice to provide good spectra. 3) Information about the electronically excited states can be obtained from the excitation profile. 4) Generally only the vibrations associated with the porphyrin skeleton are resonance enhanced and therefore the spectrum is not disturbed by the bands of peripheral substituents. Disadvantages encountered in RR spectroscopy are; 1) Fluorescence, if any, seriously interferes with the measurements of the spectrum. 2) Vibrations of axial ligands do not always appear.

Advantages of IR spectroscopy are as follows: 1) Fluorescence and undesirable photochemical effects by laser light need not to be bothered about. 2) Metal-ligand stretching and internal modes of the ligand always appear. Disadvantages are: 1) The strong IR absorption of water prevents the measurements of IR spectra of aqueous solutions. 2) A number of modes arising from the peripheral groups often complicate the spectrum and its interpretation.

## 2.3 Resonance Raman Spectra

For a general Raman scattering illustrated in Fig. 1(b), $|e\rangle$ can be a virtual state. The intensity of the scattered radiation, $I_{mn}$, due to the transition from $|m\rangle$ to $|n\rangle$ is represented by[16]

$$I_{mn} = \frac{128\,\pi^5}{9\,c^4}\,(\nu_0 \pm \nu_{mn})^4 I_0 \sum_{\varrho\sigma} |(\alpha_{\varrho\sigma})_{mn}|^2 \tag{1}$$

$$(\alpha_{\varrho\sigma})_{mn} = \frac{1}{h} \sum_e \left[ \frac{\langle n|\mu_\sigma|e\rangle\langle e|\mu_\varrho|m\rangle}{\nu_{em} - \nu_0 + i\Gamma_e} + \frac{\langle n|\mu_\sigma|e\rangle\langle e|\mu_\varrho|m\rangle}{\nu_{en} + \nu_0 + i\Gamma_e} \right] \tag{2}$$

where $\nu_0$ and $I_0$ are the frequency and intensity of the exciting light, respectively, $\nu_{mn}$ is the frequency of the vibration involved in the Raman scattering, $\alpha_{\varrho\sigma}$ ($\varrho$, $\sigma$ = x, y, z) is the $\varrho\sigma$ component of the polarizability tensor, and $\mu_\varrho$ is the $\varrho$ component of a dipole operator.

When $\langle e|\mu|m\rangle$ is finite for a real state, $|e\rangle$, the transition from $|m\rangle$ to $|e\rangle$ is accompanied by light absorption which would have the center frequency of $\nu_{em}$ and the full-width at the half-maximum of $2\,\Gamma_e$, and its maximum intensity would be proportional to $|\langle e|\mu|m\rangle|^2$. If the photon energy of Raman excitation ($= h\nu_0$) were close to the energy of the electronic transition ($= h\nu_{em}$), the first term of $(\alpha_{\varrho\sigma})_{mn}$ and thus $I_{mn}$ would be remarkably large. This is called resonance Raman scattering. In other words, when the excitation wavelength of Raman scattering comes into the absorption band (Fig. 2), RR scattering takes place. The modes which gain resonance enhancement of Raman intensity are limited to the vibrations involved in the chromophore of the absorption band. This feature is explained with Albrecht's formulation[16] below.

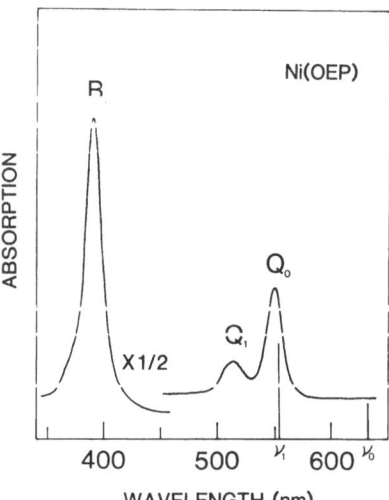

**Fig. 2.** Visible absorption spectrum of octaethyl-porphyrinatonickel(II), Ni(OEP), in $CH_2Cl_2$

Suppose that the electronic and vibrational wave functions are separated and represented as $|m\rangle = |g, i\rangle$, $|n\rangle = |g, j\rangle$ and $|e\rangle = |r, v\rangle$, where $|g\rangle$ and $|r\rangle$ are the electronic wave functions for the electronic ground and excited states, respectively, and $|i\rangle$, $|j\rangle$ and $|v\rangle$ are the vibrational wave functions for the quantum numbers of i, j, and v, respectively. Here it is postulated that the vibrational wave functions in the electronic excited state are the same as those in the electronic ground state except for the origin of the coordinate, that is, the equilibrium structure of molecules, which differs between the two states. The change of the electronic wave function due to the origin shift is incorporated with the Herzberg-Teller expansion. Then the main terms contributing to the polarizability tensor of Eq. 2 under resonance with a particular electronic excited state, $|r\rangle$, are the following two terms

$$(\alpha_{\varrho\sigma})^r_{ij} = (A_{\varrho\sigma})^r_{ij} + (B_{\varrho\sigma})^r_{ij} \tag{3}$$

$$(A_{\varrho\sigma})^r_{ij} = \frac{M^{gr}_\sigma M^{gr}_\varrho}{h} \sum_v \frac{\langle j|v\rangle\langle v|i\rangle}{\nu_{rg} + (v - i)\Delta\nu_a - \nu_0 + i\Gamma_r} \tag{4}$$

$$(B_{\varrho\sigma})^r_{ij} = \frac{M_\sigma^{gr} h_a^{rs} M_\varrho^{sg}}{h^2(v_r - v_s)} \sum_v \frac{\langle j|Q_a|v\rangle\langle v|i\rangle}{v_{rg} + (v - i)\Delta v_a - v_0 + i\Gamma_r}$$

$$+ \frac{M_\sigma^{gs} h_a^{sr} M_\varrho^{rg}}{h^2(v_r - v_s)} \sum_v \frac{\langle j|v\rangle\langle v|Q_a|i\rangle}{v_{rg} + (v - i)\Delta v_a - v_0 + i\Gamma_r} \qquad (5)$$

where $M_\sigma^{gr} = \langle g^0|\mu_\sigma|r^0\rangle$, $hv_{rg}$ is the energy separation between $|r\rangle$ and $|g\rangle$, $\Delta v_a$ is the frequency of the normal coordinate, $q_a$, and $h_a$ is a vibronic coupling operator for a normal mode $q_a$ (= $\partial H/\partial q_a$).

The A term contains the Franck-Condon factor, $\langle j|v\rangle\langle v|i\rangle$, which depends on the magnitude of the origin shift. For non-totally symmetric modes, the origin shift upon electronic excitations is zero. Therefore, the A term arises only from the totally symmetric vibrations. When atomic displacements in a given vibrational mode are along the molecular deformation produced by the electronic excitation, such vibration gains strong enhancement of Raman intensity through the A term. On the other hand, the B term is important for non-totally symmetric modes. When a given vibration helps mixing of two electronic states, $|r\rangle$ and $|s\rangle$, the B term becomes important.

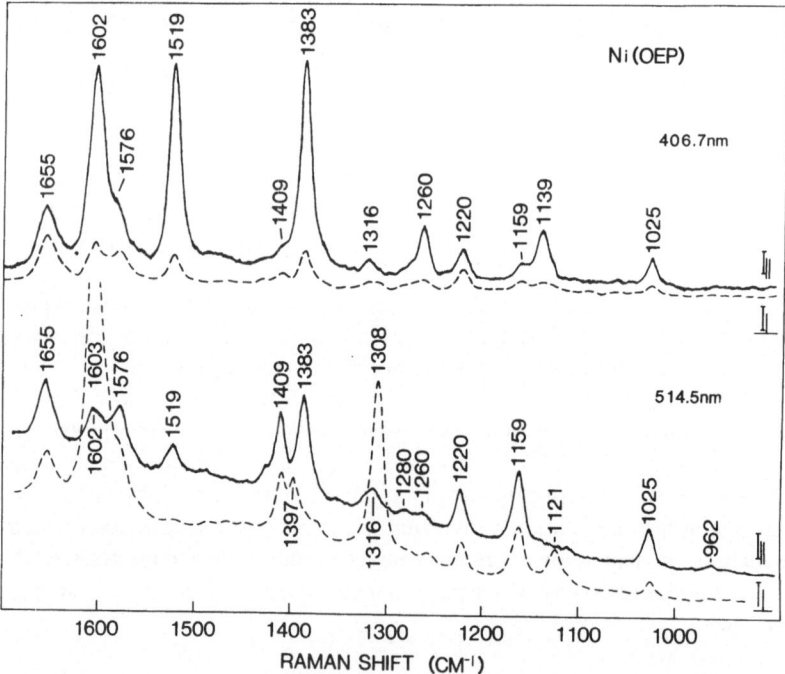

**Fig. 3.** Polarized resonance Raman spectra of Ni(OEP) in $CH_2Cl_2$ excited at 406.7 nm *(upper)* and 514.5 nm *(lower)*. Through all the polarized Raman spectra presented in this article, the solid- and broken-lines represent the parallel and perpendicular polarization components, respectively

## 2.4 Electronic Spectra of Metalloporphyrins

For a highly symmetric porphyrin with $D_{4h}$ symmetry, the highest and next highest occupied and the lowest unoccupied orbitals belong to the $a_{2u}$, $a_{1u}$, and $e_g$ species, respectively[17]. Accordingly, both of the two nearby $\pi\pi^*$ states, that is, $a_{2u}e_g$ and $a_{1u}e_g$, have $E_u$ symmetry. Due to strong configuration interaction between the two states, the transition dipole in one of the resultant transitions becomes additive and the other is subtractive. Since the magnitudes of the transition dipoles of the original two $\pi\pi^*$ transitions are similar, the subtractive component after the configuration interaction has very small transition dipole and thus results in weak absorption of light, while the additive counterpart leads to strong absorption. The weak and strong absorption bands are designated as $Q_0$ and $B_0$, respectively, as illustrated in Fig. 2, where $Q_1$ is due to vibronic transition of $Q_0$.

Since the transition dipoles of the $Q_0$ and $B_0$ bands are in the porphyrin plane ($\mu_x$ and $\mu_y$ are degenerate), the vibrations which give rise to finite values of $\langle B_0|h_a|Q_0\rangle$ of the B term in Eq. 5 are restricted to the in-plane modes with the $A_{1g}$, $A_{2g}$, $B_{1g}$, and $B_{2g}$ symmetries. Accordingly, the vibrations in these four species possibly provide strong resonance Raman bands upon excitation in the B and Q regions. Actually, however, the $A_{2g}$, $B_{1g}$, and $B_{2g}$ modes are resonance enhanced mainly in the Q region, while the $A_{1g}$ modes are noticeably enhanced in the B region as explained later. Typical RR spectra of a metalloporphyrin in resonance with the $B_0$ and $Q_0$ bands are displayed in Fig. 3.

# 3 Vibrational Spectra of Metalloporphyrins

## 3.1 General Considerations

Molecular vibrations of the four-coordinate planar metalloporphyrins are classified into the in-plane and out-of-plane modes. The vibrations of the peripheral substituents, which are included in both the in-plane and out-of-plane modes, are relatively localized, and will not be treated in this section. For the five- and six-coordinate metalloporphyrins the metal-axial ligand modes and internal vibrations of the axial ligands are added. They give essential information about the nature of the metal-ligand linkage, but are not largely coupled with the vibrations of the macrocycle. Therefore, the vibrations regarding the axial ligands will be discussed separately from the vibrations of the macrocycle later. For vibrational analysis a symmetric porphyrin is instructive. Accordingly, we first treat a planar metallo-octamethylporphyrin [M(OMP)] having $D_{4h}$ symmetry and then extend the interpretation to a less symmetric porphyrin such as protoporphyrin IX (PP). The structures of porphyrins treated in this article and their abbreviations are illustrated in Fig. 4, where the atomic numbering is also represented.

Under the assumption that the peripheral methyl group is a point mass of 15 a.m.u., the in-plane vibrations of M(OMP) are factorized into 35 gerade (9 $A_{1g}$ + 9 $B_{1g}$ + 8 $A_{2g}$ + 9 $B_{2g}$ including 1 $A_{1g}$ + 1 $B_{1g}C_m$–H stretching modes) and 18 ungerade modes (18 $E_u$ including 1 $E_u$ $C_m$–H stretching modes). The out-of-plane modes are factorized into 8 gerade (8 $E_g$) and 18 ungerade modes (3 $A_{1u}$ + 6 $A_{2u}$ + 4 $B_{2u}$ + 5 $B_{1u}$). The $A_{2u}$ and $E_u$

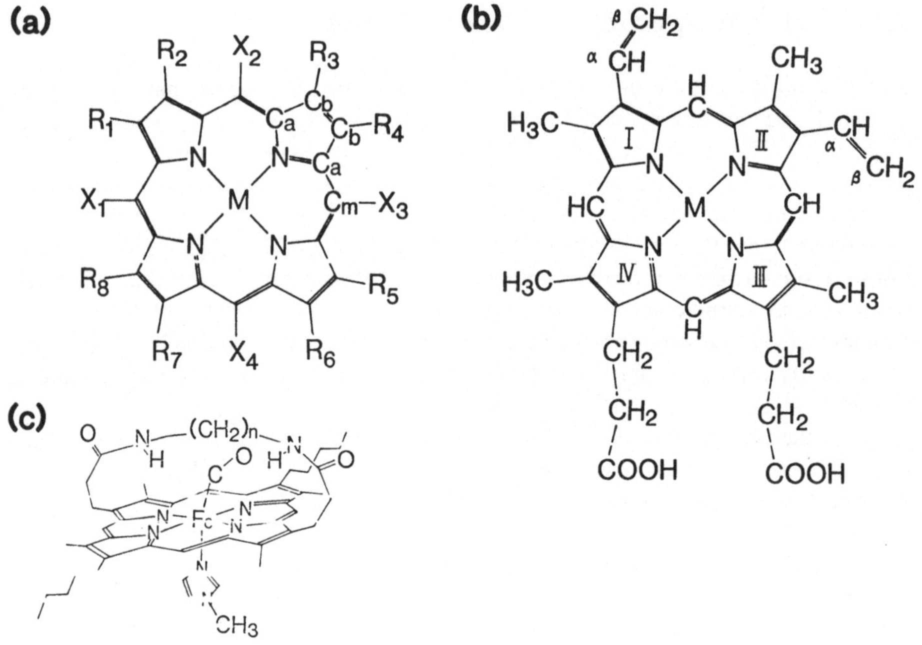

**(a)**

**(b)**

**(c)**

| Fe SP-13 | n = 5 |
|----------|-------|
| Fe SP-14 | n = 6 |
| Fe SP-15 | n = 7 |

**Fig. 4a–c.** Structural diagram of porphyrins and labellings of atoms and rings; **a)** General diagram. The substituent patterns are given in a table for several porphyrins treated in this article which are designated with abbreviations. **b)** Protoporphyrin IX. **c)** Strapped porphyrin

| Porphyrin | Abbrev | $R_1$ | $R_2$ | $R_3$ | $R_4$ | $R_5$ | $R_6$ | $R_7$ | $R_8$ | $X_1$ | $X_2$ | $X_3$ | $X_4$ |
|-----------|--------|-------|-------|-------|-------|-------|-------|-------|-------|-------|-------|-------|-------|
| Octamethylporphyrin | OMP | M | M | M | M | M | M | M | M | H | H | H | H |
| Octaethylporphyrin | OEP | E | E | E | E | E | E | E | E | H | H | H | H |
| Tetraphenylporphyrin | TPP | H | H | H | H | H | H | H | H | Ph | Ph | Ph | Ph |
| Protoporphyrin-IX | PP | M | V | M | V | M | P | P | M | H | H | H | H |
| Deuteroporphyrin-IX | DP | M | H | M | H | M | P | P | M | H | H | H | H |
| Mesoporphyrin-IX | MP | M | E | M | E | M | P | P | M | H | H | H | H |

$M=CH_3$, $E=CH_2CH_3$, $V=-CH=CH_2$, $Ph=-C_6H_5$, $P=-CH_2CH_2COOH$

modes are IR active whereas the $A_{1g}$, $B_{1g}$, $B_{2g}$, and $E_g$ modes are Raman active in an ordinary sense. However, in resonance with the $B_0$ and $Q_0$ absorption bands, the $A_{1g}$, $A_{2g}$, $B_{1g}$, and $B_{2g}$ modes are expected to gain Raman intensity. The mechanism for resonance enhancement of non-totally symmetric modes by the $Q_0$ band was discussed by Nishimura et al.[18] and Shelnutt et al.[19]. The $A_{1g}$ and $A_{2g}$ modes give the polarized (p) and anomalously polarized (ap) bands, respectively, whereas the $B_{1g}$ and $B_{2g}$ modes give the depolarized (dp) bands. Since the peripheral substituents of M(OMP) are not involved in the π conjugation system, their vibrations do not appear in RR spectra. In contrast, all vibrations of the peripheral groups can appear in the IR spectra in addition to the symmetry allowed modes of the macrocycle. This is the main reason why IR spectra are generally much more complicated than RR spectra for a given metalloporphyrin.

If the structure of the planar macrocycle were deformed, the selection rule mentioned above would be violated and the polarization properties of Raman bands would be altered. For two practical deformations of metalloporphyrins, the symmetry correlations are summarized in Table 1. Upon doming the $A_{1g}$ and $A_{2u}$ species of $D_{4h}$ are combined into the $A_1$ species of $C_{4v}$ and accordingly the RR active $A_{1g}$ species, which give rise to the polarized Raman bands, become IR active. Upon ruffling, on the other hand, the $B_{2g}$ and $A_{2u}$ species of $D_{4h}$ are combined into the $B_2$ species of $D_{2d}$ and therefore the IR inactive $B_{2g}$ species, which give rise to depolarized Raman bands, become IR active. In both cases the IR active $A_{2u}$ species become Raman active. Such Raman bands are expected to be polarized for doming and to be depolarized for ruffling. In this way the two kinds of deformations can be distinguished from the IR and Raman spectra.

**Table 1.** Symmetry correlation for deformed metalloporphyrins

|        | Ruffling $D_{2d}$ | Planar $D_{4h}$ | Domed $C_{4v}$ |
|--------|-------------------|-----------------|----------------|
|        | $A_1$             | $A_{1g}$        | $A_1$          |
|        | $A_2$             | $A_{2g}$        | $A_2$          |
|        | $B_1$             | $B_{1g}$        | $B_1$          |
|        | $B_2$             | $B_{2g}$        | $B_2$          |
|        | $E$               | $E_g$           | $E$            |
|        | $B_1$             | $A_{1u}$        | $A_2$          |
|        | $B_2$             | $A_{2u}$        | $A_1$          |
|        | $A_1$             | $B_{1u}$        | $B_2$          |
|        | $A_2$             | $B_{2u}$        | $B_1$          |
|        | $E$               | $E_u$           | $E$            |
| IR[a]  | $B_2$, $E$        | $A_{2u}$, $E_u$ | $A_1$, $E$     |
| Raman  | $A_1$, $B_1$, $B_2$, $E$ | $A_{1g}$, $B_{1g}$, $B_{2g}$, $E_g$ | $A_1$, $B_1$, $B_2$, $E$ |
| RR     | $A_2$             | $A_{2g}$        | $A_2$          |

[a]  Active species for each method. "RR" implies the species which is inactive in an ordinary Raman but active under resonance condition

## 3.2 Assignments of Vibrational Modes

Assignments of IR and RR bands are essential to make the best use of the vibrational spectroscopy. Normal coordinate treatments of metalloporphyrins have been carried out by several groups. Ogoshi et al.[20] treated the $E_u$ modes to assign the IR bands. Stein et al.[21] applied Ogoshi's force field to Raman active modes. Sunder and Bernstein[22], Susi and Ard[23], and Gladkov et al.[24] calculated the in-plane modes of Ni- and Cu-porphyrins by assuming some correlations between the force constants of the valence force field and bond-lengths. Warshel and Lappicirella[25] treated the out-of-plane modes. These theoretical treatments for the in-plane vibrations had a fatal weakness, that is, they were incompatible with the Raman data for $^{15}$N-substituted Ni(OEP) reported later[26], although for the out-of-plane vibrations no decision could be made due to insufficiency of experimental data. More experimental information particularly about classification of dp bands into the $B_{1g}$ and $B_{2g}$ species and about frequencies of unobserved fundamentals was indispensable to obtain more reliable force constants.

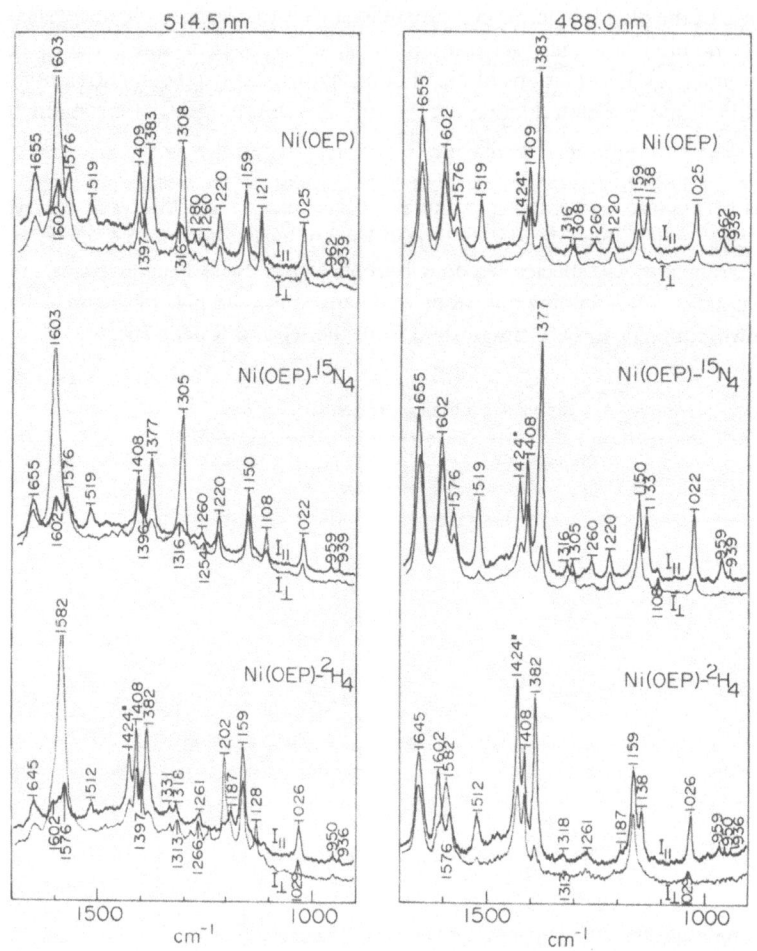

**Fig. 5.** Polarized resonance Raman spectra of Ni(OEP), Ni(OEP–$^{15}$N$_4$), and Ni(OEP-d$_4$) in CH$_2$Cl$_2$ solution excited at 514.5 *(left)* and 488.0 nm *(right)*. The Raman line at 1424 cm$^{-1}$ *(marked with an asterisk)* is due to CH$_2$Cl$_2$ (from Ref. 27)

**Table 2.** Symmetry species of the overtones and combination modes[a]

|          | A$_{1g}$ | B$_{1g}$ | A$_{2g}$ | B$_{2g}$ |
|----------|----------|----------|----------|----------|
| A$_{1g}$ | A$_{1g}$ | B$_{1g}$ | A$_{2g}$ | B$_{2g}$ |
| B$_{1g}$ |          | A$_{1g}$ | B$_{2g}$ | A$_{2g}$ |
| A$_{2g}$ |          |          | A$_{1g}$ | B$_{1g}$ |
| B$_{2g}$ |          |          |          | A$_{1g}$ |

[a]  For a point group D$_{4h}$

To overcome this problem Kitagawa et al.[27] observed the overtone and combination modes of Ni(OEP) and its meso-deuterated [Ni(OEP-d$_4$)] and $^{15}$N-substituted derivatives [Ni(OEP–$^{15}$N$_4$)]. A part of their spectra is shown in Fig. 5. About 90 non-fundamental RR bands besides fundamentals were observed for each species and analyzed. The symmetry species expected for the RR bands due to overtones and combinations are summarized in Table 2.

The key point is that if the combination of two dp bands are polarized, they belong to the same symmetry species whereas if such combinations give ap bands, individual components belong to different symmetry species. Thus the dp bands were classified into two groups. Since it was anticipated from the preliminary calculations[28] that the highest-frequency skeletal stretching mode should be considerably higher for the B$_{1g}$ species than for the B$_{2g}$ species, the assignments of each group to the B$_{1g}$ and B$_{2g}$ species were straightforward. In this way the experimental classification of the B$_{1g}$ and B$_{2g}$ species and the inference of the unobserved fundamentals were worked out. On the basis of these

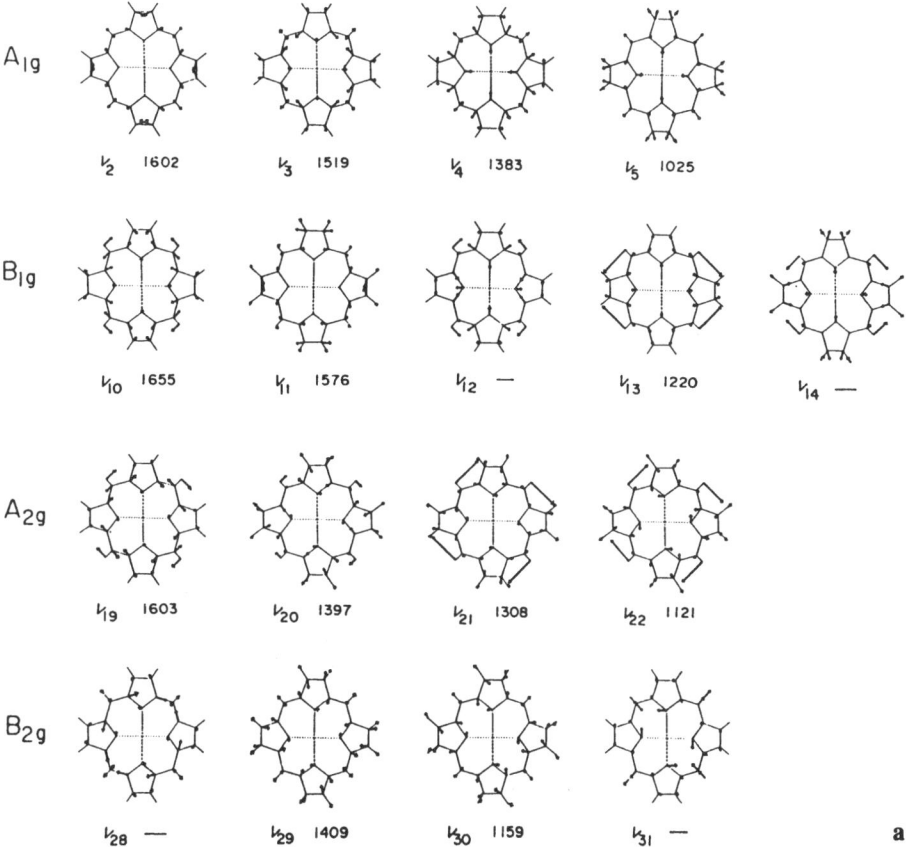

**Fig. 6. a)** Resonance Raman active vibrational modes of Ni(OEP) in the high frequency region. **b)** Resonance Raman active vibrational modes of Ni(OEP) in the low frequency region. **c)** IR active in-plane modes of Ni(OEP) (from Ref. 28 and 73)

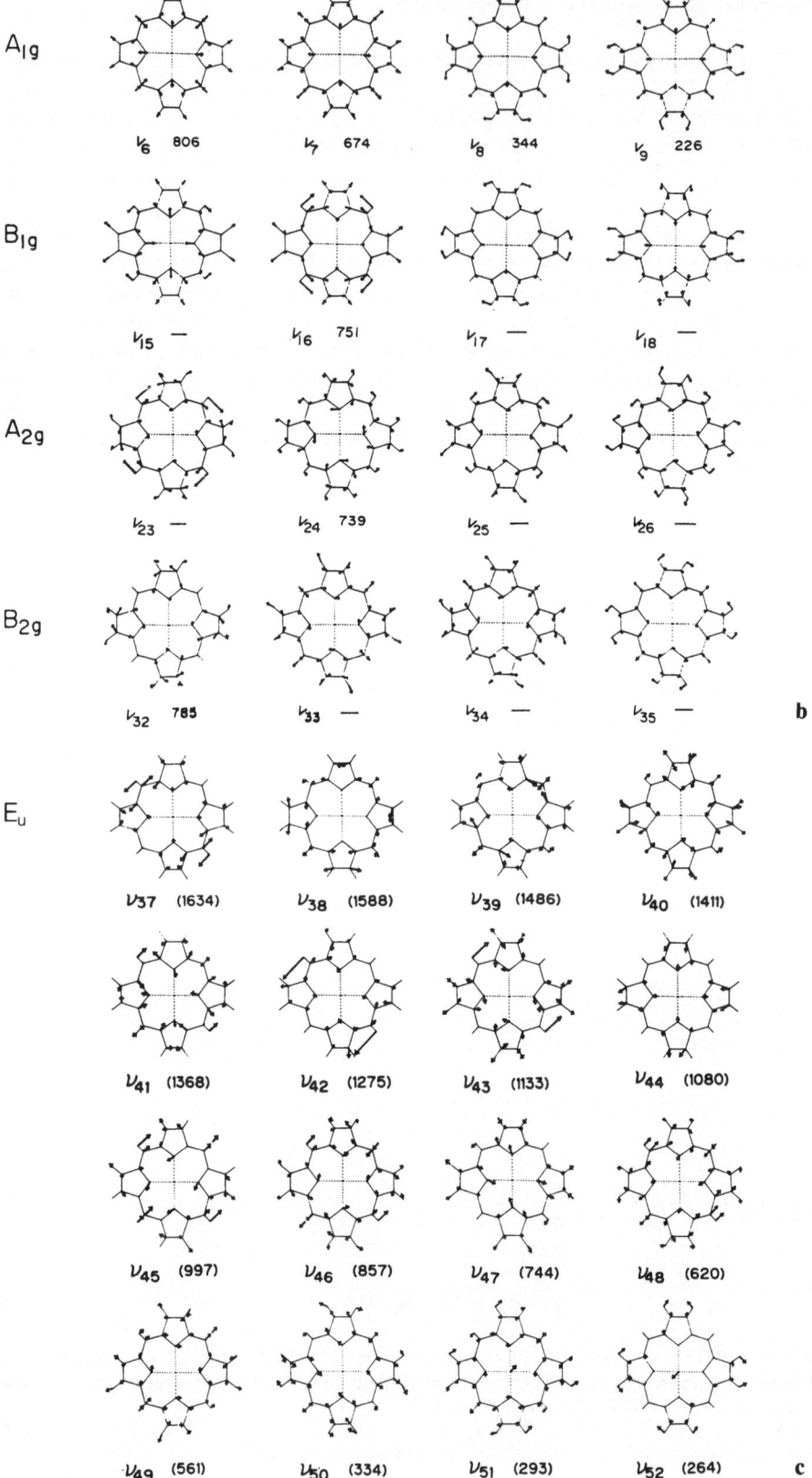

A₁g        ν₆  806        ν₇  674        ν₈  344        ν₉  226

B₁g        ν₁₅  —        ν₁₆  751        ν₁₇  —        ν₁₈  —

A₂g        ν₂₃  —        ν₂₄  739        ν₂₅  —        ν₂₆  —

B₂g        ν₃₂  785        ν₃₃  —        ν₃₄  —        ν₃₅  —        b

Eᵤ        ν₃₇  (1634)        ν₃₈  (1588)        ν₃₉  (1486)        ν₄₀  (1411)

ν₄₁  (1368)        ν₄₂  (1275)        ν₄₃  (1133)        ν₄₄  (1080)

ν₄₅  (997)        ν₄₆  (857)        ν₄₇  (744)        ν₄₈  (620)

ν₄₉  (561)        ν₅₀  (334)        ν₅₁  (293)        ν₅₂  (264)        c

**Table 3.** Calculated and observed frequencies of Ni(OEP) [cm$^{-1}$]

| Symmetry | No. | obs. | calc. | Potential Energy Distribution (%) |
|---|---|---|---|---|
| A$_{1g}$ | $\nu_1$ | – | 3072 | $\nu(C_mH)100$ |
| | $\nu_2$ | 1602 | 1591 | $\nu(C_bC_b)60$, $\nu(C_b-Et)19$ |
| | $\nu_3$ | 1519 | 1517 | $\nu(C_aC_m)41$, $\nu(C_aC_b)35$ |
| | $\nu_4$ | 1383 | 1386 | $\nu(C_aN)53$, $\delta(C_a-C_m)21$ |
| | $\nu_5$ | 1025 | 1048 | $\nu(C_b-Et)38$, $\nu(C_aC_b)23$ |
| | $\nu_6$ | 806 | 809 | $\delta(C_aC_mC_a)36$, $\nu(C_aN)27$ |
| | $\nu_7$ | 674 | 655 | $\delta(C_bC_aN)20$, $\nu(C_aC_b)19$ |
| | $\nu_8$ | 344 | 326 | $\delta(C_b-Et)57$, $\nu(C_aC_m)11$ |
| | $\nu_9$ | 264 | 230 | $\delta(C_b-Et)23$, $\nu(C_aC_m)16$ |
| B$_{1g}$ | $\nu_{10}$ | 1655 | 1656 | $\nu(C_aC_m)49$, $\nu(C_aC_b)17$ |
| | $\nu_{11}$ | 1576 | 1587 | $\nu(C_bC_b)57$, $\nu(C_b-Et)16$ |
| | $\nu_{12}$ | – | 1351 | $\nu(C_aN)63$, $\nu(C_bC_b)13$ |
| | $\nu_{13}$ | 1220 | 1262 | $\delta(C_mH)67$, $\nu(C_aC_b)22$ |
| | $\nu_{14}$ | – | 1095 | $\nu(C_aC_b)31$, $\nu(C_b-Et)30$ |
| | $\nu_{15}$ | – | 754 | $\nu(C_b-Et)25$, $\nu(C_aC_b)20$ |
| | $\nu_{16}$ | 751 | 741 | $\delta(C_aNC_a)14$, $\nu(C_b-Et)14$ |
| | $\nu_{17}$ | – | 299 | $\delta(C_b-Et)84$ |
| | $\nu_{18}$ | – | 187 | $\delta(C_a-C_m)39$, $\nu(MN)34$ |
| A$_{2g}$ | $\nu_{19}$ | 1603 | 1600 | $\nu'(C_aC_m)67$, $\nu'(C_aC_b)18$[a] |
| | $\nu_{20}$ | 1397 | 1409 | $\nu'(C_aN)29$, $\nu'(C_b-Et)24$ |
| | $\nu_{21}$ | 1308 | 1281 | $\delta'(C_mH)53$, $\nu'(C_aC_b)18$ |
| | $\nu_{22}$ | 1121 | 1118 | $\nu'(C_aN)37$, $\nu'(C_b-Et)26$ |
| | $\nu_{23}$ | – | 1022 | $\nu'(C_aC_b)26$, $\nu'(C_b-Et)20$ |
| | $\nu_{24}$ | 739 | 723 | $\delta'(C_a-C_m)43$, $\delta'(C_b-Et)29$ |
| | $\nu_{25}$ | – | 528 | $\delta'(C_aC_bC_b)39$, $\delta'(C_bC_aN)17$ |
| | $\nu_{26}$ | – | 289 | $\delta'(C_b-Et)41$, $\delta'(C_a-C_m)31$ |
| B$_{2g}$ | $\nu_{27}$ | – | 3072 | $\nu'(C_mH)100$ |
| | $\nu_{28}$ | – | 1469 | $\nu'(C_aC_m)52$, $\nu'(C_aC_b)21$ |
| | $\nu_{29}$ | 1409 | 1409 | $\nu'(C_aC_b)47$, $\nu'(C_b-Et)26$ |
| | $\nu_{30}$ | 1159 | 1157 | $\nu'(C_b-Et)49$, $\nu'(C_aN)28$ |
| | $\nu_{31}$ | – | 1016 | $\delta'(C_a-C_m)25$, $\delta'(C_aC_mC_a)23$ |
| | $\nu_{32}$ | 785 | 789 | $\delta'(C_b-Et)50$, $\delta'(C_a-C_m)22$ |
| | $\nu_{33}$ | – | 536 | $\delta'(C_aC_bC_b)28$, $\nu'(C_b-Et)15$ |
| | $\nu_{34}$ | – | 232 | $\delta'(C_aC_mC_a)25$, $\delta'(NMN)22$ |
| | $\nu_{35}$ | – | 182 | $\delta'(C_b-Et)30$, $\delta'(C_a-C_m)25$ |
| E$_u$ | $\nu_{36}$ | – | 3074 | $\nu(C_mH)100$ |
| | $\nu_{37}$ | 1604 | 1634 | $\nu(C_aC_m)34$, $\nu'(C_aC_m)24$ |
| | $\nu_{38}$ | 1557 | 1588 | $\nu(C_bC_b)56$, $\nu(C_b-Et)16$ |
| | $\nu_{39}$ | 1487 | 1486 | $\nu'(C_aC_m)36$, $\nu'(C_aN)17$ |
| | $\nu_{40}$ | 1443 | 1411 | $\nu'(C_aC_b)30$, $\nu'(C_b-Et)24$ |
| | $\nu_{41}$ | 1389 | 1368 | $\nu(C_aN)56$, $\delta(C_aC_m)14$ |
| | $\nu_{42}$ | 1268 | 1275 | $\delta(C_mH)59$, $\nu'(C_b-Et)9$ |
| | $\nu_{43}$ | 1148 | 1133 | $\nu'(C_b-Et)35$, $\nu'(C_aN)33$ |
| | $\nu_{44}$ | 1113 | 1080 | $\nu(C_b-Et)29$, $\nu(C_aC_b)26$ |
| | $\nu_{45}$ | 993 | 997 | $\nu'(C_aN)21$, $\nu'(C_aC_m)12$ |
| | $\nu_{46}$ | 924 | 857 | $\nu'(C_b-Et)30$, $\delta'(C_a-C_m)20$ |
| | $\nu_{47}$ | 726 | 744 | $\nu(C_b-Et)26$, $\nu(C_aC_b)20$ |
| | $\nu_{48}$ | 605 | 620 | $\delta'(C_b-Et)28$, $\nu(C_aC_m)13$ |
| | $\nu_{49}$ | 550 | 561 | $\delta'(C_aC_bC_b)26$, $\delta(C_aC_mC_a)15$ |
| | $\nu_{50}$ | – | 334 | $\delta(C_b-Et)25$, $\delta(C_a-C_m)15$ |
| | $\nu_{51}$ | – | 293 | $\delta(C_b-Et)23$, $\delta(NMN)16$ |
| | $\nu_{52}$ | 287[b] | 264 | $\delta(C_b-Et)41$, $\nu(MN)31$ |
| | $\nu_{53}$ | – | 175 | $\delta'(C_b-Et)29$, $\delta'(C_a-C_m)23$ |

[a] $\nu'$ and $\delta'$ represents the symmetry coordinates of antisymmetric stretching and deformation vibrations about the $C_2$ axis of pyrrole ring, respectively; [b] Ref. 33

experimental data Abe et al.[29] carried out the normal mode calculations of the in-plane modes of Ni(OMP) under $D_{4h}$ symmetry. With 37 constants of a modified Urey-Bradley force field, they assigned the 59 RR bands and 38 IR bands of Ni(OEP), Ni(OEP-$^{15}$N$_4$), and Ni(OEP-d$_4$). The magnitudes of the stretching force constants are in order of $C_bC_b >$ $C_aC_m > C_bCH_3 > C_aC_b$, which is in reasonable agreement with the order of the bond lengths. The potential energy distributions of each mode are listed in Table 3, and the individual vibrational modes are depicted in Fig. 6.

The marked difference between the force constants of Abe et al.[29] and Ogoshi et al.[20] arises from the difference in the assignment of the IR band of Ni(OEP) at 1676 cm$^{-1}$, which was considered to be $\nu_{37}$ by Ogoshi et al., but to be a combination band of two $C_m$–H out-of-plane bending modes of the $A_{2u}$ and $E_g$ species by Abe et al. The latter assignment was supported by Kincaid et al.[30] with IR spectra of Ar-matrix isolated Ni(OEP) isotope derivatives. Kincaid et al.[30] suggested to exchange the assignments of the IR bands due to the $C_aC_m$ and $C_bC_b$ stretching modes ($\nu_{37}$ and $\nu_{38}$) and of some other modes including $\nu_{40}$, $\nu_{41}$, and $\nu_{42}$. However, it is hard to do it only for the $E_u$ species without changing the assignments of the $A_{1g}$, $A_{2g}$, $B_{1g}$, and $B_{2g}$ species. On the other hand, Willems and Bocian[31, 32] examined the substituent effects at positions 2 and 4 of Ni(DP) on the IR and RR spectra, and supported Abe's assignments. Nowadays the assignments shown in Table 3 and Fig. 6 are widely accepted in the discussion of the IR and RR spectra of metalloporphyrins and heme proteins.

In porphyrin chemistry M(TPP) is rather more frequently used than M(OEP) due to its easy preparation. Therefore, the vibrational analysis of M(TPP) is important. Appearance of the meso-phenyl modes in RR spectra was first pointed out by Burke et al.[34, 35]. The IR and RR spectra of Ar-matrix isolated Co(TPP) and its deuterated derivatives were reported by Kozuka and Nakamoto[36]. There was an argument about the RR band around 1240 cm$^{-1}$, which was assigned to the $C_m$-phenyl stretching mode for Fe(TPP) by Burke et al.[34] but to a phenyl internal mode for Co(TPP) by Fuchsman et al.[37]. To determine the alternative, normal coordinate calculations were carried out for the $A_{1g}$ species of Co(TPP)[38] with the force field obtained for Ni(OEP)[29] and the RR band in question was assigned to the phenyl internal mode. This suggests the presence of appreciable mixing of the $\pi$ system of the macrocycle with that of the phenyl group at the meso positions.

# 4 Structural Information from Vibrational Spectra

What kind of information is drawn from the vibrational spectrum? This is the most important question for the practical application of vibrational spectroscopy in the study of metalloporphyrins. Indeed, apart from the vibrational assignments, a great deal of experimental data have been accumulated to establish some empirical rules of the IR and RR spectral features with regard to the coordination number, the spin- and oxidation-states, and the core-size of metal-porphyrins (i.e. the center to pyrrole nitrogen distances). *Such empirical rules are summarized here.*

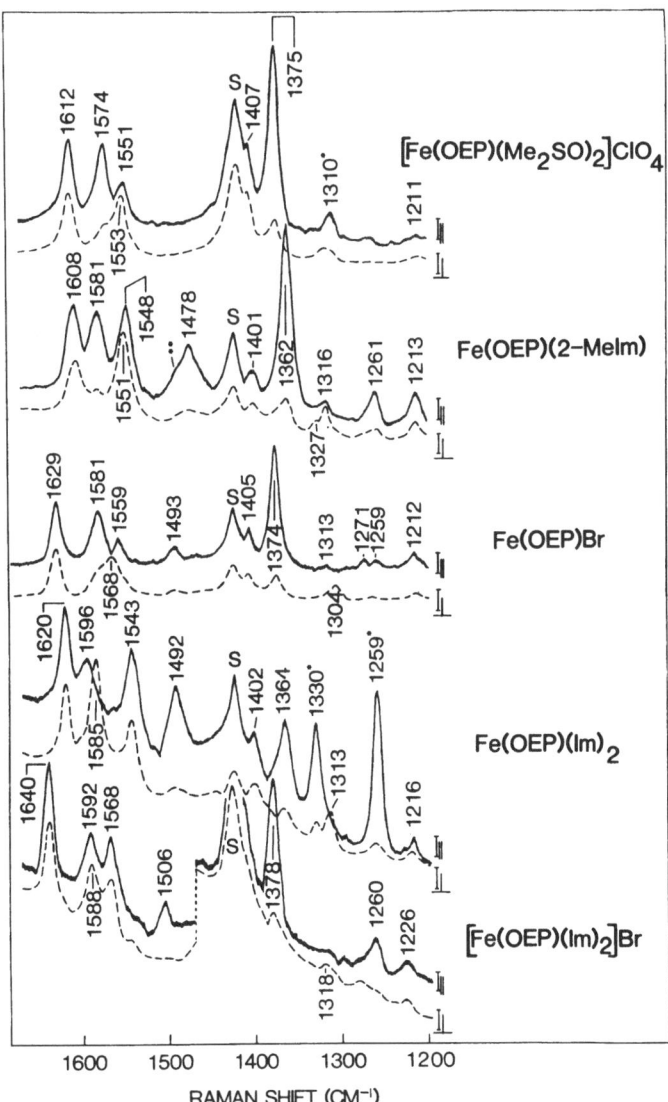

**Fig. 7.** The 488.0-nm excited polarized resonance Raman spectra of [Fe(OEP)(Me₂SO)₂]ClO₄, Fe(OEP)(2-MeIm), Fe(OEP)Br, Fe(OEP)(Im)₂, and [Fe(OEP)(Im)₂]Br in CH₂Cl₂. The bands marked with S are due to solvent and those with single and double asterisks arise from imidazole and 2-methylimidazole, respectively (from Ref. 39)

## 4.1 Markers for the Core-Size

Five typical RR spectra of Fe(OEP) derivatives[39] are displayed in Fig. 7, where the spectra are arranged in the decreasing order of the core-size. [Fe(OEP)(Me₂SO)₂]⁺ (Me₂SO: dimethylsulfoxide) and Fe(OEP)Br stand for the six- and five-coordinate ferric high-spin complexes, respectively. Fe(OEP)(Im)₂ and [Fe(OEP)(Im)₂]⁺ (Im: imidazole) are typical low-spin complexes of the ferrous and ferric porphyrins, respectively.

Fe(OEP)(2-MeIm) (2-MeIm: 2-methylimidazole) is a five-coordinate ferrous high-spin complex. These spectra resemble those of heme proteins having the corresponding oxidation-, spin-, and coordination states.

The highest-frequency band arises from the out-of-phase $C_aC_m$ stretching mode ($v_{10}$, dp), and its in-phase counterpart with regard to the $C_4$ axis gives the perpendicular peak around 1550–1600 cm$^{-1}$ ($v_{19}$, ap). The totally symmetric $C_aC_m$ stretching mode appears around 1475–1510 cm$^{-1}$ ($v_3$, p), although it is missing with [Fe(OEP)(Me$_2$SO)$_2$]$^+$. The second highest-frequency polarized peak around 1575–1595 cm$^{-1}$ is associated mainly with the in-phase $C_bC_b$ stretching modes ($v_2$, p) and its out-of-phase counterpart about the $C_4$ axis gives the second highest-frequency dp band around 1545–1570 cm$^{-1}$ ($v_{11}$, dp). The prominent peak around 1360–1375 cm$^{-1}$ is attributed to the totally symmetric $C_aN$ stretching mode ($v_4$, p). This RR band had been referred to as an oxidation state marker[40] for heme proteins, but now is explained to reflect $\pi$ delocalization; as the extent of delocalization of the $d_\pi$ electrons to the LUMO of the porphyrin increases, its frequency decreases[41, 42]. The $v_{10}$ frequency can be used for diagnosis of the coordination number for the ferric high-spin complexes in particular[43, 44]; 1615–1625 cm$^{-1}$ and 1627–1630 cm$^{-1}$ for the six- and five-coordinate complexes, respectively.

The presence of a linear correlation between the $v_{19}$ frequencies and the Ct–N distances (Ct: center of porphyrin) was first pointed out by Felton et al.[45, 46] and later similar correlations were obtained for other modes above 1450 cm$^{-1}$[43, 47]. In Fig. 8 the $v_{10}$, $v_2$, $v_{19}$, $v_{11}$, and $v_3$ frequencies of Fe(OEP) derivatives and Ni(OEP) are plotted against the Ct–N distances[39]. They appear to satisfy the relation of $v = K(A-d)$, in which d is the Ct–N distance and A is an arbitrary constant. The coefficients, K, of the straight lines shown in Fig. 8, which are 576, 495, 414, 288, and 322 [cm$^{-1}$ Å$^{-1}$] for $v_{19}$, $v_{10}$, $v_3$, $v_{11}$, and $v_2$, respectively, seem to parallel the amount of contribution of the $C_aC_m$ stretching

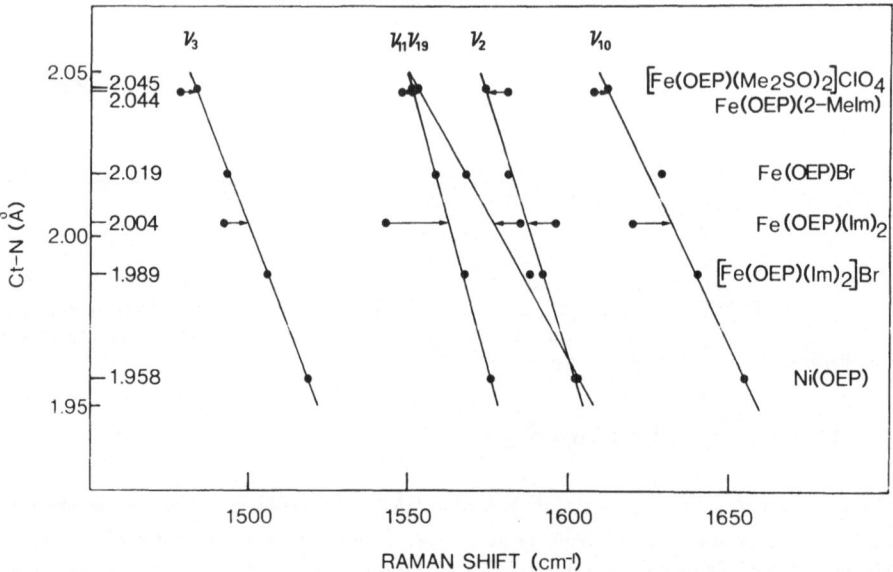

**Fig. 8.** Correlations between the $v_{10}$, $v_2$, $v_{19}$, $v_{11}$, and $v_3$ frequencies and the porphyrin core-size for the indicated complexes of OEP (from Ref. 39)

coordinate to the individual modes. This implies that when the core-size becomes larger the $C_aC_m$ bonds bear the corresponding deformation without bringing about deformation of the pyrrole rings[44]. In Fig. 8 the points of $Fe(OEP)(Im)_2$ considerably deviate from the linear correlation. Similar deviations were also noted for $Fe(PP)(Im)_2$ and ascribed to increased $\pi$-back donation[47]. It still remains to be elucidated theoretically how each bond-strength is altered upon a change of the $\pi$ delocalization. Since new ab initio M.O. calculations on metalloporphyrins are in progress[48], more basic interpretation of Raman spectra might be provided from a theoretical point of view.

The frequencies of the vibrations described above depend on the metal species coordinated. The RR spectra of Co, Ni, Cu, and Zn(OEP) were investigated and their $v_2$, $v_3$, $v_{10}$, $v_{11}$, and $v_{19}$ frequencies were plotted against the transition energy of the $Q_0$ band[49]. As shown in Fig. 9, straight lines are obtained. This strongly suggests that the electronic absorption reflects the core expansion in the electronic ground state rather than a change in the excited state. The metal dependence of the RR spectra of metal-substituted chlorophylls was investigated by Fujiwara et al.[50] who obtained linear correlations between Raman frequencies and the Ct–N distances. Consequently it can be concluded that the stronger the $M–N_{pyrr}$ bond, the shorter the Ct–N distance and the higher the ring stretching frequencies. Such metal dependence would result from a balance of the attractive conjugation interaction between the $4p_z$(metal) and $a_{2u}$(porphyrin) orbitals[49, 51] with the repulsive interaction between the $d_{x^2-y^2}$(metal) and $2p_{x,y}$(N) orbitals[52]. The latter interaction becomes important only when the $d_{x^2-y^2}$ orbital is occupied.

## 4.2 Markers for Delocalization of $\pi$ Electrons

With regard to the interaction with an axial ligand (L), orbital overlapping between $d_{z^2}$(Fe) or $d_\pi$(Fe) (= $d_{xz}$, $d_{yz}$) and $\Pi$(L) or $\sigma$(L) yields two kinds of bonding interactions; $d_{z^2}$(Fe)–$\sigma$(L) and $d_\pi$(Fe)–$\Pi$(L). The former and latter are conveniently called the $\sigma$- and $\pi$-type interactions, respectively. Since the $d_{z^2}$(Fe) orbital belongs to the $a_{1g}$ species under

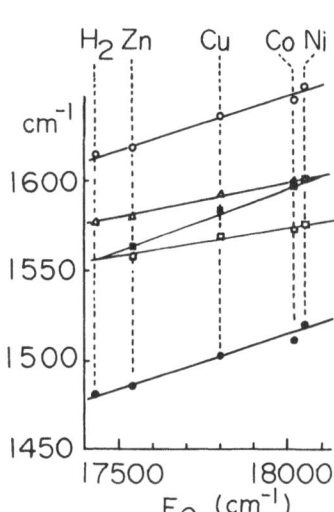

**Fig. 9.** Plots of the wavenumbers of a few selected Raman bands *vs* the wavenumbers of the $Q_0$ band. ○: $v_{10}$, △: $v_2$, ▨: $v_{19}$, □: $v_{11}$, ●: $v_3$ (from Ref. 49)

$D_{4h}$ symmetry, the interaction involving the $d_{z^2}(Fe)$ orbital does not very much affect a $\pi$-state of the porphyrin. By contrast, the $\pi$-interaction causes a change in the extent of the delocalization of $\pi$ electrons in the porphyrin plane, because the $d_\pi(Fe)$ orbitals belong to the $e_g$ species as well as the LUMO of porphyrin, $\Pi^*(ring)$. Since they have the same symmetry, there would be a $\pi$-type molecular orbital represented by Eq. 6.

$$\Psi_\pi = c_1 d_\pi(Fe) + c_2 \Pi^*(ring) + c_3 \Pi^*(L) \tag{6}$$

For the σ-type complex, $c_3$ is zero. Since the low-spin complexes adopt the planar structure regardless of the oxidation states, the coefficients would not be very different between the ferrous and ferric states. There should be 3 and 4 $d_\pi(Fe)$ electrons for the ferric and ferrous states, respectively. Then the amount of electrons delocalized to $\Pi^*(ring)$ is $3c_2^2$ and $4c_2^2$ for the ferric and ferrous low-spin states, respectively. No matter what the value of $c_2^2$ is, the delocalization of electrons from $d_\pi(Fe)$ to $\Pi^*(ring)$ is larger

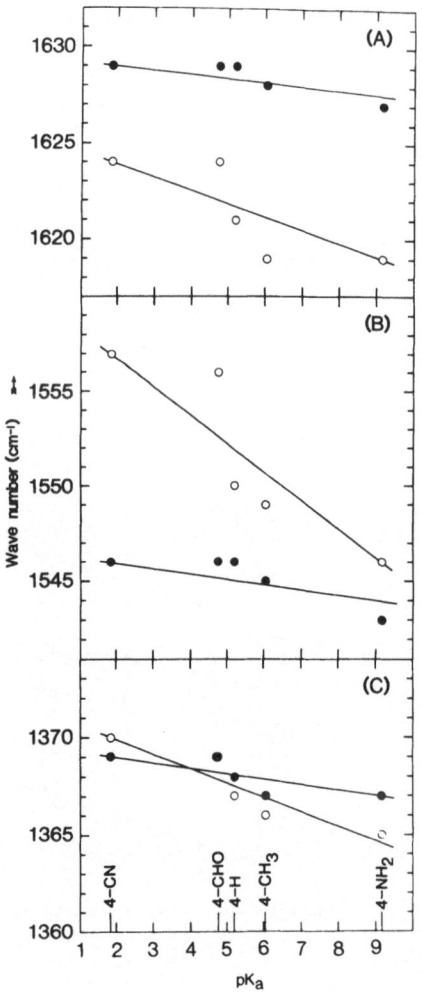

**Fig. 10.** The frequencies of $\nu_{10}$ (A), $\nu_{11}$ (B), and $\nu_4$ (C) modes of Fe(OEP)(4-YPy)$_2$ (Y = CN, CHO, H, CH$_3$, and NH$_2$) and Fe(OEC)(4-YPy)$_2$ in CH$_2$Cl$_2$ are plotted against pK$_a$ values of 4-CNPy, 4-CHOPy, Py, γ-Pic (4-CH$_3$Py), and 4-NH$_2$Py, respectively. ○: OEP complexes, ●: OEC complexes (from Ref. 39)

for ferrous low-spin state than for the ferric low-spin state, although the coefficient would be changed for the high-spin state with a pyramidal structure.

According to recent ab initio MO calculations[53], the $\Pi^*$(ring) (= 5 $e_g$ orbital) is antibonding with regard to the $C_a$–N bond. Therefore, the larger the number of electrons occupying this orbital, the weaker the $C_a$–N bond and thus the greater the reduction of the $C_a$–N stretching force constant. As a result, the $\nu_4$ frequency is lower for the ferrous low-spin state than for the ferric low-spin state. If the axial ligand has $\pi$ acidity, the empty or half occupied $\Pi^*$(L) orbital withdraws electrons in proportion to $c_3^2$ and therefore $c_2^2$ becomes smaller than that for the case of $c_3^2 = 0$. Then the $\nu_4$ frequency would become higher. This is applied to the $O_2$, CO, and NO complexes of reduced hemoproteins[42b]. Conversely, if the axial ligand has $\pi$ basicity, electrons would be more donated to $\Pi^*$(ring) and then the $\nu_4$ frequency would become lower. This is the case of reduced cytochrome P-450[54]. In this way the $\nu_4$ frequency reflects delocalization of $\pi$ electrons[41, 42].

Due to the metal-axial ligand interactions other RR bands are also affected by the basicity of axial ligands. Although the sensitivity of the $\nu_{10}$ frequency to axial ligands had been pointed out[44], this was partly caused by the repulsive interaction between the axial ligand and pyrrole nitrogens. To abstract the pure electronic effects, Ozaki et al.[39] examined the RR spectra of p-substituted pyridine complexes. In Fig. 10 the $\nu_{10}$, $\nu_{11}$, and $\nu_4$ frequencies of Fe(OEP)(4-YPy)$_2$ are plotted against $pK_a$ of 4-YPy (Y = CN, CHO, H, CH$_3$, and NH$_2$). These frequencies become lower as the $pK_a$ of the axial ligand increases and simultaneously they deviate more from the straight line shown in Fig. 8. It is interesting that the $\nu_{11}$ mode involving the $C_bC_b$ stretching character is most sensitive to $pK_a$, while the $\nu_4$ frequency is only moderately sensitive.

# 5 Vibrations of Less Symmetric Metalloporphyrins

## 5.1 Effects of Peripheral Substituents

A naturally abundant metalloporphyrin with an asymmetric disposition of the peripheral groups is M(PP) (Fig. 4b). It is important to examine to what extent the assignments established for M(OEP) can be applied to M(PP). Willems and Bocian[31, 32] investigated the effects of the substituents at positions 2, and 4 of PP by placing acetyl and formyl groups there, pointing out that the assignments of the $A_{1g}$, $A_{2g}$, $B_{1g}$, and $B_{2g}$ modes of Ni(OEP) are generally applicable to Ni(PP) except for a few modes including $\nu_2$, $\nu_8$, $\nu_{13}$, and $\nu_{20}$. Spiro and coworkers[47, 55] analyzed the vinyl mode contribution to the RR spectra of Ni(PP) by observing the frequency shift due to vinyl deuteration. Lee et al.[56] observed the RR spectra for specifically deuterated Ni(PP), which are displayed in Fig. 11. The RR bands of Ni(PP) were assigned on the basis of the normal coordinate calculations for appropriate model systems. The observed deuteration shifts and plausible assignments are summarized in Table 4.

The RR bands of Ni(PP) at 1633 and 1610 cm$^{-1}$ are assigned to the vinyl C=C stretching modes as proposed by Choi et al.[55]. Although Choi et al. assumed a large splitting (90 cm$^{-1}$) between the in-phase and out-of-phase modes of the CH$_2$ scissoring

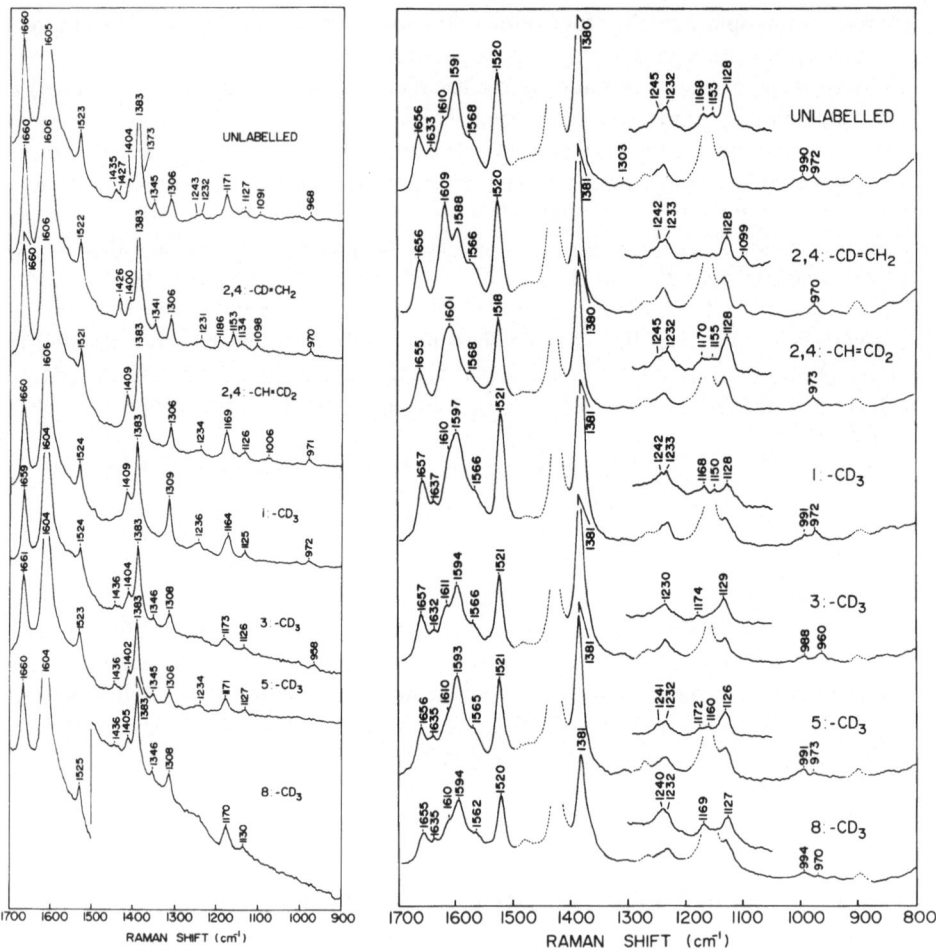

**Fig. 11.** Resonance Raman spectra of the dimethylester of unlabelled Ni(PP) (top) and its deuter-ated isotopomers. The label positions are specified at the right side of each spectrum; *left*) the 514.5-nm excitation spectra for the $CH_2Cl_2$ solution, *right*) the 406.7-nm excitation spectra for the benzene solution (the inserted spectra were observed for the $CCl_4$ solution) (from Ref. 56)

modes of the two vinyl groups at positions 2 and 4, the normal mode calculations indi-cated a much smaller splitting, and bands at 1435 and 1427 cm$^{-1}$ are assigned to the out-of-phase and in-phase modes, respectively[56]. The $CH_2$ rocking mode at 1091 cm$^{-1}$ is expected to be considerably coupled with the skeletal modes such as $\nu_{20}$ and $\nu_{29}$, and these might be affected by vinyl deuteration. Therefore, the $\nu_{20}$ band of M(PP) [1345 cm$^{-1}$ for Ni(PP)], which has been used as diagnostic for the presence of protoheme in heme proteins[57], apparently seems to disappear upon substitution of the vinyl group with other groups.

For the porphyrin with the formyl substituents, the $-CH=O$ stretching mode ($\nu_{CH=O}$) also gives a prominent RR band. It is noted that the $\nu_{CH=O}$ frequency of mono-formyl porphyrin is independent of the substituted position in an organic solution[32] but when

**Table 4.** Observed frequencies and isotope shifts of Raman bands of Ni(PP) in the spectral region above 800 cm$^{-1}$

| Unlabelled [cm$^{-1}$] | ϱ | Isotope shift [cm$^{-1}$] | | | | | | Assignment |
|---|---|---|---|---|---|---|---|---|
| | | C$_\alpha$D[a] | C$_\beta$D$_2$[b] | 1-CD$_3$ | 3-CD$_3$ | 5-CD$_3$ | 8-CD$_3$ | |
| 1656[c] | dp | 0 | −1 | +1 | +1 | 0 | −1 | ν$_{10}$ |
| 1633[c] | p | − | − | +4 | −1 | +2 | +2 | ν(C=C)[f] |
| 1610[c] | p | −1 | − | 0 | +1 | 0 | 0 | ν(C=C)[f] |
| 1605[d] | ap | +1 | +1 | +1 | −1 | −1 | −1 | ν$_{19}$ |
| 1591[c] | p | −3 | − | +6 | +3 | +2 | +3 | ν$_2$ |
| 1568[c] | dp | −2 | 0 | −2 | −2 | −3 | −4 | ν$_{11}$ |
| 1520[c] | p | 0 | −2 | +1 | +1 | +1 | 0 | ν$_3$ |
| 1435[d] | dp | −9 | − | − | +1 | +1 | +1 | β$_s$(CH$_2$)[f] |
| 1427[d] | dp | − | − | − | − | − | − | β$_s$(CH$_2$)[f] |
| 1404[d] | dp | −4 | +5 | +5 | 0 | −2 | +1 | ν$_{29}$ |
| 1383[d] | p | 0 | 0 | 0 | 0 | 0 | 0 | ν$_4$ |
| 1345[d] | ap | −4 | − | − | +1 | 0 | +1 | ν$_{20}$ |
| 1306[d] | ap | 0 | 0 | +3 | +2 | 0 | +2 | ν$_{21}$ |
| 1245[e] | p | −3 | 0 | −3 | − | −4 | −5 | |
| 1232[d] | dp | −1 | +2 | +4 | − | +2 | − | ν$_{13}$ |
| 1232[e] | p | +1 | 0 | +1 | −2 | 0 | 0 | |
| 1171[d] | dp | +15 | −2 | −7 | +2 | 0 | −1 | ν$_{30}$ |
| 1168[e] | p | − | +2 | 0 | +6 | +4 | +1 | |
| 1153[e] | p | − | +2 | −3 | − | +7 | − | |
| 1128[e] | p | 0 | 0 | 0 | +1 | −2 | −1 | |
| 1127[d] | | +7 | −1 | −2 | −1 | 0 | +3 | ν$_{22}$ |
| 1091[d] | | +7 | − | − | − | − | − | ϱ(CH$_2$)[f] |
| 990[c] | | − | − | +1 | −2 | +1 | +4 | ν$_5$ |
| 972[c] | | −2 | +1 | 0 | −12 | +1 | −2 | ν$_{31}$ |

a   2,4-CD=CH$_2$

b   2,4-CH=CD$_2$

c   Frequencies and isotope shifts are obtained from the spectra excited at 406.7 nm for the CH$_2$Cl$_2$ solution

d   Frequencies and isotope shifts are obtained from the spectra excited at 514.5 nm for the CCl$_4$ solution

e   Frequencies and isotope shifts are obtained from the spectra excited at 406.7 nm for the CCl$_4$ solution

f   The coordinates for the vinyl group are as follows; ν(C=C), C=C stretching, β$_s$(CH$_2$), CH$_2$ scissoring, ϱ(CH$_2$), CH$_2$ rocking

the corresponding iron-porphyrin is incorporated into a protein, the 2-formyl and 4-formyl porphyrins give rise to different ν$_{CH=O}$ frequencies[58].

The formyl substituent causes a characteristic red shift of the Q bands for the heme $a$ groups of cytochrome oxidase. Babcock and coworkers[59–62] investigated the RR spectra of heme $a$ derivatives, and found that the ν$_{CH=O}$ frequency depends on the spin- and oxidation-states of the heme iron. The assignments of the RR bands of heme $a$ derivatives are discussed by Choi et al.[63] and Kitagawa et al.[64]. Since a porphyrin with formyl substituents is unstable to laser irradiation in the Soret region, although it is stable in the protein, careful examination of the sample after the measurements of RR spectra is desirable upon excitation around 400–440 nm.

## 5.2 Assignments of Low Frequency Vibrations

In the lower frequency region, the RR bands due to the metal ligand stretching modes, and the out-of-plane and in-plane deformation modes are expected. Since they are sensitive to structure, assignments of the low frequency modes are important. As seen in Table 3 the low frequency vibrations of metallooctaalkylporphyrins involve significant contribution from the in-plane deformation modes of the peripheral groups and thus should depend not only on disposition and identity of the peripheral substituents but also on the interaction of the peripheral groups with their vicinity. For example, the RR spectrum of an intestinal peroxidase is quite similar to that of myoglobin in the region above 1200 cm$^{-1}$ but is distinctly different from it in the region below 450 cm$^{-1}$ [65]. This implies that the low frequency RR spectra sensitively reflect the protein-heme interaction through the peripheral groups.

The RR spectra of Ni(PP) and its specifically deuterated derivatives in the 200–800 cm$^{-1}$ region are shown in Fig. 12. Table 5 summarizes the polarization properties and the observed isotope shifts for Ni(PP) together with the observed frequencies for Ni(OEP). To interpret these results a new idea was introduced as explained below. When the deformation modes of the peripheral groups are strongly coupled with the deformation modes of pyrrole rings for the asymmetrically substituted porphyrins, each pyrrole ring would have different vibrational frequencies in principle. Since pyrrole rings I and II of PP in Fig. 4 have a pair of methyl and vinyl substituents, they are expected to have similar frequencies. Due to small interactions between the two rings, the in-phase and out-of-phase modes of the two pyrrole rings may slightly split. The approximate values of the splitting can be deduced from the calculated frequency separation between the $A_g$ and $B_g$ species of tetra methyl-vinyl porphyrin having $C_{4h}$ symmetry. When the phase difference between the two adjacent pyrrole rings is $\pi/2$, the modes fall on Raman inactive but IR active $E_u$ modes under $D_{4h}$ symmetry. The frequency of the $E_u$ species

**Table 5.** Observed frequencies and isotope shifts of Raman bands of Ni(PP) in the spectral region below 800 cm$^{-1}$

| Ni(OEP)[a] [cm$^{-1}$] | Unlabelled [cm$^{-1}$] | $\varrho$ | $C_\alpha D$[b] | $C_\beta D_2$[c] | 1-CD$_3$ | 3-CD$_3$ | 5-CD$_3$ | 8-CD$_3$ | Assignment |
|---|---|---|---|---|---|---|---|---|---|
| 751 | 755 | dp | 0 | 0 | 0 | 0 | −2 | −1 | $\nu_{16}$ |
| 712 | 710 |    | +1 | 0 | −1 | 0 | +1 | −1 | $\nu_{24}$ |
| 673 | 681 | p | −2 | −5 | −1 | −4 | −2 | −3 | $\nu_7$ |
|     | 422 | p | −2 | −19 | −10 | −2 | −1 | +1 | $\delta$[d] (I, II)[e] |
| 359 | 375 | p | +1 | +2 | −3 | −2 | −14 | −1 | $2\nu_{35}$ (III, IV) |
| 343 | 350 | p | 0 | +1 | 0 | +1 | −9 | +2 | $\nu_8$ (III, IV) |
|     | 330 |    | 0 | 0 | 0 | −3 | +2 | 0 | $2\nu_{18}$ |
|     | 320 |    | 0 | 0 | − | − | − | − | $2\nu_{18}$ |
| 260 | 269 | p | 0 | −1 | 0 | 0 | −6 | +2 | $\nu_9$ (III, IV) |

[a]  The frequencies are obtained from the spectrum excited at 406.7 nm under the same conditions as those for Ni(PP)
[b]  2,4-CD=CH$_2$
[c]  2,4-CH=CD$_2$
[d]  $\delta$, the C$_b$–CH=CH$_2$ bending mode of the vinyl group
[e]  The ring number

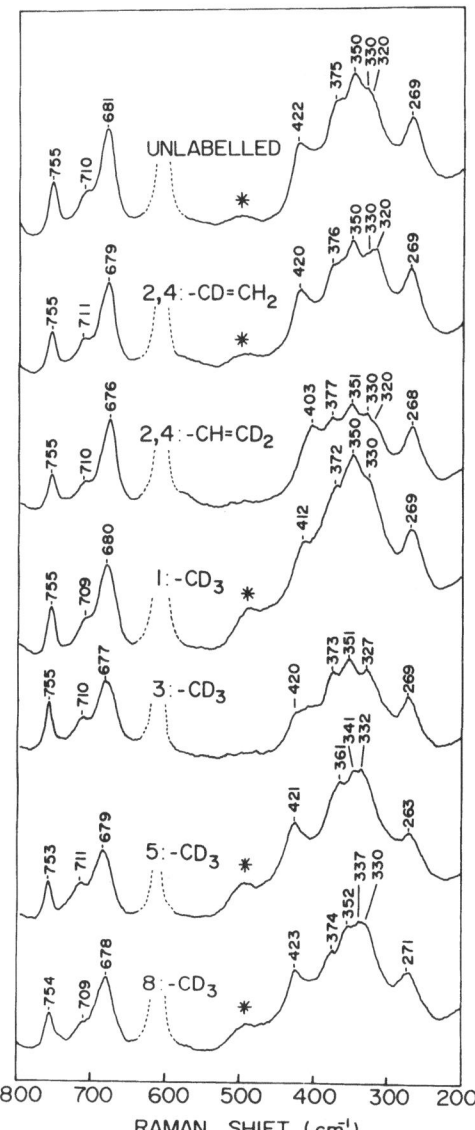

**Fig. 12.** Resonance Raman spectra of the dimethylester of Ni(PP) (top) and its deuterated isotopomers in CH$_2$Cl$_2$ solution. The labelled positions are specified at the right side of each spectrum. Solvent peaks are depicted by *dotted lines*. Asterisks denote the contribution from the quartz cell. Excitation, 406.7 nm (from Ref. 56)

might be closest to its intrinsic frequency, because the magnitude of the vibrational coupling between the adjacent pyrrole rings would be smaller than those of the in-phase and out-of-phase modes. Quite similar considerations can be applied to rings III and IV having the methyl and propionic side chains.

When a group on ring I or II is deuterated, it is assumed in this idea that the non-deuterated pyrrole rings keep their Raman bands and the deuterated ring gives rise to additional RR bands at frequencies slightly shifted from those of the non-deuterated rings, although the amount of the shift depends on the extent of the vibrational displacement of the hydrogen atom to be substituted with a deuteron. The same idea is applied to the deuteration of one of the rings III or IV. The amount of the deuteration shift was estimated from normal coordinate calculations for tetramethyl-tetravinyl Ni-porphyrin. In this way the low frequency RR bands of Ni(PP) were assigned as shown in Table 5[56]. It is noteworthy that the contribution of the four pyrrole rings to intensity of a given RR band is not always equal even if their frequencies are close to one another.

## 5.3 Metal-Pyrrole Stretching Modes

Since the M–N$_{pyrr}$ stretching mode ($\nu_{M-N}$) most directly reflects the M–N$_{pyrr}$ bond-strength, identification of the mode is an interesting subject. Under D$_{4h}$ symmetry the $\nu_{M-N}$ modes are factorized into A$_{1g}$ + B$_{1g}$ + E$_u$, and $\nu_9$(A$_{1g}$), $\nu_{18}$(B$_{1g}$), and $\nu_{52}$(E$_u$) are mainly responsible for them, although they are considerably mixed with other skeletal vibrations. In the expected frequency region, Ni(OEP) gave two bands at 222 and 260 cm$^{-1}$ but Ni(PP) in THF, CH$_2$Cl$_2$ or benzene gave only one band at 269 cm$^{-1}$[56]. The $\nu_9$ and $\nu_{52}$ modes were calculated at 230 and 264 cm$^{-1}$, respectively[29]. For Ni(OEP) Ogoshi et al.[33] found the IR band attributable to the E$_u$ species at 287 cm$^{-1}$. Considering the calculated frequency difference between the $\nu_9$ and $\nu_{52}$ modes and absence of the 222 cm$^{-1}$ band for Ni(PP), it seems most likely to assign the RR band of Ni(OEP) and Ni(PP) at 260 and 269 cm$^{-1}$, respectively, to the $\nu_9$ mode[56], although the 222 cm$^{-1}$ band of Ni(OEP) was previously assigned to the $\nu_9$ mode[28].

Sunder et al.[66] observed the IR active $\nu_{M-N}$ mode at 264 cm$^{-1}$ for Cu(OMP). Boucher and Katz[67] carried out a systematic IR study on divalent M(PP) complexes and the corresponding hematoporphyrin derivatives. They found a strong band near 350 cm$^{-1}$, which was deduced to arise from a coupling of the M–N$_{pyrr}$ stretching with a porphyrin skeletal deformation mode. The frequencies of IR bands at 970–920 cm$^{-1}$ and 530–500 cm$^{-1}$ are metal dependent and the frequency shifts can be linked to the order of the M–N bond strength[67]; 522(Co) > 521(Ni) > 512(Cu) > 509(Ag) > 506(Zn) $\approx$ 506(Cd) > 504(Mg). On the other hand, Ogoshi et al.[33] and Bürger et al.[68] found a few metal-sensitive IR bands for M(OEP) around 970–990, 910–925, 335–350, 200–280 and 120–160 cm$^{-1}$. The second lowest band is metal-isotope sensitive for Zn(OEP)[20], but the lowest one is not. Consequently, the bands at 203(Zn), 214(Mg), 234(Cu), 264(Co), 275(Pd) and 287(Ni) cm$^{-1}$ of M(OEP) were assigned to the E$_u$ species of the M–N$_{pyrr}$ stretching modes. Interestingly, the frequencies of all the metal sensitive IR bands change in parallel with the shifts of the Q and B bands.

Oshio et al.[69] searched structure-sensitive bands among twenty Fe(TPP) derivatives, finding two spin-state marker bands at 1333–1349 and 432–469 cm$^{-1}$ and an oxidation state marker at 790–906 cm$^{-1}$. It is noted that the spin-state marker bands are also metal-sensitive. For Fe(OEP) derivatives, Ogoshi et al.[70] found an IR marker band to distinguish between the pyramidal and planar structures of the FeN$_4$ core; the Fe–N$_{pyrr}$ stretching mode gives rise to apparently a doublet around 270 cm$^{-1}$ and singlet around 310 cm$^{-1}$ for the pyramidal and planar structures, respectively.

## 5.4 Out-of-Plane Modes

Although Raman intensities of out-of-plane modes are not expected to be enhanced in resonance with the Q and B bands of a metalloporphyrin, Choi and Spiro[71] demonstrated the appearance of such Raman bands below 1000 cm$^{-1}$ for Fe(PP) derivatives:

a $C_m$–H out-of-plane deformation at $840 \text{ cm}^{-1}$ ($E_g$), pyrrole folding modes at $425$–$510 \text{ cm}^{-1}$ ($A_{2u}$), a $C_m$–$C_a$ deformation mode at $300 \text{ cm}^{-1}$ ($A_{2u}$), and a pyrrole tilting mode at $\sim 255 \text{ cm}^{-1}$. These modes are sensitive to protein-heme interactions in the protein. The appearance of the former three bands were confirmed by Willems and Bocian[31] for Ni(PP) derivatives. In the IR spectra, on the other hand, the $A_{2u}$ species is expected to appear, and in fact Ogoshi and Yoshida[72] reported a metal-sensitive $C_m$–H out-of-plane deformation mode for M(OEP); $834(\text{Mg})$, $836(\text{Zn})$, $837(\text{Cu})$, $837(\text{Co})$, $837(\text{Ni})$, $839(\text{Pd}) \text{ cm}^{-1}$. Kincaid et al.[30] observed the IR spectra of matrix isolated Mn, Fe, Co, Ni, Cu, and Zn complexes of OEP at 15 K and pointed out that the metal sensitive bands involve the vibrations of the $C_a$–$C_m$ bonds, and accordingly that the metal dependent frequency changes arise from different extent of ruffling. The vibrational modes of the IR($A_{2u}$) and Raman($E_g$) active out-of-plane vibrations, which were calculated with appropriate force constants, are depicted elsewhere[73].

# 6 Axial Ligand Modes

Since the axial coordination site of metalloporphyrin is the catalytic site for chemical as well as biological systems, we are curious to know an interaction mode between the metal and an axial ligand (L). Vibrational spectroscopy is the most powerful means to probe it as long as the M–L stretching ($\nu_{M-L}$) and internal modes of L can be identified.

## 6.1 Hemoproteins

In relation to the strain model for hemoglobin (Hb) cooperativity[74], the assignment of the Fe-histidine stretching ($\nu_{Fe-His}$) RR band of Hb has attracted general attention. Hori and Kitagawa[75] studied the $Fe^{2+}$–N(2-MeIm) stretching mode for the five-coordinate Fe($T_{piv}$PP)(2-MeIm) ($T_{piv}$PP: picket-fence porphyrin) and Fe(PP)(2-MeIm). Figure 13 shows the RR spectra of the five-coordinate Fe(PP)(2-MeIm) and Fe(PP)(2-MeIm-$d_5$) excited at 441.6 nm. Previously the RR bands at 377, 410, and $206 \text{ cm}^{-1}$ had been proposed as the $\nu_{Fe-His}$ mode of Mb by Kincaid et al.[76a], Desbois et al.[77], and Kitagawa et al.[78], respectively. However, as shown in Fig. 13, neither the band at 377 nor that at $\sim 410 \text{ cm}^{-1}$ exhibited a frequency shift with the deuterated 2-MeIm(2-MeIm-$d_5$), but the RR band at $206 \text{ cm}^{-1}$ was shifted to a lower frequency with 2-MeIm-$d_5$. This band exhibited a small shift to a higher frequency upon $^{54}$Fe substitution, and therefore was assigned to the $\nu_{Fe-(2\text{-MeIm})}$ mode of Fe(PP)(2-MeIm). This assignment was later supported by Kincaid et al.[76b]. The conclusive evidence for the assignment of the $\nu_{Fe-His}$ mode was obtained by Rousseau and coworkers[79], who supported Kitagawa et al.'s assignment. From the $\nu_{Fe-His}$ RR band, Nagai et al.[80] demonstrated the presence of strain in the Fe–His bond of low affinity Hb, particularly in the α subunit[81], and more recently Matsukawa et al.[82] observed continuous distribution of the $\nu_{Fe-His}$ frequencies under some correlation with the oxygen affinity. The $\nu_{Fe-His}$ frequency of the photo-dissociated carbonmonoxy Hb is distinctly different from that of the equilibrium state[83].

The $\nu_{Fe-L}$ mode of Fe(PP)(2-MeIm) is very solvent sensitive; its frequency becomes lower as the solvent becomes more hydrophobic[84], but in contrast, it becomes higher

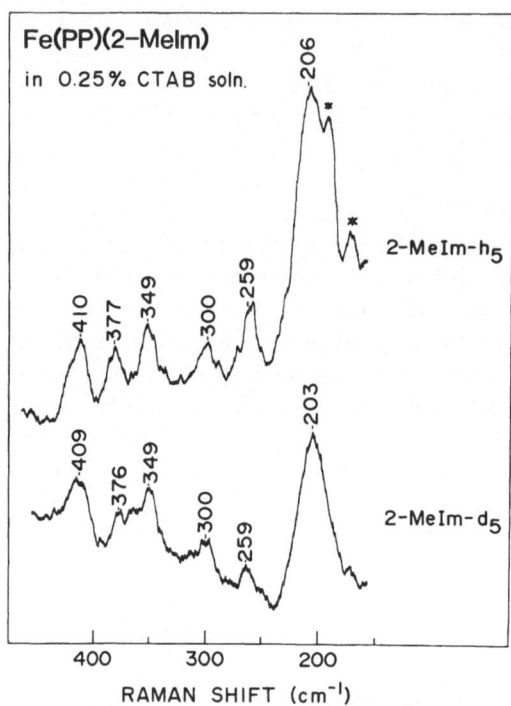

**Fig. 13.** Resonance Raman spectra of Fe(PP)(2-MeIm) and Fe(PP)(2-MeIm-$d_5$) in 0.25% (CTAB (w/w) aqueous solutions; excitation, 457.9 nm (from Ref. 75)

upon ionization of the coordinated imidazole by ca. 27 cm$^{-1}$, and based on this observation the higher $v_{Fe–His}$ frequency for peroxidase than for an oxygen carrier was attributed to strong hydrogen bonding of the proximal histidine and resultant strong σ basicity of an imidazolate-like state[84, 85]. In this way the Fe–L bond in the fifth coordination site of heme proteins plays an essential role for their functions and the Fe–L stretching RR band serves as a sensitive probe for it.

The manner in which $O_2$ is bound to metalloporphyrins and hemoproteins has been a matter of chemist's concern, since it is directly related to the reactivity of the ligand and thus the function of the protein. The internal modes of the ligand (L) were examined more deeply with IR spectroscopy[86]. Brunner[87] first observed the RR band attributable to the Fe–$O_2$ stretching mode at 567 cm$^{-1}$ for oxyHb. Unexpectedly, the Fe–$O_2$ stretching frequency remained unchanged upon the T–R transition of the Hb quaternary structure[80]. The RR band due to the O–O stretching mode ($v_{O–O}$) was observed at 1107, 1137, and 1152 cm$^{-1}$ for Co-substituted oxyHb by Tsubaki and Yu[88]. The former two bands were interpreted as a Fermi doublet arising from the porphyrin mode at 1123 cm$^{-1}$ and the unperturbed $v_{O–O}$ mode at 1122 cm$^{-1}$. With IR spectroscopy only the 1107 cm$^{-1}$ band was observed by Maxwell and Caughey[89]. Since the other counterpart of the Fermi doublet was not observed in the IR spectra and furthermore it is not clear whether there is a vibrational interaction between the O–O stretching mode and a porphyrin mode, this interpretation should be examined with appropriate model compounds. The probability of the Fermi resonance between the O–O stretch and an overtone of the Co–$O_2$ stretch seems low, because the latter band was identified at 538 cm$^{-1}$.

It is noteworthy that the O–O stretching Raman band of oxygenated cytochrome P-450 was identified at 1140 cm$^{-1}$ (1074 cm$^{-1}$ for $^{18}O_2$)[90] by using a spinning cell at

$-60\,^{\circ}\text{C}$. This is the first observation of the $\nu_{OO}$ mode in the RR spectra of the iron-containing hemeproteins, although unexpectedly, this frequency is close to the $\nu_{OO}$ frequency of Co-substituted oxyHb.

The $Fe^{3+}-F^-$ ($\nu_{Fe-F}$) and $Fe^{3+}-OH^-$ stretching ($\nu_{Fe-OH}$) modes of Hb were investigated by Asher et al.[91-93] and found at 471 and 495 cm$^{-1}$, respectively, upon excitation around 550–650 nm. The $\nu_{Fe-F}$ frequency is shifted to 443 cm$^{-1}$ when $F^-$ is hydrogen bonded to distal histidine. The $\nu_{Fe-F}$ frequencies for Hb and Mb are considerably lower than that of Fe(OEP)F shown later in Table 8. With regard to the $N_3^-$ complex, controversial assignments had been proposed partly due to its characteristic spin-equilibrium. The $N_3^-$ complexes of Hb and Mb have a low-spin ground state and exhibit a measurable spin-equilibrium at room temperature[94]. IR bands due to the antisymmetric stretching mode of bound azide of $MbN_3$ at 2023 and 2046 cm$^{-1}$ were assigned to the low- and high-spin states, respectively[95]. It had been suggested that a broad z-polarized absorption band of $MbN_3$ around 650 nm arised from a CT band of the low-spin component[96]. By excitation into the 640 nm absorption region, Asher et al.[91] observed selective enhancement of a Raman band at 413 cm$^{-1}$ in $MbN_3$ and assigned it to the $Fe^{3+}-N_3$ stretching mode. On the other hand, Desbois et al.[77] observed a $^{15}N$ isotope sensitive RR band at 570 cm$^{-1}$ for $MbN_3$ upon excitation at the Soret band. Later Asher and Schuster[92] suggested that the 413 cm$^{-1}$ mode arised from the high-spin form and the 570 cm$^{-1}$ mode from the low-spin form. Contrary to it, Tsubaki et al.[97] demonstrated that the intensity of the 413 cm$^{-1}$ mode increases with decreasing temperature in spite of the decrease in population of the high-spin component. They concluded from studies on temperature dependence, isotope substitution, depolarization measurements, and normal coordinate calculations that the 413 and 570 cm$^{-1}$ modes are the $Fe^{3+}-N_3$ stretching and the azide internal bending modes, respectively, both in the low-spin state.

For the $CN^-$ complex of Hb, two isotope-sensitive bands were observed. With $^{12}C^{14}N$, $^{13}C^{14}N$, $^{12}C^{15}N$ and $^{13}C^{15}N$ adducts, the bands were observed at 453, 450, 450, and 446 cm$^{-1}$ for one series and at 412, 414, 412, and 414 cm$^{-1}$ for the other, respectively[98]. From the pattern of the isotope shift, the former and latter series were assigned to the Fe–CN stretching ($\nu_{Fe-CN}$) and Fe–C–N bending modes, respectively. These modes were found to be insensitive to a change of the quaternary structure of Hb[99].

For the NO complex of metMb, two isotope-sensitive bands were observed. With $^{14}N^{16}O$, $^{15}N^{16}O$, and $^{14}N^{18}O$ adducts, the bands were observed at 595, 589, and 587 cm$^{-1}$ for one series and at 573, 562, and 569 cm$^{-1}$ for the other one, respectively. The former and latter series were assigned to the $Fe^{3+}-NO$ stretching ($\nu_{Fe-NO}$) and the $Fe^{3+}-N=O$ bending modes ($\delta_{FeNO}$), respectively[100]. With the NO complex of deoxyMb, only the $\delta_{FeNO}$ band was observed. The NO stretching mode ($\nu_{NO}$) can be observed with IR spectroscopy. The $\nu_{NO}$ frequency depends on the coordination number; 1660 cm$^{-1}$ for Fe(PPDME)NO (PPDME: protoporphyrin-dimethylester) and 1676 cm$^{-1}$ for Fe(PPDME)NO(1-MeIm)[101]. The corresponding IR band of HbNO was identified with $^{15}N$ isotope at 1615 cm$^{-1}$, which was split into two bands at 1615 and 1668 cm$^{-1}$ in the presence of inositol hexaphosphate, suggesting coexistence of the five- and six-coordinate NO complexes in the T-state HbNO[101]. There are cases in which the $\nu_{Fe-L}$ mode is not observed but internal modes of L are found. For $NO_2^-$ and $SCN^-$ complexes of metHb, $NO_2$ symmetric stretching and SCN bending modes were observed at 1324 and 510 cm$^{-1}$, respectively[99].

## 6.2 Metalloporphyrins

The M–L stretching and internal modes of L have been investigated for a wide variety of metalloporphyrins. In the case of L = $O_2$, the ligand mode is identified by $^{18}O_2$ isotope labelling. The frequency of the O–O stretching mode ($\nu_{O-O}$) is located at 1555, 1145, and 842 cm$^{-1}$ for $O_2$, $O_2^-(KO_2)$[102], and $O_2^{2-}(Na_2O_2)$[103], respectively. It is also generally accepted that the asymmetric end-on and symmetric side-on structures give rise to $\nu_{O-O}$ around 1140–1280 and 800–1100 cm$^{-1}$, respectively[104]. The observed frequencies for the $\nu_{O-O}$ bands of metalloporphyrins are summarized in Table 6.

With the IR matrix isolation technique, Nakamoto et al.[105] first succeeded in observing the $\nu_{O-O}$ band of a $Fe^{2+}$ porphyrin. Fe(TPP)$O_2$ exhibited the bands at 1195 and 1106 cm$^{-1}$. The former was stable at 100–240 K whereas the latter was stable below 100 K, and the two bands were assigned to structural isomers. The $\nu_{O-O}$ frequency depends little on the type of porphyrin; 1143 cm$^{-1}$ for Co(TPP)(Py)$O_2$ (Py: pyridine), 1145 cm$^{-1}$ for Co(OEP)(Py)$O_2$, 1143 cm$^{-1}$ for Co(PPDME)(Py)$O_2$, 1143 cm$^{-1}$ for Co(DFDPDME)(Py)$O_2$ (DFDPDME: diformyldeuteroporphyrin-dimethylester), and 1151 cm$^{-1}$ for Co(T$_{piv}$PP)(Py)$O_2$ all in CH$_2$Cl$_2$[106]. However, it depends on the coordination number; 1278 cm$^{-1}$ for Co(TPP)$O_2$[36], 1142 cm$^{-1}$ for Co(TPP)(1-MeIm)$O_2$[107], and 1143 cm$^{-1}$ for Co(TPP)(Py)$O_2$[108]. As shown in Table 6, the $\nu_{O-O}$ frequencies of the free base adducts are in the order of Co(TPP)$O_2$ (1278 cm$^{-1}$)[36] > Fe(TPP)$O_2$ > Mn(TPP)$O_2$ (983 cm$^{-1}$)[112]. Co–$O_2$ adducts are of the end-on type whereas Mn–$O_2$ and Ti–$O_2$ adducts are of the side-on type. The Fe–$O_2$ adducts usually adopt the end-on type as in hemopro-

**Table 6.** $O_2$ stretching frequencies (cm$^{-1}$) of dioxygen adducts of metalloporphyrins

| Molecule | Structure | $^{16}O_2$ | $^{16}O^{18}O$ | $^{18}O_2$ | Reference |
|---|---|---|---|---|---|
| Fe(OEP)$O_2$ | end-on<br>side-on | 1190<br>1104 | a | 1124<br>1042 | 109 |
| Fe(TPP)$O_2$ | end-on<br>side-on | 1195<br>1106 | b | 1127<br>1043 | 109 |
| Co(OEP)$O_2$ | end-on | 1275 | – | 1202 | 110 |
| Co(TPP)$O_2$ | end-on | 1278 | 1252, 1241 | 1209 | 36 |
| Mn(OEP)$O_2$ | side-on | 991 | – | 934 | 111 |
| Mn(TPP)$O_2$ | side-on | 983 | 957 | 933 | 112 |
| Ti(OEP)$O_2$ | side-on | 898 | – | – | 113 |
| Co(TPP)(1-MeIm)$O_2$ | end-on | 1142 | – | 1071 | 107 |
| Co(Cap)(1-MeIm)$O_2$[c] | end-on | 1176 | – | 1084 | 107 |
| Co(TpivPP)(1-MeIm)$O_2$ | end-on | 1150 | – | 1065 | 114 |
| Fe(Cap)(1-MeIm)$O_2$ | end-on | 1172 | – | 1097 | 107 |
| Fe(TpivPP)(1-MeIm)$O_2$ | end-on | 1159 | – | 1075 | 114 |
| Cr(TPP)(Py)$O_2$ | end-on | 1142 | – | – | 115 |
| [Fe$^{3+}$(OEP)$O_2$]$^-$ | side-on | 806 | – | 759 | 116 |
| HbO$_2$ | end-on | 1107 | – | 1065 | 117 |
| MbO$_2$ | end-on | 1103 | – | 1065 | 117 |

[a]  Hidden under the strong Fe(OEP) bands
[b]  The $\nu_{16O18O}$ mode for the end-on structure was observed at 1162 cm$^{-1}$ while that for the side-on structure was hidden under the strong Fe(TPP) band at 1076 cm$^{-1}$
[c]  Cap: capped porphyrin

teins, but $[Fe(OEP)O_2]^-$, in which the iron ion is in a ferric high-spin state, assumes the side-on geometry[116].

The $Co-O_2$ stretching mode ($\nu_{Co-OO}$) changes with oxygen affinity but its behavior is opposite to ordinary expectation. $Co(T_{piv}PP)(1,2-Me_2Im)O_2$ having low affinity gives the $\nu_{Co-OO}$ RR band at 527 cm$^{-1}$, which is appreciably higher than the 516 cm$^{-1}$ of $Co(T_{piv}PP)(1-MeIm)O_2$ with higher affinity[118]. The Fe–O–O bending vibration has never been identified with IR and RR spectroscopies for model compounds as well as hemoproteins. Nevertheless, two $^{18}O$ isotope sensitive RR bands were observed at 486 and 279 cm$^{-1}$ for oxyphthalocyaninato-iron(II)$[Fe(Pc)O_2]$ in an oxygen matrix at 15 K[119]. The former shifts to 477 and 466 cm$^{-1}$ with $^{17}O_2$ and $^{18}O_2$, respectively, but the latter shifts to 275 and 271 cm$^{-1}$. The low and high frequency components were assigned to the Fe–O–O bending and Fe–$O_2$ stretching modes, respectively, although both are mixed with each other[119].

The $Fe^{4+}=O$ stretching mode, $\nu_{Fe=O}$, was observed at 852 cm$^{-1}$ (818 cm$^{-1}$ for $^{18}O$) by Nakamoto et al. for Fe(TPP) in an oxygen matrix using Raman spectroscopy[120, 121]. The $Fe^{4+}=O$ derivative was produced by photolysis of $Fe(TPP)O_2$. A similar band was observed at 843 cm$^{-1}$ for a toluene solution of FeO(TMP) (TMP: tetramesityl porphyrin)[122], which was generated by thermal decomposition of the μ-peroxo-bridged $Fe^{3+}$ porphyrin complex[123]. These are for the five-coordinate complexes, but the $\nu_{Fe=O}$ mode for the six-coordinate complexes was observed at lower frequencies[124]; 829 cm$^{-1}$ for $FeO(T_{piv}PP)(THF)$ and 807 cm$^{-1}$ for $FeO(T_{piv}PP)(1-MeIm)$. The iron(IV) porphyrin is an important intermediate in a catalytic cycle of peroxidase, for which the RR band due to the $\nu_{Fe=O}$ mode is identified at 787 cm$^{-1}$ [125-127]. Interestingly, the oxygen atom of the $Fe^{4+}=O$ heme in the enzymic intermediate is exchanged with bulk water only when the bound oxygen is hydrogen bonded to the adjacent amino acid residue[128].

In the case of the $Fe^{4+}$ heme, the extra oxidizing equivalent is fairly localized to the iron ion. However, when the oxidizing equivalent is delocalized to the porphyrin ring, a characteristic marker band is observed in the IR spectra at ~ 1280 cm$^{-1}$ for $M(TPP\cdot)^+$ and at ~ 1550 cm$^{-1}$ for $M(OEP\cdot)^+$ [129]. The RR and IR spectra of M(EP) dications (EP: etioporphyrin) were treated by Aleksandrov et al.[130].

The CO molecule usually binds to the site where dioxygen is bound, and a binding mode of CO is often investigated to probe the binding site of $O_2$. The CO molecule binds to the metal ion in a linear or bent fashion, and the CO stretching ($\nu_{CO}$) observed with IR spectroscopy and the M–CO stretching ($\nu_{M-CO}$) and the M–C–O bending ($\delta_{MCO}$) modes observed with RR spectroscopy, reflect the structure. Table 7 lists the frequencies of the $\nu_{CO}$ modes for CO adducts of some metalloporphyrins.

Yu and coworkers investigated extensively the RR spectra of the CO complexes. For $Fe(T_{piv}PP)CO(1-MeIm)$ the isotope sensitive bands were observed at 489, 485, 481, and 477 cm$^{-1}$ for the $^{12}C^{16}O$, $^{13}C^{16}O$, $^{12}C^{18}O$, and $^{13}C^{18}O$ adducts, respectively, and were assigned to $\nu_{Fe-CO}$[137]. The $\nu_{Fe-CO}$ frequencies of $Fe(T_{piv}PP)COL$ were 486, 489, 496, and 527 cm$^{-1}$ for L = Py, 1-MeIm, 1,2-MeIm, and THF, respectively[137]. This suggests that the weaker $Fe^{2+}$–L bond at the trans position provides a stronger Fe–CO bond. The cis- and trans-effects on the CO stretching frequencies in iron porphyrins were studied with IR spectroscopy[138] and a comprehensive review on this subject for iron, ruthenium, and osmium porphyrins has been published[133]. If there were steric hindrance for the CO binding site, the Fe–C–O linkage might adopt a bent structure. Yu et al.[135] examined such a possibility with the strapped porphyrin illustrated in Fig. 4. The RR bands due to

**Table 7.** CO stretching frequencies ($cm^{-1}$) of carbon monooxide adducts of metalloporphyrins

| Molecule | CO | $^{13}CO$ | Reference |
|---|---|---|---|
| Co(TPP)(CO)$_2$ | 2078 | 2032[a] | 36 |
| Co(TPP)CO | 2073 | 2032[a] | 36 |
| Co(TPP)(CO)(O$_2$) | 2089 | | 36 |
| Co(TPP)(CO)(NO) | ~2078 | | 36 |
| Fe(TPP)(CO)$_2$ | 2042 | | 131 |
| Fe(TPP)CO | 1973 | | 131 |
| Ru(TPP)(CO)$_2$[b] | 2005 | | 132 |
| Fe(OEP)(1-MeIm)CO | 1970 | | 134 |
| Fe(TpivPP)(1-MeIm)CO | 1969 | | 114 |
| Fe(Cap)(1-MeIm)CO | 2002 | | 107 |
| Fe(SP-15)CO[c] | 1945 | 1901 | 135 |
| Fe(SP-14)CO[c] | 1939 | 1894 | 135 |
| Fe(SP-13)CO[c] | 1933 | 1888 | 135 |
| heme 5 (CO)[d] | 1954 | 1910 | 135 |
| Ru(OEP)(CO)(SCH$_2$CH$_2$CO$_2$CH$_3$)$^-$ | 1917 | | 136 |
| Ru(OEP)(CO)(PPh$_3$) | 1966 | | 136 |
| Fe(OEP)(CO)(SCH$_2$CH$_2$CO$_2$CH$_3$)$^-$ | 1950 | | 136 |

[a] Broad band
[b] The CO stretching frequencies of various Ru and Os complexes are summarized in Ref. 133, but the Ru complexes listed in this table are not contained in Ref. 133
[c] SP: strapped porphyrin, see Fig. 4
[d] Heme 5 has similar peripheral substituents to those of the strapped porphyrin, but it has no hydrocarbon strap which hinders the CO binding

$\nu_{Fe-CO}$ were observed at 509, 512, and 514 $cm^{-1}$ for n = 7, 6, and 5, respectively, which are significantly higher than 495 $cm^{-1}$ of the corresponding unstrapped heme. Accordingly, Yu et al.[135] suggested that the increase of steric hindrance increases the $\nu_{Fe-CO}$ frequency, but lowers the CO affinity. The $\nu_{CO}$ frequencies are 1945, 1939, and 1932 $cm^{-1}$ for n = 7, 6, and 5, respectively, which are lower than 1954 $cm^{-1}$ for the unstrapped species. The $\delta_{FeCO}$ band was identified from the characteristic zigzag behavior for the isotope shift in contrast with the monotonous pattern of the $\nu_{Fe-CO}$ band. For instance, the $\delta_{FeCO}$ band for n = 6 appeared at 578, 563, 575, and 561 $cm^{-1}$ for $^{12}C^{16}O$, $^{13}C^{16}O$, $^{12}C^{18}O$, and $^{13}C^{18}O$ adducts, respectively[135]. The $\delta_{FeCO}$ mode of HbCO was observed at 577 $cm^{-1}$ [139], which was close to the frequency observed for the strapped porphyrin. It was pointed out that the Raman band due to $\delta_{FeCO}$ is observable only when binding of CO is sterically hindered for the model system[135]. On the basis of the $\nu_{Fe-CO}$, $\delta_{FeCO}$ and $\nu_{CO}$ frequencies, the bond angle of the Fe–C–O linkage was estimated for proteins[98].

The results shown in Table 7 indicate that the $\nu_{CO}$ frequencies of the 1:2 adducts are always higher than those of 1:1 adducts. This may result from the decreased $\pi$-back donation from metal to CO which is caused by competition between two CO groups in the 1:2 adduct[36]. For 1:2 adducts, $\nu_{CO}$ frequencies are in order of $Co^{2+} > Fe^{2+} > Ru^{2+} > Os^{2+}$. Thus, the degree of $\pi$-back donation to CO increases in the order of $Co^{2+} < Fe^{2+} < Ru^{2+} < Os^{2+}$ [36, 133].

The $\nu_{M-L}$ band had originally been looked for with $Fe^{3+}$ porphyrin complexes. Nakamoto and coworkers first observed the $^{54}Fe$ isotopic frequency shift for various

$Fe^{3+}(OEP)L$ ($L = F^-$, $N_3^-$, $NCS^-$) and assigned the $\nu_{Fe-L}$ bands[70]. Table 8 summarizes the M–L stretching frequencies observed with IR spectroscopy so far. Although the IR data were obtained for a solid matrix[70], the $\nu_{Fe-L}$ frequencies observed for a solution with RR spectroscopy (606, 364, and 279 $cm^{-1}$ for $L = F^-$, $Cl^-$, and $Br^-$, respectively)[142] were in good agreement with them. For $Mn^{3+}$-porphyrins, the ligand sensitive RR bands were reported at 495, 285, 245, and 186 $cm^{-1}$ for $L = F^-$, $Cl^-$, $Br^-$, and $I^-$, respectively, and they are resonance enhanced upon excitation at the CT band around 460–490 nm[143].

It is of particular interest to compare the $Fe^{3+}$–$SR^-$ and $Fe^{3+}$–$SR_2$ stretching frequencies in connection with cytochrome P-450 and cytochrome $c$ having the thiolate and thioether ligands, respectively. Oshio et al.[140] observed IR bands attributable to the $\nu_{Fe-SR^-}$ and $\nu_{Fe-SR_2}$ modes. The results are also included in Table 8. The $\nu_{Fe-SR^-}$ band appears around 333–341 $cm^{-1}$, which are appreciably lower than 351 $cm^{-1}$ of the $\nu_{Fe-SR^-}$ band of oxidized cytochrome P-450 observed with RR spectroscopy[144].

Regarding the six-coordinate complexes of the type L–Fe–L, the symmetric and anti-symmetric L–Fe–L stretching modes are expected to appear in Raman and IR spectra, respectively. The anti-symmetric stretching mode of the $RS^-$–$Fe^{3+}$–$SR^-$ linkage was observed at 345 $cm^{-1}$ which is higher by ca. 20 $cm^{-1}$ than the corresponding mode of the $R_2S$–$Fe^{3+}$–$SR$ linkage. This implies that the Fe–$SR^-$ bond is stronger than the Fe–$SR_2$ bond by 12% in terms of the force constant. For $[Fe(OEP)(Im)_2]^+$, the symmetric and anti-symmetric L–$Fe^{3+}$–L stretching modes were reported at 290 (RR)[142] and 377 (IR)[70] $cm^{-1}$, respectively. Wright et al.[145] analyzed the RR spectra of $Fe(MP)(Py)_2$, locating the Py–$Fe^{2+}$–Py symmetric stretching mode at 179 $cm^{-1}$ and the $d_\pi(Fe) \rightarrow \Pi^*(Py)$ CT band at 496.5 nm on the basis of the Raman excitation profiles. The RR spectra of

**Table 8.** IR frequencies ($cm^{-1}$) of metal-axial ligand stretching modes of iron porphyrins

| Molecule | Mode | Frequency | Isotope shift ($^{54}Fe$) | Reference |
|---|---|---|---|---|
| Fe(OEP)F | Fe–F | 605.5 | 3 | 70 |
| Fe(OEP)Cl | Fe–Cl | 357 | – | 70 |
| Fe(OEP)Br | Fe–Br | 270 | – | 70 |
| Fe(OEP)I | Fe–I | 246 | – | 70 |
| Fe(OEP)NCS | Fe–NCS | 315 | 1.5 | 70 |
| Fe(OEP)N$_3$ | Fe–N$_3$ | 421 | 3 | 70 |
| Fe(OEP)SC$_6$H$_5$ | Fe–S$^-$ | 341.0 | 1.5 | 140 |
| Fe(TPP)SC$_6$H$_5$ | Fe–S$^-$ | 335.5 | 2.5 | 140 |
| [Fe(TPP)(SC$_6$H$_5$)$_2$]$^-$ | S$^-$–Fe–S$^-$ | 345.0 | – | 140 |
| Fe(OEP)(SC$_6$H$_4$-p-NO$_2$) | Fe–S$^-$ | 338.0 | – | 140 |
| Fe(PPDME)(SC$_6$H$_4$-p-NO$_2$) | Fe–S$^-$ | 333.5 | – | 140 |
| [Fe(TPP)(THT)$_2$]ClO$_4$$^a$ | S–Fe–S | 328.0 | 2.5 | 140 |
| [Fe(TPP)(PMS)$_2$]ClO$_4$$^b$ | S–Fe–S | 323.5 | 2.0 | 140 |
| [Fe(OEP)(γ-Pic)$_2$]ClO$_4$ | γ-Pic–Fe–γ-Pic | 373 | 2.5 | 70 |
| [Fe(OEP)(Im)$_2$]ClO$_4$ | Im–Fe–Im | 376.5 | 3.5 | 70 |
| [Fe(OEP)(BIm)$_2$]ClO$_4$$^c$ | BIm–Fe–BIm | 333 | – | 70 |
| [Fe(TPP)]$_2$O | Fe–O–Fe | 892, 878$^d$ | – | 141 |

$^a$ THT; tetrahydrothiophene
$^b$ PMS; pentamethylene sulfide
$^c$ BIm; benzimidazole
$^d$ Observed as a doublet

Fe(OEP)(Py)$_2$, Ru(OEP)(Py)$_2$ and Os(OEP)(Py)$_2$ were investigated by Schick and Bocian[146]. From the analysis of the Raman excitation profile for the bound-Py modes, they explored the CT bands from metal to Py at 514, 472, and 503 nm for the Fe$^{2+}$, Ru$^{2+}$, and Os$^{2+}$ complexes, respectively. The IR frequencies for the L–Fe–L type complexes are also included in Table 8.

# 7 Vibrational Spectra of Reduced Metalloporphyrins

When a metalloporphyrin is reduced without hydrogenation, it becomes a porphyrin anion, but with hydrogenation, it gives a metallochlorin or a metallobacteriochlorin. Vibrational spectra for the two cases are described separately below.

## 7.1 Metalloporphyrin Anions

Physicochemical properties of so-called Fe$^{1+}$ and Fe$^0$ porphyrins have been reviewed by Reed[147], who pointed out a possibility of partial migration of the reducing equivalents to the macrocycle. Vibrations of the porphyrin are expected to reflect the extent of delocalization of the reducing equivalents. The RR spectra of mono- and di-anions of Zn(TPP) and VO(EP) were reported by Ksenofontova et al.[148] and Yamaguchi et al.[149] and appreciable low frequency shifts of RR bands compared with those of neutral porphyrins were pointed out. Contrary to it, Srivatsa et al.[150] did not find any difference between the RR spectra of Fe(TPP) and [Fe(TPP)]$^-$ both in DMF (DMF: dimethylformamide).

To clarify this contradiction, Teraoka et al.[151] produced Fe(OEP), [Fe(OEP)]$^-$ and [Fe(OEP)]$^{2-}$ with a sodium-mirror contact reduction technique and observed the absorption, EPR, and RR spectra for an identical preparation. Figure 14 shows the RR spectra of [Fe(OEP)]$^-$, [Fe(OEP–$^{15}$N$_4$)]$^-$, and [Fe(OEP–d$_4$)]$^-$ in THF. The inset shows the EPR spectrum of the sample. The signals characteristic of Fe$^{1+}$ are seen at $g_\perp$ = 2.26 and $g_\parallel$ = 1.93. [Fe(OEP)]$^-$ gave the $\nu_4$ band at the same frequency as the Fe(OEP) but $\nu_{10}$ at extremely lower frequency and the overall spectral pattern of [Fe(OEP)]$^-$ in THF was distinctly different from that of Fe(OEP) in THF. Upon further reduction, it became EPR silent, and gave $\nu_4$ at considerably lower frequency but $\nu_{10}$ and other C$_a$C$_m$ stretching bands at the similar frequencies to those of the Fe(OEP) in THF. In this state, the $\nu_2$ and $\nu_{11}$ modes were shifted to lower frequencies. Consequently, Teraoka et al.[151] suggested that the [Fe(OEP)]$^-$ in THF has no significant contribution from π-anion radical but [Fe(OEP)]$^{2-}$ has such character.

## 7.2 Comparison of Metallochlorins and Metalloporphyrins

Metallochlorins, in which one of four C$_b$C$_b$ bonds of metalloporphyrins is saturated, serve as models of chlorophylls and heme $d$. Finding of various heme $d$ proteins in particular raised the question, what kind of differences between iron-porphyrins and iron-chlorins made the latter complexes more suitable for a prosthetic group of some

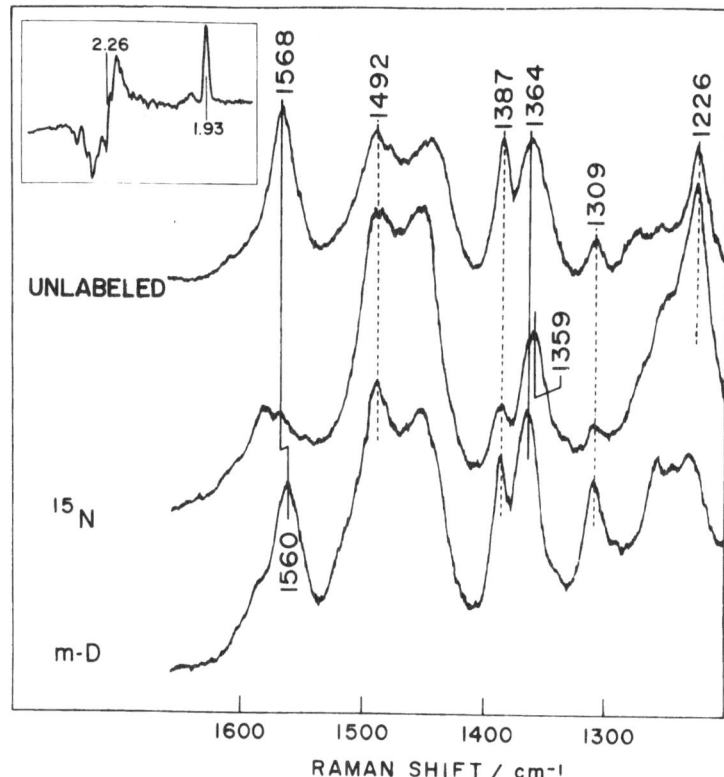

**Fig. 14.** Resonance Raman spectra of $[Fe(OEP)]^-$ (upper), $[Fe(OEP-^{15}N_4)]^-$ (middle) and $[Fe(OEP-d_4)]^-$ (bottom) in a THF solution at 285 K excited at 441.6 nm. The inset figure shows the EPR spectra (77 K) of the sample used for the Raman measurements (from Ref. 151)

heme proteins. RR spectroscopy is expected to tell the effects on the Fe–L bond as well as the structure of the macrocycle brought about by saturation of one $C_bC_b$ bond of the Fe-porphyrin.

RR spectra of Cu(OEC) were first reported by Ozaki et al.[152]. Figure 15 compares the polarized RR spectra of Cu(OEC) with those of Cu(OEP). The RR bands of Cu(OEC) in the 900–1700 cm$^{-1}$ region were assigned on the basis of frequency shifts upon 15,20-deuteration of meso-carbons and $^{15}N$ substitution of pyrrolic nitrogens[152]. Anderson et al.[153] and Hanson et al.[154] also reported the RR spectra of metallo-chlorin derivatives. Ozaki et al.[39, 155] extended the comparative study of Fe(OEP) and Fe(OEC) to the Fe–L stretching and ligand internal modes. Here the vibrational spectra of M(OEC) are explained on the basis of M(OEP) and then the difference in the Fe–L interaction between Fe(OEC) and Fe(OEP) will be discussed.

M(OEC) can be assumed to have $C_{2v}$ symmetry. Symmetry properties of the in-plane vibrations of M(OEC) are correlated with those of M(OEP) having $D_{4h}$ symmetry in Table 9. The $A_{1g}$ and $B_{1g}$ species of $D_{4h}$ group are combined into the $A_1$ species and the $A_{2g}$ and $B_{2g}$ species of $D_{4h}$ group are combined into the $B_2$ species of $C_{2v}$ group, whereas the $E_u$ species of $D_{4h}$ group split into the $A_1$ and $B_2$ species of the $C_{2v}$ group. Both modes are IR active. In the RR spectra the $B_2$ modes are expected to give dp or ap bands while

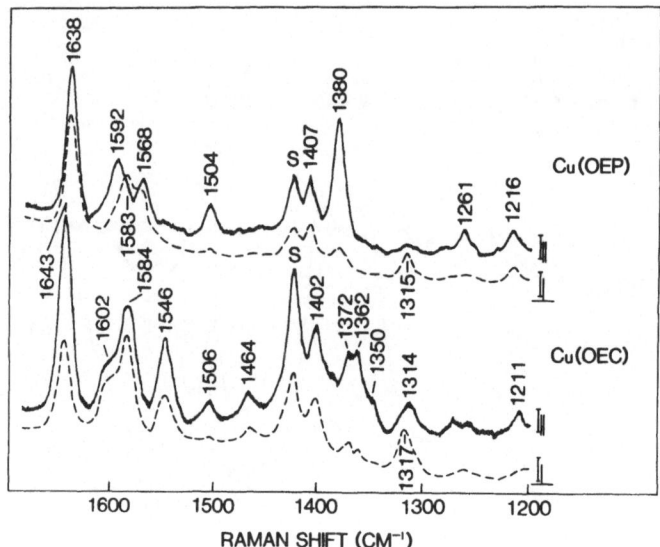

**Fig. 15.** The 488.0-nm excited polarized resonance Raman spectra of Cu(OEP) and Cu(OEC) in CH$_2$Cl$_2$

**Table 9.** Symmetry correlation between D$_{4h}$ and C$_{2v}$ groups for the in-plane vibrations

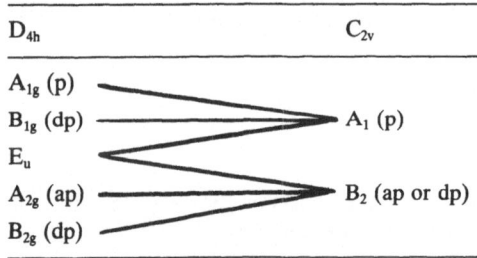

| D$_{4h}$ | C$_{2v}$ |
|---|---|
| A$_{1g}$ (p) | |
| B$_{1g}$ (dp) | A$_1$ (p) |
| E$_u$ | |
| A$_{2g}$ (ap) | B$_2$ (ap or dp) |
| B$_{2g}$ (dp) | |

the A$_1$ modes are expected to give p bands. The modes corresponding to the B$_{1g}$ modes of M(OEP) would give p bands with M(OEC) whereas those corresponding to the B$_{2g}$ modes of M(OEP) would give ap or dp bands with M(OEC). Accordingly, it is anticipated that the number of bands and their polarization properties would be altered in the RR spectra of M(OEC) from those of M(OEP). In fact, as shown in Fig. 15, extra bands are observed at 1464, 1362, and 1350 cm$^{-1}$ for Cu(OEC), although the general patterns of the RR spectra of Cu(OEP) and Cu(OEC) are similar.

The RR bands of Cu(OEC) at 1643(p), 1584(ap), and 1506(p) cm$^{-1}$ are assigned to the $\nu_{10}$-, $\nu_{19}$-, and $\nu_3$-like modes of M(OEP), respectively, and are accordingly associated with C$_a$C$_m$ stretching modes. The RR bands of Cu(OEC) at 1602(dp) and 1546(p) cm$^{-1}$ are assigned to the $\nu_2$- and $\nu_{11}$-like modes of M(OEP) and associated with the C$_b$C$_b$ stretching modes. Separation of the $\nu_2$ and $\nu_{11}$ bands is considerably larger for Cu(OEC) than for Cu(OEP), while the $\nu_{10}$ and $\nu_3$ frequencies of Cu(OEC) are close to those of Cu(OEP). These features were retained by all Fe(OEC) derivatives[39]. The two kinds of

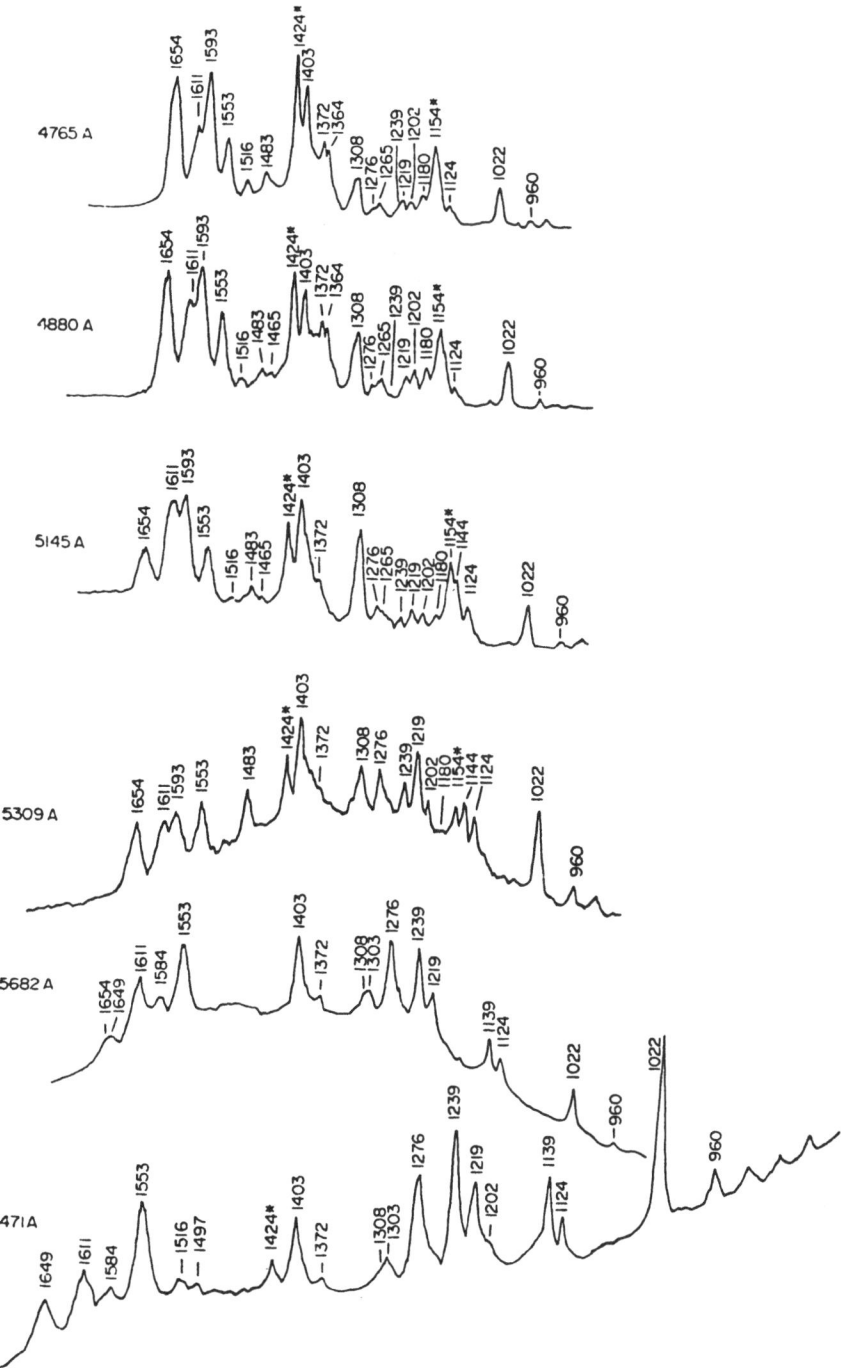

**Fig. 16.** Resonance Raman spectra of Ni(OEC) in $CH_2Cl_2$ excited at various wavelengths. The bands at 1424 and 1154 $cm^{-1}$ with an asterisk are due to $CH_2Cl_2$, but the latter contains some contribution from Ni(OEC)

$C_bC_b$ stretching modes belong to different symmetry species with Cu(OEP) whereas to identical species ($A_1$) with Cu(OEC) and accordingly their mutual interaction takes place. The Raman bands of Cu(OEC) at 1372(p), and 1362(p) cm$^{-1}$ exhibit a $^{15}$N isotopic frequency shift and therefore are assigned to the $\nu_4$- and $\nu_{12}$-like modes of M(OEP), respectively.

## 7.3 Characteristic Features of Metallochlorins

Metallochlorins have generally strong absorption in the red region and each absorption band of M(OEP) splits into two components because of symmetry lowering. Although the RR band positions of M(OEC) are not largely different from those of M(OEP),

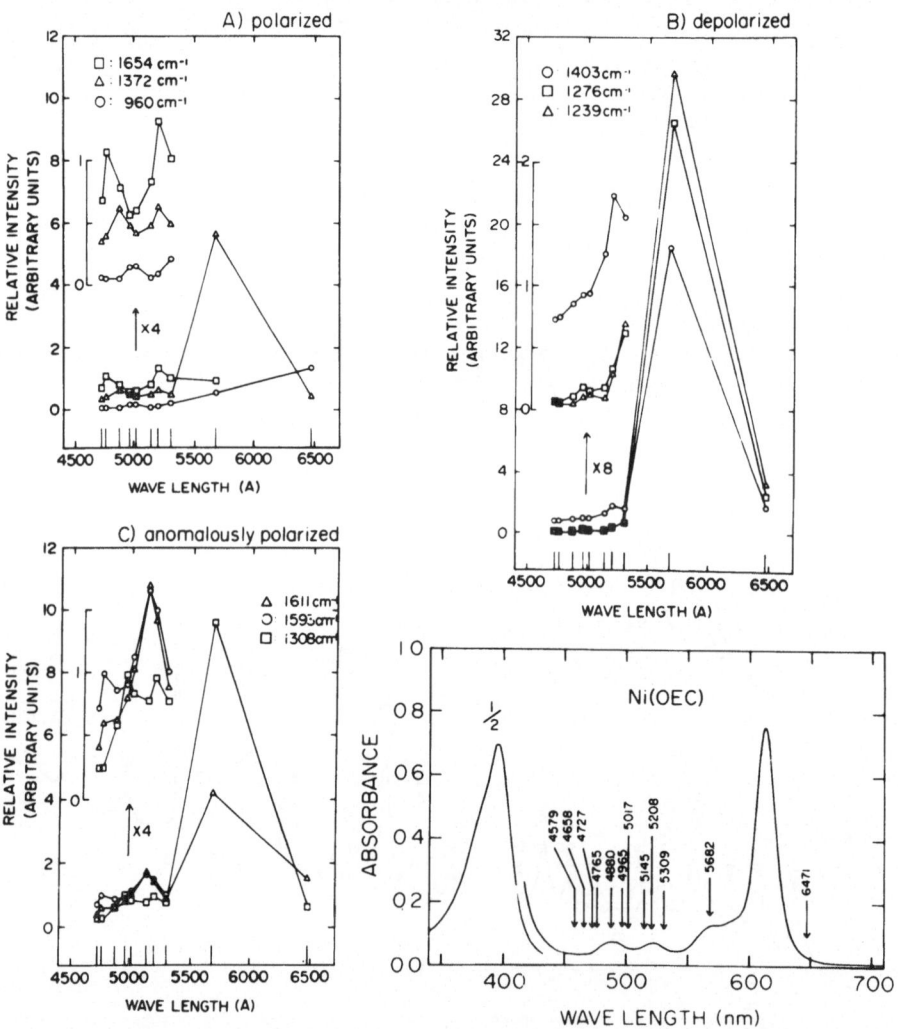

**Fig. 17 A–C.** Excitation profiles of the resonance Raman bands of Ni(OEC) in CH$_2$Cl$_2$ and its absorption spectrum in which the *arrows* indicate the excitation wavelengths used for Raman measurements; **A)** polarized bands, **B)** depolarized bands, **C)** anomalously polarized bands

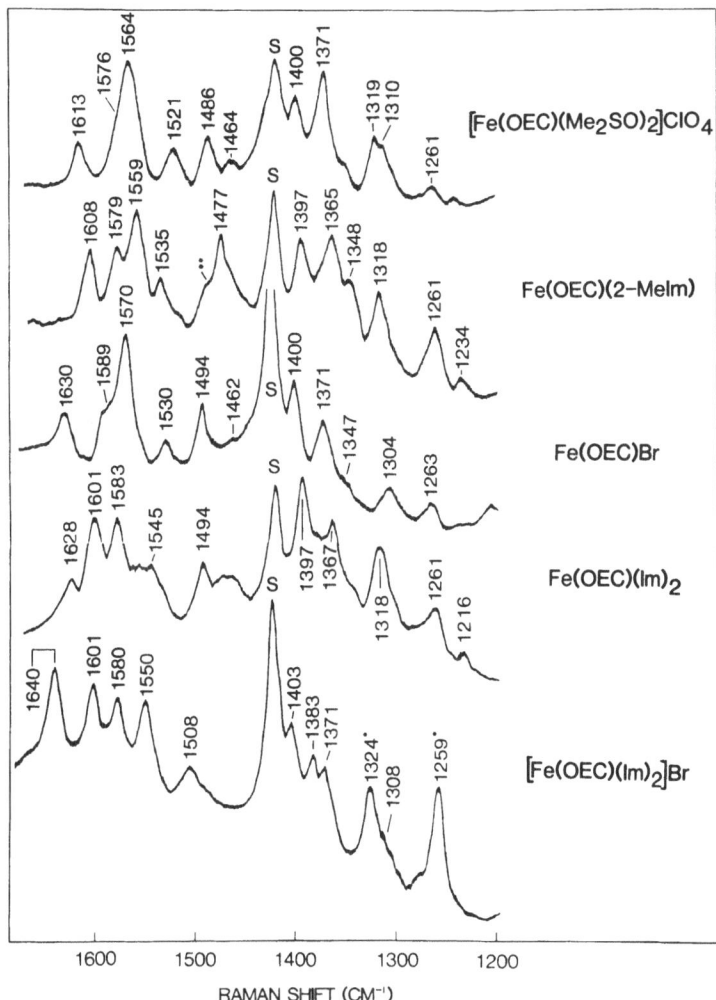

**Fig. 18.** The 441.6-nm excited resonance Raman spectra of [Fe(OEC)(Me$_2$SO)$_2$]ClO$_4$, Fe(OEC) (2-MeIm), Fe(OEC)Br, Fe(OEC)(Im)$_2$, and [Fe(OEC)(Im)$_2$]Br in CH$_2$Cl$_2$ (from Ref. 39)

excitation profiles of the RR bands are greatly different from those of M(OEP). Figure 16 shows the RR spectra of Ni(OEC) in CH$_2$Cl$_2$ solution excited at 476.5, 488.0, 514.5, 530.9, 568.2, and 647.1 nm. Although the concentration of Ni(OEC) is unaltered, RR intensities increase as the excitation wavelength becomes longer.

The intensities of individual RR bands of Ni(OEC) relative to that of the solvent at 1424 cm$^{-1}$ are plotted against the excitation wavelength in Fig. 17. It is apparent that all the dp and ap bands are strongly resonance enhanced around 570 nm. It is noteworthy that RR bands in the 1300–1000 cm$^{-1}$ region are strongly resonance enhanced in the Q band region. This is distinct from M(OEP) for which RR bands above 1500 cm$^{-1}$ are intensified in the Q band region. Such difference was confirmed with Fe(OEC)Cl, [Fe(OEC)(Im)$_2$]$^+$ and Cu(OEC), and therefore considered to be characteristic of metallochlorins[152]. This suggests that electronic excitation in the Q band region involves a change of the $C_aC_b$ bond for M(OEC) but little so for M(OEP).

Figure 18 shows the RR spectra of representative Fe(OEC) derivatives. The RR spectra of Fe(OEC) derivatives change with the oxidation-, spin- and coordination states similar to those of Fe(OEP) derivatives shown in Fig. 7. When the frequencies of the RR bands above 1450 cm$^{-1}$ were plotted against the Ct–N distances, straight lines similar to those shown in Fig. 8 for M(OEP) were obtained[39]. Regarding the relation between the $v_{10}$ and $v_{11}$ frequencies and pK$_a$ of axial ligands shown in Fig. 10, Fe(OEC) is less sensitive to a basicity of axial ligands. It is also noted that $v_4$ frequency of Fe$^{3+}$(OEC) derivatives is generally lower than that of the corresponding Fe$^{3+}$(OEP) derivatives, whereas the $v_4$ frequency of Fe$^{2+}$(OEC) derivatives is higher than that of Fe$^{2+}$(OEP) derivatives. As a result, a change of the $v_4$ frequency upon the redox change is smaller for Fe(OEC) derivatives than for Fe(OEP) derivatives. These facts may imply that π-back donation from Fe to the macrocycle is much less important with Fe(OEC) than with Fe(OEP).

## 7.4 Axial Ligand Modes

Elucidation of the difference between the Fe–L interactions of Fe(OEC)L and Fe(OEP)L, if any, is an interesting subject to think about the significance of chlorin chromophore in some heme proteins. For such purpose, comparison of the M–L stretching frequencies of Fe(OEC) derivatives with those of Fe(OEP) derivatives is instructive. As shown in Table 10, the Fe–L stretching frequencies of the Fe(OEC) and Fe(OEP) derivatives are remarkably close to each other irrespective of the oxidation and spin states[155].

In Fig. 19 the RR spectra of Fe(OEP)(Py)$_2$ and Fe(OEC)(Py)$_2$ are compared. The RR bands marked by b are due to the internal modes of bound pyridine which were confirmed with Py–d$_5$[155]. Appearance of such bands for Fe(MP)(Py)$_2$ (MP: mesoporphy-

**Table 10.** RR frequencies of Fe–L stretching modes of Fe(OEP) and Fe(OEC) complexes

| Mode | Molecule[a] | Frequencies (cm$^{-1}$) | | Solvent |
|---|---|---|---|---|
| | | OEP | OEC | |
| Fe$^{3+}$–F | Fe(P)F | 607 (605.5[c]) | 608[b] (589[d]) | THF |
| Fe$^{3+}$–Cl | Fe(P)Cl | 364 (357[c]) | 361[b] (352[d]) | THF |
| Fe$^{3+}$–Br | Fe(P)Br | 279 (270[c]) | 272 (270[d]) | THF |
| Fe$^{3+}$–I | Fe(P)I | – (246[c]) | 248 (240.5[d]) | THF |
| Fe$^{2+}$–(2-MeIm) | Fe(P)(2-MeIm) | 206 | 209 | CH$_2$Cl$_2$ |
| Fe$^{2+}$–(1,2-Me$_2$Im) | Fe(P)(1,2-Me$_2$Im) | 188 | 187 | CH$_2$Cl$_2$ |
| (Py)–Fe$^{2+}$–(Py) | Fe(P)(Py)$_2$ | 178 | 179 | CH$_2$Cl$_2$ |
| (γ-Pic)–Fe$^{2+}$–(γ-Pic) | Fe(P)(γ-Pic)$_2$ | 161 | 161 | CH$_2$Cl$_2$ |
| (Pip)–Fe$^{2+}$–(Pip) | Fe(P)(Pip)$_2$ | 164 | 163 | Pip |

a   P refers to OEP or OEC
b   Data from Ref. 142
c   Infrared data (KBr disc) from Ref. 70
d   Infrared data (KBr disc) from Ref. 156

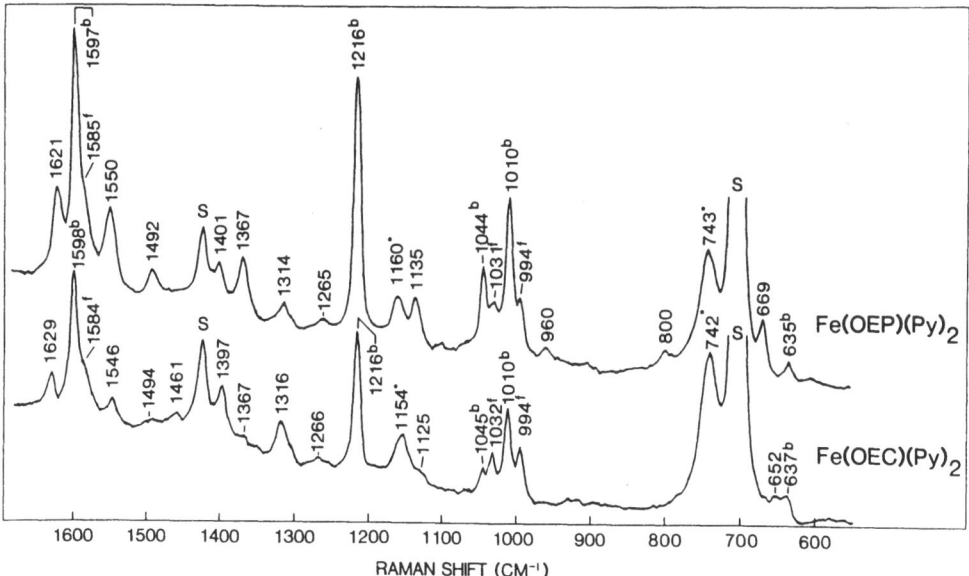

**Fig. 19.** Resonance Raman spectra of Fe(OEP)(Py)$_2$ and Fe(OEC)(Py)$_2$ in CH$_2$Cl$_2$ excited at 488.0 nm. The bands marked with b and f are due to bound- and free-pyridine, respectively. The bands with an asterisk contain contributions from both sample and solvent (from Ref. 155)

rin) was found by Spiro and Burke[157] and the vibrational assignments were carried out[145]. The proximity of the bound Py modes of Fe(OEC)(Py)$_2$ to those of Fe(OEP)(Py)$_2$ in addition to the proximity of the Fe–L stretching frequencies strongly suggests that saturation of one C$_b$–C$_b$ bond of the porphyrin little affects the nature of the iron-ligand interaction.

# 8 Metallo-Phthalocyanines

Metallo-phthalocyanines [M(Pc)] have attracted attention from view of their practical application to material science, but the vibrational spectra have been much less studied. This is partly due to low solubility in any solvent and difficulty in spectral analysis. The molecular vibrations of M(Pc) with D$_{4h}$ symmetry are factorized into 14 A$_{1g}$ + 13 A$_{2g}$ + 14 B$_{1g}$ + 14 B$_{2g}$ + 13 E$_g$ + 6 A$_{1u}$ + 8 A$_{2u}$ + 7 B$_{1u}$ + 28 E$_u$ and their selection rules are the same as those described for M(OEP).

The RR spectra of VO(Pc)[158], Mg, Cu, and Zn(Pc)[159] in a solid film were measured with the excitation lines near 600 nm. The breathing vibration of the 16-membered ring corresponding to $\nu_7$ of M(OEP) was observed at 682 cm$^{-1}$ for Mg(Pc) and 678 cm$^{-1}$ for Cu(Pc) and Zn(Pc). The RR bands at 221, 236, and 232 cm$^{-1}$ for Mg(Pc), Cu(Pc), and Zn(Pc), respectively, might be attributed to M–N$_{pyrr}$ stretching modes. Melendres et al.[160] observed the RR spectra of Fe(Pc) on Cu and Ag electrodes and also carried out normal coordinate calculations[161]. Data on isotopic frequency shifts are very desirable for detailed band assignments. Huang et al.[162] investigated excitation profiles of RR bands of Pt(Pc) in the 570–660 nm region, showing that the third absorption maximum of Pt(Pc) in the visible region (587 nm) is not the 0–2 as previously interpreted, but is the 0–1 transition.

Homborg and coworkers extensively investigated the vibrational spectra of metal-lophthalocyanines[163]. The $Fe^{3+}$–X stretching RR bands of Fe(Pc)X were resonance enhanced upon excitation within the CT band near 500 nm, and indeed identified at 485, 308, 225, and 195 $cm^{-1}$ for X = F, Cl, Br, and I, respectively, while the corresponding IR bands were observed at 475, 303, 221, and 193 $cm^{-1}$, for X = F, Cl, Br, and I, respectively. These frequencies are considerably lower than those of Fe(OEP)X (see Table 8). Kalz and Homborg[164] reported the symmetric and antisymmetric X–$Co^{3+}$–X stretching vibrations of $Co(Pc)X_2$ and $[Co(Pc)X_2]^{-1}$. The symmetric mode was strongly resonance enhanced and identified up to the fourth overtone: 280 and 168 $cm^{-1}$ for X = Cl and Br, respectively, of $Co(Pc)X_2$ and 270 and 165 $cm^{-1}$ for X = Cl and Br, respectively, of $[Co(Pc)X_2]^{-1}$. The antisymmetric mode was located at 326 and 250 $cm^{-1}$ for X = Cl and Br, respectively, of the former complexes and at 546, 333, and 243 $cm^{-1}$ for X = F, Cl, and Br, respectively, of the latter complexes.

Aleksandrov et al.[165] measured the RR spectra of M(Pc) anions, that is, $Co(Pc)^m$ (m = 0, − 1, and − 2), $Fe(Pc)^m$ (m = O, − 1, − 2, and − 4) and $Mg(Pc)^m$ (m = 0, − 2, and − 4). They found that the RR spectrum of $[Co(Pc)]^{-1}$ retained the spectral pattern of Co(Pc) with appreciable low frequency shifts of all RR bands, but that the RR spectrum of $[Fe(Pc)]^{-1}$ was distinctly different from that of Fe(Pc). Such a difference must arise from whether the extra electron is localized to the metal ion or delocalized to the macrocycle. The RR spectra of free radicals of Pc were reported by Homborg and Kalz[166] and Sugimoto et al.[167]. The former authors also studied the π-cation radical of Pc and pointed out that IR bands at 1350 and 1450 $cm^{-1}$ can be used as diagnostic markers for the π-cation radicals[168]. This is interesting because the porphyrin π-cation radicals also provide a characteristic IR band: 1280 $cm^{-1}$ for $M(TPP·)^+$ and 1550 $cm^{-1}$ for $M(OEP·)^{+}$ [129].

With regard to IR spectra, polymeric $[M(Pc)L]_n$ (M = Fe, Co; L = bidentate bridging ligand) were investigated by Metz et al.[169] who pointed out that the 1589 $cm^{-1}$ band of 4,4-bipyridine can distinguish between the monodentate and bidentate bridging states and that its intensity enables to determine a chain length by comparison with the spectrum of the monomeric compound. The complex-forming ability of M(Pc) with formic acid was examined by Sauvage et al.[170]. Interaction of Fe(Pc) with $O_2$ usually leads to the formation of a dimeric μ-oxo species[171], but Nakamoto and coworkers[109] succeeded in observing the O–O stretching IR band at 1207 $cm^{-1}$ with a matrix isolation technique. The O–O stretching mode for $Mn(Pc)O_2$ was located at 992 $cm^{-1}$ from the IR spectrum[111].

# 9 Concluding Remarks

Vibrational spectroscopy provides detailed structural information about metalloporphyrins. Recent progress in Fourier transform IR spectroscopy allows the measurements of

IR spectra with higher sensitivity and higher resolution. On the other hand, the development of pulse lasers and array detectors allows to measure RR spectra of short-lived species such as reaction intermediates and electronically excited species using time resolved RR spectroscopy. These new techniques will increasingly contribute to the characterization of metalloporphyrins. Elucidation of the data obtained with those advanced techniques may need basic knowledge about the vibrational spectra of metalloporphyrins. We wish this article would be helpful to analyze these fascinating spectra.

# 10 References

1. Strekas, T. C., Spiro, T. G.: Biochim. Biophys. Acta *263*, 830 (1972)
2. Spiro, T. G., Strekas, T. C.: Proc. Natl. Acad. Sci. USA *69*, 2622 (1972)
3. Brunner, H., Mayer, A., Sussner, H.: J. Mol. Biol. *70*, 153 (1972)
4. Lutz, M.: C. R. Hebd. Seances Acad. Sci. Ser. *B275*, 497 (1972)
5. Kitagawa, T., Ozaki, Y., Kyogoku, Y.: Adv. Biophys. *11*, 153 (1978)
6. Warshel, A.: Annu. Rev. Biophys. Bioeng. *6*, 273 (1977)
7. Felton, R. H., Yu, N.-T.: In The Porphyrins Vol. 3, Part A (Dolphin, D., ed.) p. 347, Academic Press, New York 1978.
8. Asher, S. A.: Methods Enzymol. *76*, 371 (1981)
9. Rousseau, D. L., Ondrias, M. R.: Annu. Rev. Biophys. Bioeng. *12*, 357 (1983)
10. Spiro, T. G. (ed.): Biological Applications of Raman Spectroscopy Vol. 3, John Wiley & Sons, New York, in press
11. Kitagawa, T.: in Spectroscopy of Biological Systems (Clark, R. J. H., Hester, R. E., eds.) p. 443, John Wiley & Sons, London 1986
12. Lutz, M.: in Advances in Infrared and Raman Spectroscopy (Clark, R. J. H., Hester, R. E. eds.) Vol. 11, p. 211, John Wiley & Sons, New York 1984
13. Bürger, H.: in Porphyrins and Metalloporphyrins (Smith, K. M., ed.) p. 525, Elsevier, Amsterdam 1975
14. Spiro, T. G.: in Iron Porphyrins (Lever, A. B. P., Gray, H. B., eds.) Vol. 2, p. 91, Addison-Wesley, Reading 1983
15. Solovyov, K. N., Gladkov, L. L., Starukhin, A. S., Shkirman, S. F.: Spectroscopy of Porphyrins: Vibrational State, Nauka I Tekhnika 1985
16. Tang, J., Albrecht, A. C.: in Raman Spectroscopy (Szymanski, H. A., ed.) Vol. 2, p. 33, Plenum, New York 1970
17. Gouterman, M.: in The Porphyrins (Dolphin, D., ed.) Vol. 3, p. 1, Academic Press, New York 1978
18. Nishimura, Y., Hirakawa, A. Y., Tsuboi, M.: J. Mol. Spectrosc. *68*, 335 (1977)
19. Shelnutt, J. A., O'Shea, D. C., Yu, N.-T., Cheung, L. D., Felton, R. H.: J. Chem. Phys. *64*, 1154 (1976)
20. Ogoshi, H., Saito, Y., Nakamoto, K.: ibid. *57*, 4194 (1972)
21. Stein, P., Burke, J. M., Spiro, T. G.: J. Am. Chem. Soc. *97*, 2304 (1975)
22. Sunder, S., Bernstein, H. J.: J. Raman Spectrosc. *5*, 351 (1976)
23. Susi, H., Ard, S.: Spectrochim. Acta, Teil A *33*, 561 (1977)
24. Gladkov, L. L., Gradyushko, A. T., Shulga, A. M., Solovyov, K. N., Starukhin, A. S.: J. Mol. Struct. *47*, 463 (1978)
25. Warshel, A., Lappicirella, A.: J. Am. Chem. Soc. *103*, 4664 (1981)
26. Kitagawa, T., Abe, M., Kyogoku, Y., Ogoshi, H., Sugimoto, H., Yoshida, Z.: Chem. Phys. Lett. *48*, 55 (1977)
27. Kitagawa, T., Abe, M., Ogoshi, H.: J. Chem. Phys. *69*, 4516 (1978)
28. Abe, M., Kitagawa, T., Kyogoku, Y.: Chem. Lett. 249 (1976)
29. Abe, M., Kitagawa, T., Kyogoku, Y.: J. Chem. Phys. *69*, 4526 (1978)
30. Kincaid, J. R., Urban, M. W., Watanabe, T., Nakamoto, K.: J. Phys. Chem. *87*, 3096 (1983)

31. Willems, D. L., Bocian, D. F.: J. Am. Chem. Soc. *106*, 880 (1984)
32. Willems, D. L., Bocian, D. F.: J. Phys. Chem. *89*, 234 (1985)
33. Ogoshi, H., Masai, N., Yoshida, Z., Takemoto, J., Nakamoto, K.: Bull. Chem. Soc. Jpn. *44*, 49 (1971)
34. Burke, J. M., Kincaid, J. R., Spiro, T. G.: J. Am. Chem. Soc. *100*, 6077 (1978)
35. Burke, J. M., Kincaid, J. R., Peters, S., Gagne, R. R., Collman, J. P., Spiro, T. G.: ibid. *100*, 6083 (1978)
36. Kozuka, M., Nakamoto, K.: ibid. *103*, 2162 (1981)
37. Fuchsman, W. H., Goldberg, J. M., Levy, D. D., Smith, Q. R.: Bioinorg. Chem. *9*, 461 (1978)
38. Kozuka, M., Iwaizumi, M.: Bull. Chem. Soc. Jpn *56*, 3165 (1983)
39. Ozaki, Y., Iriyama, K., Ogoshi, H., Ochiai, T., Kitagawa, T.: J. Phys. Chem. *90*, 6105 (1986)
40. Yamamoto, T., Palmer, G., Gill, D., Salmeen, I. T., Rimai, L.: J. Biol. Chem. *248*, 5211 (1973)
41. Spiro, T. G., Strekas, T. C.: J. Am. Chem. Soc. *96*, 338 (1974)
42. a) Kitagawa, T., Kyogoku, Y., Iizuka, T., Ikeda-Saito, M.: Chem. Lett. 849 (1975); b) idem J. Am. Chem. Soc. *98*, 5169 (1976)
43. Spiro, T. G., Stong, J. D., Stein, P.: J. Am. Chem. Soc. *101*, 2648 (1979)
44. Teraoka, J., Kitagawa, T.: J. Phys. Chem. *84*, 1928 (1980)
45. Felton, R. H., Yu, N.-T., O'Shea, D. C., Shelnutt, J. A.: J. Am. Chem. Soc. *96*, 3675 (1974)
46. Spaulding, L. D., Chang, C. C., Yu, N.-T., Felton, R. H.: ibid. *97*, 2517 (1975)
47. Choi, S., Spiro, T. G., Langry, K. C., Smith, K. M., Budd, D. L., LaMar, G. N.: ibid. *104*, 4345 (1982)
48. Saito, M., Kashiwagi, H.: J. Chem. Phys. *82*, 848 (1985)
49. Kitagawa, T., Ogoshi, H., Watanabe, E., Yoshida, Z.: J. Phys. Chem. *79*, 2629 (1979)
50. Fujiwara, M., Tasumi, M.: ibid. *90*, 250 (1986)
51. Gouterman, M.: J. Chem. Phys. *30*, 1139 (1959)
52. Kincaid, J., Nakamoto, K.: J. Inorg. Nucl. Chem. *37*, 85 (1975)
53. Kashiwagi, H., Obara, S.: Int. J. Quant. Chem. *20*, 843 (1981)
54. Ozaki, Y., Kitagawa, T., Kyogoku, Y., Imai, Y., Hashimoto-Yutsudo, C., Sato, R.: Biochemistry *17*, 5826 (1978)
55. Choi, S., Spiro, T. G., Langry, K. C., Smith, K. M.: J. Am. Chem. Soc. *104*, 4337 (1982)
56. Lee, H., Kitagawa, T., Abe, M., Pandey, R. K., Leung, H.-K., Smith, K. M.: J. Mol. Structr. *146*, 329 (1986)
57. Adar, F.: Arch. Biochem. Biophys. *181*, 5 (1977)
58. Tsubaki, M., Nagai, K., Kitagawa, T.: Biochemistry *19*, 379 (1980)
59. Babcock, G. T., Salmeen, I.: ibid. *18*, 2493 (1979)
60. Steelandt-Frentrup, J. V., Salmeen, I., Babcock, G. T.: J. Am. Chem. Soc. *103*, 5981 (1981)
61. Callahan, P. M., Babcock, G. T.: Biochemistry *20*, 952 (1981)
62. Babcock, G. T., Callahan, P. M.: ibid. *22*, 2314 (1983)
63. Choi, S., Lee, J. J., Wei, Y. H., Spiro, T. G.: J. Am. Chem. Soc. *105*, 3692 (1983)
64. Kitagawa, T., Kyogoku, Y., Orii, Y.: Arch. Biochem. Biophys. *181*, 228 (1977)
65. Kimura, S., Yamazaki, I., Kitagawa, T.: Biochemistry *20*, 4632 (1981)
66. Sunder, S., Mendelsohn, R., Bernstein, H. J.: Spectrochim. Acta, Teil A *33*, 715 (1977)
67. Boucher, L. J., Katz, J. J.: J. Am. Chem. Soc. *89*, 1340 (1967)
68. Bürger, H., Burczyk, K., Fuhrhop, J. H.: Tetrahedron *27*, 3257 (1971)
69. Oshio, H., Ama, T., Watanabe, T., Kincaid, J., Nakamoto, K.: Spectrochim. Acta *40A*, 863 (1984)
70. Ogoshi, H., Watanabe, E., Yoshida, Z., Kincaid, J., Nakamoto, K.: J. Am. Chem. Soc. *95*, 2845 (1973)
71. Choi, S., Spiro, T. G.: ibid. *105*, 3683 (1983)
72. Ogoshi, H., Yoshida, Z.: Bull. Chem. Soc. Jpn. *44*, 1722 (1971)
73. Abe, M.: in Spectroscopy of Biological Systems (Clark, R. J. H., Hester, R. E., eds.) p. 347, John Wiley & Sons, London 1986
74. Perutz, M. F.: Nature (London) *273*, 495 (1972)
75. Hori, H., Kitagawa, T.: J. Am. Chem. Soc. *102*, 3608 (1980)
76. a) Kincaid, J., Stein, P., Spiro, T. G.: Proc. Natl. Acad. Sci. USA *76*, 549 (1979); b) ibid. p. 4156
77. Desbois, A., Lutz, M., Banerjee, R.: Biochemistry *18*, 1510 (1979)

78. Kitagawa, T., Nagai, K., Tsubaki, M.: FEBS Lett. *104,* 376 (1979)
79. Argade, P. V., Sassaroli, M., Rousseau, D. L., Inubushi, T., Ikeda-Saito, M., Lapidot, A.: J. Am. Chem. Soc. *106,* 6593 (1984)
80. Nagai, K., Kitagawa, T., Morimoto, H.: J. Mol. Biol. *136,* 271 (1980)
81. Nagai, K., Kitagawa, T.: Proc. Natl. Acad. Sci. USA 77, 2033 (1980)
82. Matsukawa, S., Mawatari, K., Yoneyama, Y., Kitagawa, T.: J. Am. Chem. Soc. *107,* 1108 (1985)
83. Friedman, J. M., Rousseau, D. L., Ondrias, M. R.: Annu. Rev. Phys. Chem. *33,* 471 (1982)
84. Teraoka, J., Kitagawa, T.: J. Biol. Chem. *256,* 3969 (1981)
85. Stein, P., Mitchell, M., Spiro, T. G.: J. Am. Chem. Soc. *102,* 7795 (1980)
86. Alben, J. O.: in The Porphyrins (Dolphin, D., ed.) Vol. 3, p. 334, Academic Press, New York 1978
87. Brunner, H.: Naturwiss. *61,* 129 (1974)
88. Tsubaki, M., Yu, N.-T.: Proc. Natl. Acad. Sci. USA 78, 3581 (1981)
89. Maxwell, J. C., Caughey, W. S.: Biochem. Biophys. Res. Commun. *60,* 1309 (1974)
90. Bangcharoenpaurpong, O., Rizos, A. K., Champion, P. M., Jollie, D., Sligar, S. G.: J. Biol. Chem. *261,* 8089 (1986)
91. Asher, S. A., Vickery, L. E., Schuster, T. M., Sauer, K.: Biochemistry *16,* 5849 (1977)
92. Asher, S. A., Schuster, T. M.: ibid. *18,* 5377 (1979)
93. Asher, S. A., Schuster, T. M.: ibid. *20,* 1866 (1981)
94. Iizuka, T., Yonetani, T.: Adv. Biophys. *1,* 157 (1970)
95. Alben, J. O., Fajer, L. Y.: Biochemistry *11,* 842 (1972)
96. Eaton, W. A., Hochstrasser, R. M.: J. Chem. Phys. *49,* 985 (1968)
97. Tsubaki, M., Srivastava, R. B., Yu, N.-T.: Biochemistry *20,* 946 (1981)
98. Yu, N.-T., Benko, B., Kerr, E. A., Gersonde, K.: Proc. Natl. Acad. Sci. USA *81,* 5106 (1984)
99. Henry, E. R., Rousseau, D. L., Hopfield, J. J., Noble, R. W., Simon, S. R.: Biochemistry *24,* 5907 (1985)
100. Benko, B., Yu, N.-T.: Proc. Natl. Acad. Sci. USA *80,* 7042 (1983)
101. Maxwell, J. C., Caughey, W. S.: Biochemistry *15,* 388 (1976)
102. Creighton, J. A., Lippencott, E. R.: J. Chem. Phys. *40,* 1779 (1964)
103. Evance, J. C.: J. Chem. Soc. D, 682 (1969)
104. Jones, R. D., Summerville, D. A., Basolo, F.: Chem. Revs. *79,* 139 (1979)
105. Nakamoto, K., Watanabe, T., Ama, T., Urban, M. W.: J. Am. Chem. Soc. *104,* 3744 (1982)
106. Bajdor, K., Kincaid, J. R., Nakamoto, K.: ibid. *106,* 7741 (1984)
107. Jones, R. D., Budge, J. R., Ellis, P. E., Linard, J. E., Summerville, D. A., Basolo, F.: J. Organomet. Chem. *181,* 151 (1979)
108. Bajdor, K., Nakamoto, K., Kincaid, J.: J. Am. Chem. Soc. *105,* 678 (1983)
109. Watanabe, T., Ama, T., Nakamoto, K.: J. Phys. Chem. *88,* 440 (1984)
110. Urban, M. W., Nakamoto, K., Kincaid, J.: Inorg. Chim. Acta *61,* 77 (1982)
111. Watanabe, T., Ama, T., Nakamoto, K.: Inorg. Chem. *22,* 2470 (1983)
112. Urban, M. W., Nakamoto, K., Basolo, F.: Inorg. Chem. *21,* 3406 (1982)
113. Guilard, R., Fontesse, M., Fournari, P., Lecomte, C., Protas, J.: J. Chem. Soc. Chem. Commun. 161 (1976)
114. Collman, J. P., Brauman, J. I., Halbert, T. R., Suslick, K. S.: Proc. Natl. Acad. Sci. USA 73, 3333 (1976)
115. Cheung, S. K., Grimes, C. J., Wong, J., Reed, C. A.: J. Am. Chem. Soc. *98,* 5028 (1976)
116. McCandlish, E., Miksztal, A. R., Nappa, M., Sprenger, A. Q., Valentine, J. S., Stong, J. D., Spiro, T. G.: J. Am. Chem. Soc. *102,* 4268 (1980)
117. Maxwell, J. C., Volpe, J. A., Barlow, C. H., Caughey, W. S.: Biochem. Biophys. Res. Commun. *58,* 166 (1974)
118. Mackin, H. C., Tsubaki, M., Yu, N.-T.: Biophys. J. *41,* 349 (1983)
119. Bajdor, K., Oshio, H., Nakamoto, K.: J. Am. Chem. Soc. *106,* 7273 (1984)
120. Bajdor, K., Nakamoto, K.: ibid. *106,* 3045 (1984)
121. Proniewicz, L. M., Bajdor, K., Nakamoto, K.: J. Phys. Chem. *90,* 1760 (1986)
122. Hashimoto, S., Tatsuno, Y., Kitagawa, T.: Proc. Intl. Conf. Raman Spectrosc. (Peticolas, W. L., Hudson, B., eds.) p. 1–28 (1986)
123. Chin, D.-H., LaMar, G. N., Balch, A. L.: J. Am. Chem. Soc. *102,* 4344 (1982)
124. Schappacher, M., Chottard, G., Weiss, R.: J. Chem. Soc. Chem. Commun. 93 (1986)

125. Hashimoto, S., Tatsuno, Y., Kitagawa, T.: Proc. Japan Acad. 60 Ser. B, 345 (1984)
126. Terner, J., Sitter, A. J., Reczek, C. M.: Biochim. Biophys. Acta 828, 73 (1985)
127. Sitter, A. J., Reczek, C. M., Terner, J.: J. Biol. Chem. 260, 7515 (1985)
128. Hashimoto, S., Tatsuno, Y., Kitagawa, T.: Proc. Natl. Acad. Sci. USA 83, 2417 (1986)
129. Shimomura, E. T., Phillippi, M. A., Goff, H. M., Scholz, W. F., Reed, C. A.: J. Am. Chem. Soc. 103, 6778 (1981)
130. Aleksandrov, I. V., Yeletskii, N. P., Sidorov, A. N.: Biophysics 25, 389 (1980)
131. Wayland, B. B., Mehne, L. F., Swartz, J.: J. Am. Chem. Soc. 100, 2379 (1978)
132. Eaton, G., Eaton, S.: ibid. 97, 235 (1975)
133. Buchler, J. W., Kokisch, W., Smith, P. D.: Struct. Bonding 34, 79 (1978)
134. Rougee, M., Brault, D.: Biochemistry 14, 4100 (1975)
135. Yu, N.-T., Kerr, E. A., Ward, B., Chang, C. K.: Biochemistry 22, 4534 (1983)
136. Ogoshi, H., Sugimoto, H., Yoshida, Z.: Bull. Chem. Soc. Jpn. 51, 2369 (1978)
137. Kerr, E. A., Mackin, H. C., Yu, N.-T.: Biochemistry 22, 4373 (1983)
138. Alben, J. O., Caughey, W. S.: Biochemistry 7, 175 (1968)
139. Tsubaki, M., Srivastava, R. B., Yu, N.-T.: Biochemistry 21, 1132 (1982)
140. Oshio, H., Ama, T., Watanabe, T., Nakamoto, K.: Inorg. Chim. Acta 96, 61 (1985)
141. Fleischer, E., Srivastava, T. S.: J. Am. Chem. Soc. 91, 2403 (1969)
142. Kitagawa, T., Abe, M., Kyogoku, Y., Ogoshi, H., Watanabe, E., Yoshida, Z.: J. Phys. Chem. 80, 1181 (1976)
143. Asher, S., Sauer, K.: J. Chem. Phys. 64, 4115 (1976)
144. Champion, P. M., Gunsalus, I. C., Wagner, G. C.: J. Am. Chem. Soc. 100, 3743 (1978)
145. Wright, P. G., Stein, P., Burke, J. M., Spiro, T. G.: ibid. 101, 3531 (1979)
146. Schick, G. A., Bocian, D. F.: ibid. 106, 1682 (1984)
147. Reed, C. A.: Adv. Chem. Ser. (Kadish, K. M., ed.) 201, 333 (1982)
148. Ksenofontova, N. M., Maslov, V. G., Sidorov, A. N., Bobovich, Ya. S.: Opt. Spektrosk. 40, 809 (1976)
149. Yamaguchi, H., Soeta, A., Toeda, H., Itoh, K.: J. Electroanal. Chem. 159, 347 (1983)
150. Srivatsa, G. S., Sawyer, D. T., Boldt, N. J., Bocian, D. F.: Inorg. Chem. 24, 2123 (1985)
151. Teraoka, J., Hashimoto, S., Sugimoto, H., Mori, M., Kitagawa, T.: J. Am. Chem. Soc. 109, 180 (1987)
152. Ozaki, Y., Kitagawa, T., Ogoshi, H.: Inorg. Chem. 18, 1772 (1979)
153. Andersson, L. A., Loehr, T. M., Chang, C. K., Mauk, A. G.: J. Am. Chem. Soc. 107, 182 (1985)
154. Hanson, L. K., Chang, C. K., Ward, B., Callahan, P. M., Babcock, G. T., Head, J. D.: ibid. 106, 3950 (1984)
155. Ozaki, Y., Iriyama, K., Ogoshi, H., Ochiai, T., Kitagawa, T.: J. Phys. Chem. 90, 6113 (1986)
156. Ogoshi, H., Watanabe, E., Yoshida, Z., Kincaid, J., Nakamoto, K.: Inorg. Chem. 14, 1344 (1975)
157. Spiro, T. G., Burke, J. M.: J. Am. Chem. Soc. 98, 5482 (1976)
158. Aroca, R., Loutfy, R. O.: Spectrochim. Acta 39A, 847 (1983)
159. Jennings, C., Aroca, R., Hor, A.-M., Loutfy, R. O.: J. Raman Spectrosc. 15, 34 (1984)
160. Melendres, C. A., Rios, C. B., Feng, X., McMasters, R.: J. Phys. Chem. 87, 3526 (1983)
161. Melendres, C. A., Maroni, V. A.: J. Raman Spectrosc. 15, 319 (1984)
162. Huang, T.-H., Rieckhoff, K. E., Voigt, E.-M.: Can. J. Chem. 56, 976 (1978)
163. Kalz, W., Homborg, H.: Z. Naturforsch. 38b, 470 (1983)
164. Homborg, H., Kalz, W.: ibid. 39b, 1490 (1984)
165. Aleksandrov, I. V., Bovovich, Ya. S., Maslov, V. G., Sidorov, A. N.: Opt. Spektrosk. 37, 265 (1974)
166. Homborg, H., Kalz, W.: Z. Naturforsch. 33b, 1067 (1978)
167. Sugimoto, H., Higashi, T., Mori, M.: J. Chem. Soc. Chem. Commun. 622 (1983)
168. Homborg, H.: Z. Anorg. Chem. 507, 35 (1983)
169. Metz, J., Schneider, O., Hanack, M.: Spectrochim. Acta 38A, 1265 (1982)
170. Sauvage, F. X., De Backer, M. G., Stymne, B.: ibid. 38A, 803 (1982)
171. Ercolani, C., Gardini, M., Monacelli, F., Penness, G., Rossi, G.: Inorg. Chem. 22, 2584 (1983)

# Synthesis and Structure of Biomimetic Porphyrins

**Brian Morgan and David Dolphin**

Department of Chemistry, University of British Columbia, Vancouver, B.C., Canada V6T 1Y6

Structure and Bonding 64
© Springer-Verlag Berlin Heidelberg 1987

# 1. Introduction

Because of their ubiquitousness and the variety of their natural functions, heme proteins have been investigated on multi- and interdisciplinary levels. These proteins, all containing an iron porphyrin as the prosthetic group, are responsible for oxygen transport and storage (hemoglobin and myoglobin)[1], electron transport (cytochromes b, c)[2], oxygen reduction (cytochrome oxidase)[3], hydrogen peroxide utilization and destruction (peroxidases and catalases)[4], and hydrocarbon oxidation (cytochrome P450)[5]. The active site in each case contains an iron porphyrin (usually protoporphyrin IX), the nitrogens of the porphyrin ring occupying four essentially planar coordination sites of the metal. Therefore their diversity of function must be dictated by the number and nature of the axial ligands, the spin and oxidation state of the iron and the nature of the polypeptide chain.

A basic tenet of bioinorganic chemistry is that the structure and function of large biomolecules may be mimicked using simpler inorganic complexes to model the active sites. Obviously to fully understand the mechanisms of heme protein function, a study of iron porphyrins must be undertaken in which the characteristics of the metal (spin state, oxidation state and coordination number) and the steric and electronic effects of the porphyrin and other ligands are systematically varied.

Historically, much of the research on metalloporphyrins has focussed on the mechanism of reversible oxygen binding to myoglobin and hemoglobin. Oxygen binding heme proteins are five coordinate high spin $(S = 2)$ iron(II) species, which upon oxygenation become six-coordinate low spin $(S = 0)$. The difficulty in reproducing this behaviour is dominated by two problems:
(i)   the irreversible oxidation of iron(II) porphyrins on exposure to oxygen, and
(ii)  the difficulty in obtaining well-defined five-coordinate iron porphyrins.

Simple iron(II) porphyrins cannot reversibly bind oxygen, except at low temperature. At room temperature, and in the absence of a large excess of a sixth ligand, formation of the six-coordinate iron(II) dioxygen species is immediately followed by attack of a second five-coordinate iron(II) complex to give the μ-peroxo bisiron(III) complex. This rapidly breaks down, presumably via a ferryl intermediate to give a μ-oxo bisiron(III) complex in which the iron has been irreversibly oxidized to the ferric form (Scheme 1)[8-11].

**Table 1.** Axial ligands of selected heme proteins[6, 7]

| Protein | Ligand | |
|---|---|---|
| Hemoglobin | | |
|     deoxy | His-F8 | a |
|     oxy | His-F8 | $O_2$ |
|     carbonmonoxy | His-F8 | CO |
| Myoglobin – deoxy | His-F8 | a |
| Cytochrome P450 | Cysteine | a |
| Cytochrome c – tuna | His-18 | Met-80 |
| Cytochrome $b_5$ – calf liver | His-39 | His-63 |
| Peroxidase – horseradish | His | a |
| Catalase – beef liver | Tyr-357 | a |

a Sixth ligand site vacant or occupied by water

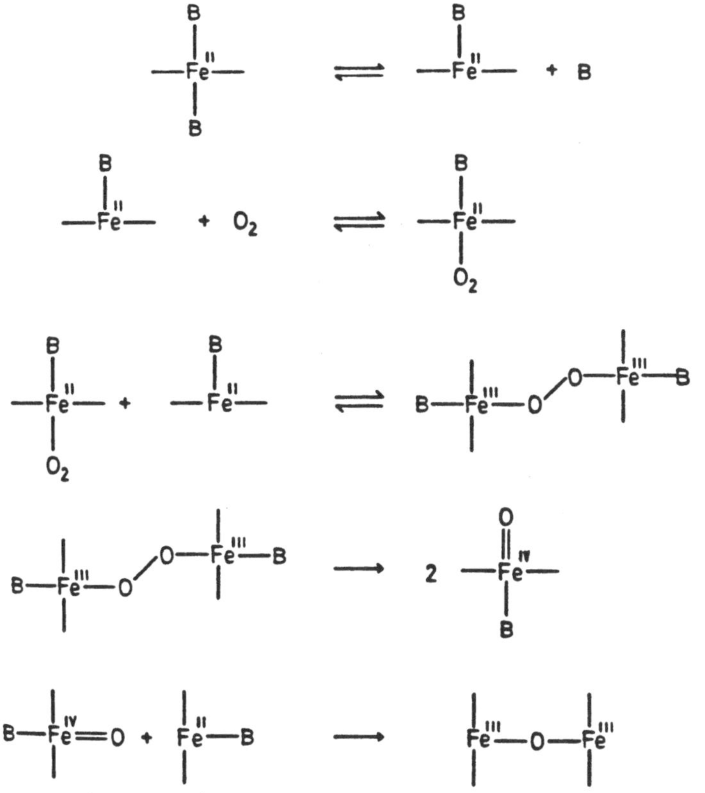

**Scheme 1**

Therefore, a major role for the polypeptide backbone of heme proteins is to sheath the oxygen binding site, preventing the close approach of two heme rings and consequent irreversible oxidation via the μ-peroxo complex. That irreversible oxidation of iron(II) porphyrins is possible by another mechanism is demonstrated by the fact that the body must provide an enzyme to reduce methemoglobin (the oxidized iron(III) hemoglobin being incapable of oxygen transport) to the functional iron(II) form[12]. Even so, hemoglobin exists in the body in the ferric form to the extent of about 3%. This alternative oxidation pathway occurs in aqueous acid or under conditions where μ-peroxo complex formation is inhibited, and is believed to involve proton assisted formation of protonated superoxide[13]. Similar oxidations occur with the production of superoxide, for the cytochromes P450[14] and cytochrome oxidase[3]. The peptide chain also stabilizes the $Fe^{II}-O_2$ species by

$$Fe^{II}-O_2 + H^+ \rightarrow Fe^{III} + HO_2 \qquad (6)$$

$$Fe^{II} + HO_2 + H^+ \rightarrow Fe^{III} + H_2O_2 \qquad (7)$$

$$2\,Fe^{II} + H_2O_2 + 2\,H^+ \rightarrow 2\,Fe^{III} + 2\,H_2O \qquad (8)$$

enclosing the porphyrin in a hydrophobic pocket to which access by protons is inhibited. In addition, recent neutron[15] and X-ray diffraction[16] studies have indicated that stabilization of the iron-oxygen bond in oxyhemoglobin and oxymyoglobin may in part be due to hydrogen bonding between the terminal oxygen atom and the imidazole of the distal histidine (His-E7).

The influence of the protein backbone is more pervasive than simply providing a barrier to oxidation. Conformational changes upon oxygen binding at the active site are believed to be responsible for the remarkable cooperativity exhibited by hemoglobin[17]. Similarly, the arrangement of certain residues on the protein has been postulated to provide an avenue along which electron transfer may occur in the cytochromes. But it is the protein's role in maintaining the coordination sphere of the iron porphyrin which determines the functions of the various heme proteins.

The second major problem in studying simple iron porphyrins is the preference of the metal for six-coordination. For example, in solution containing strongly coordinating N-donor ligands, six-coordination is favoured over five-coordination, i.e., for the equilibria in Eq. 9, $K_2 > K_1$ ($K_2/K_1 = 10$–$30$ in aprotic solvent at $25\,°C$)[18]. In benzene at $25\,°C$, the

$$\text{Fe}^{\text{II}}(\text{Por}) + \text{B} \underset{}{\overset{K_1}{\rightleftharpoons}} \text{Fe}^{\text{II}}(\text{Por})\text{B} + \text{B} \underset{}{\overset{K_2}{\rightleftharpoons}} \text{Fe}^{\text{II}}(\text{Por})(\text{B})_2 \tag{9}$$

binding constants of pyridine to $\text{Fe}^{\text{II}}(\text{TPP})$ have been estimated at $K_1 \sim 1.5 \times 10^3\ \text{M}^{-1}$ and $K_2 \sim 1.9 \times 10^4\ \text{M}^{-1}$. The size of $K_1$ and $K_2$ is obviously controlled by the spin state of the iron. The four-coordinate iron porphyrin is in an intermediate spin ($S = 1$) state. Addition of one ligand gives the high spin ($S = 2$) five-coordinate complex which adds a second ligand to form the low spin ($S = 0$) six-coordinate species with a gain in crystal field stabilization energy[19, 20]. In contrast, for $\text{Co}^{\text{II}}$, no stabilization is gained on going from five- to six-coordinate since $\text{Co}^{\text{II}}$ is low spin in both cases, and $K_1 \gg K_2$[21].

A further consequence of this is the difficulty of preparing mixed-ligand six-coordinate iron porphyrins. A typical example is the preparation of models for cytochrome c in which the iron is coordinated by an imidazole (His-18) and a thioether (Met-80). The greater ligating power of imidazole coupled with its tendency to form six-coordinate bis(imidazole) complexes makes self assembly of the mixed ligand system, $\text{Im–Fe–SR}_2$, difficult. Strategies which control coordination are essential for preparing a range of heme protein model porphyrins.

Numerous approaches have been used to control oxidation and coordination in model porphyrin systems.

(i) *Excess Ligand:* The presence of excess base (imidazole, pyridine) will minimize the concentration of five-coordinate heme and reduce μ-peroxo complex formation.

$$
\begin{array}{c}
\text{B} \\
| \\
-\text{Fe}^{\text{II}}- \\
| \\
\text{B}
\end{array}
\rightleftharpoons
\begin{array}{c}
\text{B} \\
| \\
-\text{Fe}^{\text{II}}- \\
|
\end{array}
+
\begin{array}{c}
\text{B} \\
| \\
-\text{Fe}^{\text{II}}- \\
| \\
\text{O}_2
\end{array}
\rightleftharpoons
\begin{array}{c}
| \\
\text{B}-\text{Fe}^{\text{III}}-\text{O} \\
|
\end{array}
\diagup
\begin{array}{c}
| \\
\text{O}-\text{Fe}^{\text{III}}-\text{B} \\
|
\end{array}
\tag{10}
$$

However in this case one is limited to studying competitive oxygen binding to six-coordinate hemes.

$$
\begin{array}{c}
\text{B} \\
| \\
-\text{Fe}^{\text{II}}- \\
| \\
\text{B}
\end{array}
+ \text{O}_2 \rightleftharpoons
\begin{array}{c}
\text{B} \\
| \\
-\text{Fe}^{\text{II}}- \\
| \\
\text{O}_2
\end{array}
+ \text{B} \tag{11}
$$

(ii) *Low Temperatures*[22-25]: Iron(II)–$O_2$ porphyrin complexes are stable at low temperatures ($\sim -60\,°C$), where the irreversible oxidation reactions are slowed down. Again one is reduced to studying competitive oxygen binding as $K_2/K_1$ increases as temperature decreases.

$$-\overset{}{Fe^{II}}- \; + \; B \; \underset{}{\overset{K_1}{\rightleftharpoons}} \; -\overset{B}{\underset{}{Fe^{II}}}- \; + \; B \; \underset{}{\overset{K_2}{\rightleftharpoons}} \; -\overset{B}{\underset{B}{Fe^{II}}}- \; \underset{}{\overset{O_2}{\rightleftharpoons}} \; -\overset{B}{\underset{O_2}{Fe^{II}}}- \qquad (12)$$

(iii) *Kinetic Measurements:* Fast spectroscopic methods may be used to observe reversible oxygen binding even under conditions where irreversible oxidation will occur. Traylor has exploited the stability of imidazole-heme-CO complexes towards oxidation[26]. A solution of Im–Hm–CO, equilibrated with a mixture of oxygen and carbon monoxide, is subjected to a short laser pulse which dissociates the carbon monoxide. The deoxy heme then reacts preferentially with oxygen at a fast but measurable rate ($k_B^{O_2} > 10^7\,M^{-1}s^{-1}$). In $10^3 - 10\,s$, the Im–Hm–$O_2$ complex dissociates and returns to the Im–Hm–CO complex.

$$-\overset{B}{\underset{CO}{Fe^{II}}}- \; \underset{k_B^{CO}}{\overset{h\nu\ k_B^{-CO}}{\rightleftharpoons}} \; -\overset{B}{\underset{}{Fe^{II}}}- \; \underset{-O_2\ k_B^{-O_2}}{\overset{+O_2\ k_B^{O_2}}{\rightleftharpoons}} \; -\overset{B}{\underset{O_2}{Fe^{II}}}- \qquad (13)$$

The kinetics are described by:[27]

$$1/k_{return} = 1/k_B^{O_2} + K_B^{O_2}[O_2]/k_B^{CO}[CO] \qquad (14)$$

Since $k_B^{CO}[CO]$ is the rate of return before $O_2$ is added, $k_B^{CO}[CO]$ may be accurately determined in the experiment and $k_B^{O_2}$ and $k_B^{-O_2}$ may be calculated from a plot of $1/k_{return}$ vs $O_2$ (pressure).

(iv) *Metal Substitution:* Replacement of iron with cobalt[28] or ruthenium[29] leads to metalloporphyrins which are more inert to oxidation and possess different coordination properties. Such an approach is applicable since apoproteins may be reconstituted with Co and Ru porphyrins. In the case of Co, reconstituted Co hemoglobin exhibits cooperative oxygen binding although to a diminished extent.

(v) *Immobilization:* This approach attempts to prevent irreversible oxidation by anchoring the porphyrin to a solid support. In Wang's classic experiment[30] a heme diethyl ester was embedded in a matrix of polystyrene and 1-(2-phenylethyl)imidazole. The matrix not only prevented close approach of two hemes but also provided a hydrophobic environment. Reversible oxygen uptake was observed. Alternatively, either the porphyrin or the ligand may be covalently attached to a rigid support. Basolo and his colleagues have undertaken the latter approach and prepared a silica gel support which contained 3-imidazolylpropyl groups attached to the surface[31]. Reaction with $Fe^{II}(TPP)(B)_2$ coordinated the porphyrin and heating in flowing helium removed the sixth axial base (pyridine or piperidine) to give the five-coordinate iron(II) porphyrin. However, attempts to observe reversible oxygen binding was obscured by the physisorption of oxygen by the silica support.

(vi) *Steric encumbrance:* By sterically blocking one or both faces of the porphyrin, close approach of two porphyrin rings and therefore μ-oxo bridge formation may be prevented. The approach most closely mimicking the natural system is to enfold the porphyrin ring in a polymer chain. This approach has been vigorously pursued in an attempt to prepare compounds capable of reversible oxygen binding in aqueous solution at room temperature. However, doubts about the number and nature of the active sites and the reversibility of oxygenation have made this approach less fruitful.

In contrast, porphyrins have been synthesized in which one or both faces of the ring are obstructed by some group(s) covalently bonded to the ring. The function of the steric hindrance is two-fold:

(i)   to direct base binding to the open face, ensuring five-coordination, and
(ii)  to allow $O_2$ to bind on the hindered face, steric hindrance preventing μ-oxo bridge formation.

Five-coordination may also be ensured in these systems by using bulky axial bases which cannot bind on the protected face.

This approach has been used by many groups to produce a wide variety of architecturally different model porphyrins, e.g. picket-fence, capped, cyclophane, crowned, strapped, basket-handle, etc., which are discussed below.

(vii) *Chelated Hemes:* Covalent attachment of the ligand to the porphyrin periphery allows one to control the extent of coordination. For poor ligands such as thiolate, thioether, phenoxide, etc., covalent attachment increases the local concentration and the likelihood of coordination to the metal without the necessity of a large excess of external ligand (Eq. 15). As long as displacement does not occur, addition of a second ligand allows formation of six-coordinate mixed ligand systems.

On the other hand, for strong ligands e.g. imidazole, pyridine, chelation provides a built-in 1:1 base/porphyrin stoichiometry. As long as dimerization, to form mixtures of six- and four-coordinate (Eq. 16), is prevented this approach produces five-coordinate complexes.

$$\tag{15}$$

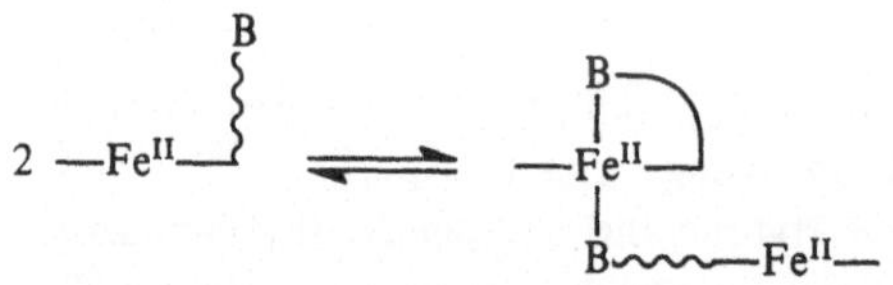

$$\tag{16}$$

In the following sections we examine those porphyrins which employ steric encumbrance and chelation as models for heme proteins. We also review a series of biomimetic porphyrins that have been synthesized in order to create complexes with quinones at fixed distances from the porphyrin ring. Using these models, important information on energy transfer between various components of the photosynthetic apparatus can be gained. Several excellent reviews exist which discuss the ligand binding properties of these models and their congruency with the natural systems[20, 32–37]. Instead, we will concentrate here on the strategy and synthetic details of model porphyrin preparations.

To this end, the compounds have been grouped together more in terms of structure than of function.

The reader is directed to reference 56 for a discussion on porphyrin dimers and strati bisporphyrins which are not reviewed here.

## 2. Porphyrins with Appended Peptides

Perhaps the most obvious approach to the synthesis of heme protein models is to reproduce the local environment of the heme active site by covalently attaching various peptide fragments to a suitable porphyrin. If the peptide fragments contain suitable amino acids (e.g. histidine, methionine), reproduction of the coordination sphere of the heme protein may be possible.

An early example of this approach was that of Lautsch et al.[38], who coupled various histidine-containing tripeptides to the propionic acid side-chains of mesoporphyrin IX **1a** (Scheme 2). After metal insertion into the porphyrin, intramolecular coordination by the histidyl imidazole was possible depending on the length of the peptide chain. Similarly, histidine-containing peptides were attached to the ethyl side chains of mesoporphyrin IX via sulfide linkages to give **2(b–d)**, a situation similar to that in cytochrome c. Losse and Müller[39] coupled L-histidine methylester and protohemin **3** with dicyclohexylcar-

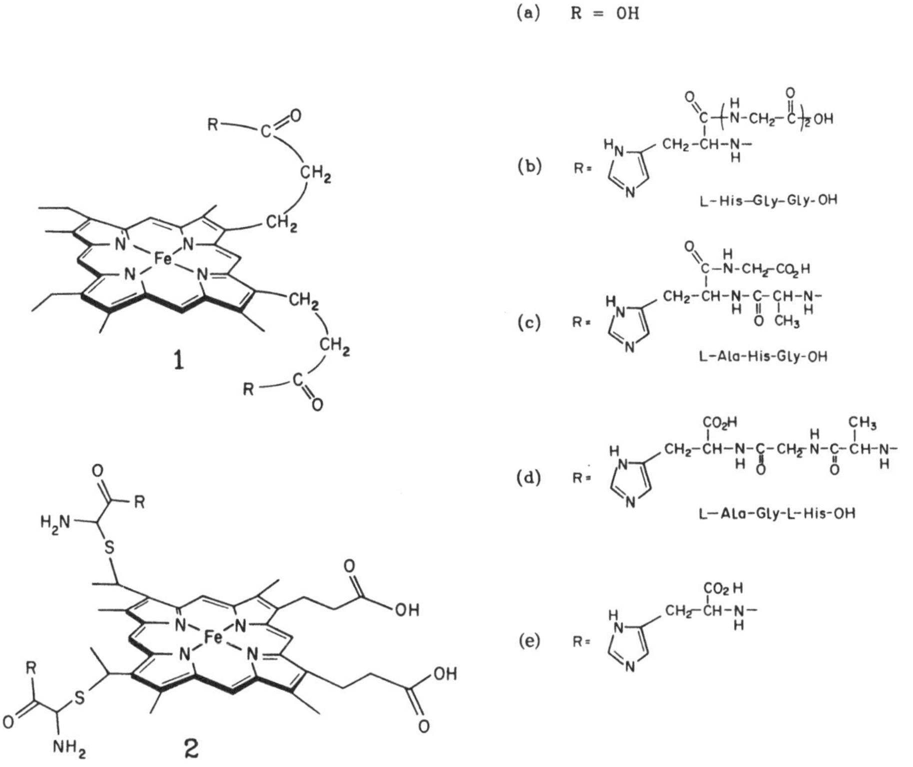

**Scheme 2**

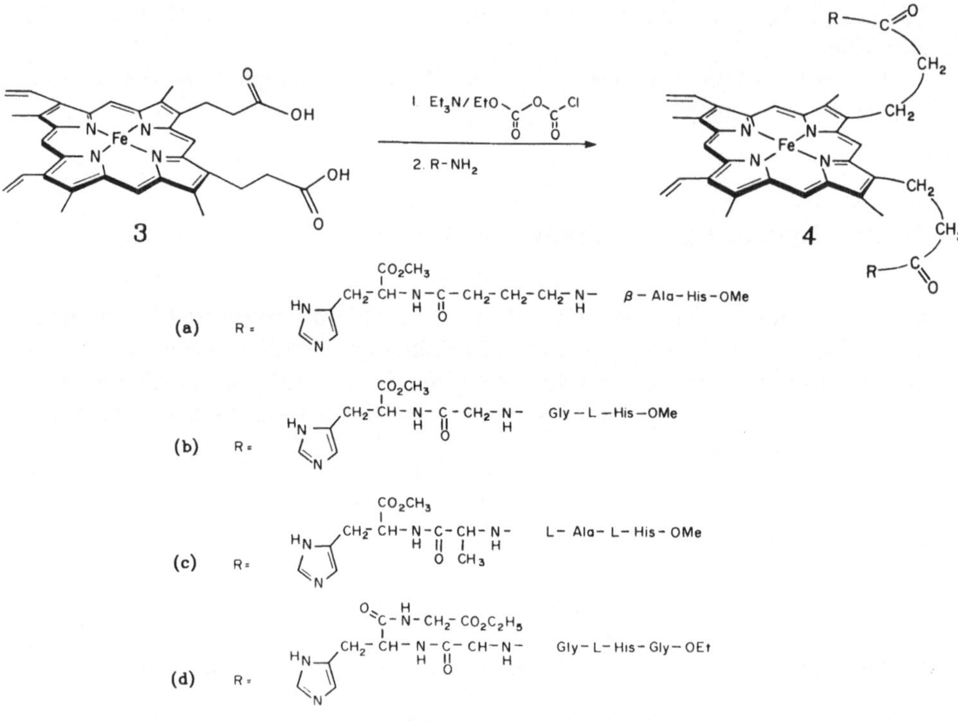

**Scheme 3**

bodiimide in N,N-dimethylformamide. However, models indicated that, with a single histidine bound to the porphyrin propionic acid, the length of the side chain was too short for coordination of the imidazole to the iron centre. Van der Heijden et al.[40], coupled various di- and tripeptide fragments (e.g. β-Ala–His; Gly–L–His; L-Ala–L–His; Gly–L–His–Gly–OEt) to protohemin **3** in DMF in the presence of triethylamine and ethyl chlorocarbonate to yield the bispeptidyl derivatives **4a–d** (Scheme 3). At the same time Warme and Hager[41] prepared porphyrins containing appended histidine and methionine groups. The reaction of mesohemin **6** with a $SO_3$/DMF complex **5** yielded the mesohemin sulfuric anhydride, in which one or both of the propionic acid side chains had been converted into a sulfonic anhydride. Subsequent reaction with amino acids such as histidine or methionine methyl ester yielded either mono- or disubstituted hemins **7a–e** (Scheme 4). A potential cytochrome c model **7e**, containing both histidine and methionine covalently attached to the porphyrin was also prepared. Unfortunately preparation and isolation of the products was quite tedious and yields were low. It was also recognized that the peptide-containing side arms were too short to allow unstrained intramolecular coordination. Momenteau et al.[42], have synthesized a five-coordinate iron porphyrin and characterized the coordinating properties. Treatment of deuterohemin **8** with equimolar quantities of ethylchloroformate and triethylamine, followed by L-histidine methyl ester dihydrochloride and more triethylamine gave, after purification, a mixture of three compounds **9a–c**. The desired deuterohemin 6(7)-mono-(histidine methyl ester **9c**) was separated from unreacted deuterohemin and deuterohemin 6,7-bis(histidine methyl ester) **9b**, and was obtained in 16% yield as a mixture of isomers (Scheme 5). The reduced iron(II) species was capable of binding oxygen reversibly at low

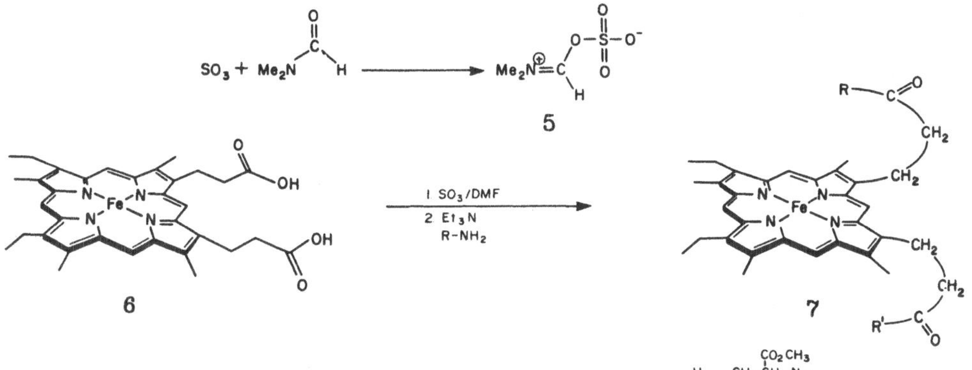

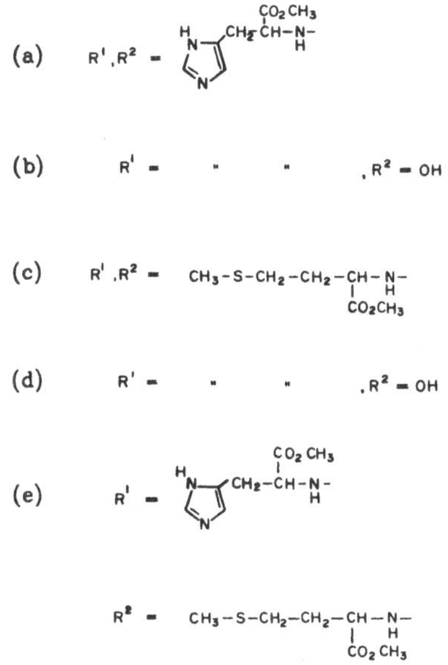

**Scheme 4**

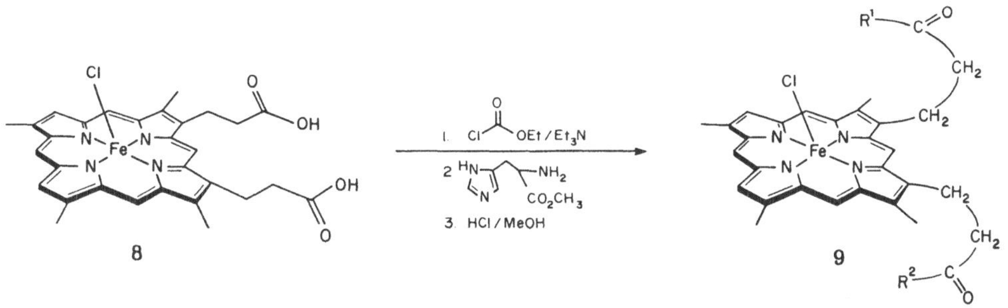

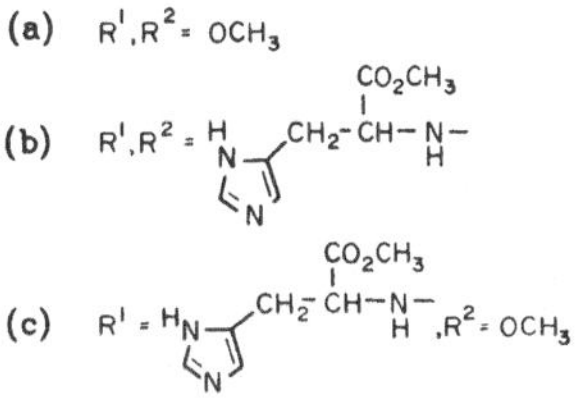

(a) $R^1, R^2 =$ [imidazole-$CH_2CH-N-$ with $CO_2CH_3$]

(b) $R^1 =$ " " , $R^2 = OH$

(c) $R^1, R^2 = CH_3-S-CH_2-CH_2-CH-N-$ with $CO_2CH_3$

(d) $R^1 =$ " " , $R^2 = OH$

(e) $R^1 =$ [imidazole-$CH_2-CH-N-$ with $CO_2CH_3$]

$R^2 = CH_3-S-CH_2-CH_2-CH-N-$ with $CO_2CH_3$

**Scheme 5**

(a) $R^1, R^2 = OCH_3$

(b) $R^1, R^2 = H$ [imidazole-$CH_2-CH-N-$ with $CO_2CH_3$]

(c) $R^1 = H$ [imidazole-$CH_2-CH-N-$ with $CO_2CH_3$] , $R^2 = OCH_3$

temperature but oxidized irreversibly at room temperature. Furthermore, extensive dimerization of the 5-coordinate species occurred at low temperature ($-60\,°C$), complicating oxygen binding studies.

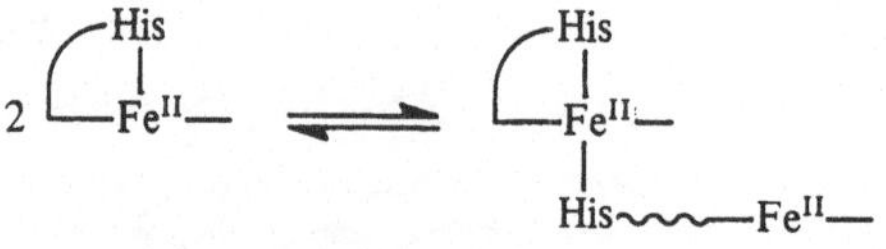

$$\tag{17}$$

A similar dimerization was observed for the iron(III) species at room temperature in concentrated solutions.

To this stage all the syntheses have yielded model systems which are 6-coordinate, or mixtures of the 5- and 6-coordinate species, separation of which could be tedious. Castro prepared porphyrin derivatives having two covalently attached imidazoles, by heating deuterohemin **8** or mesohemin **1a** with excess histamine in vacuo in the absence of solvent for three hours[43]. The bis-chelated product **10** was obtained in up to 50% yield after purification. Controlled hydrolysis with 2 M hydrochloric acid gave a 20% yield of the monochelated hemin **11**, again as a mixture of isomers (Scheme 6).

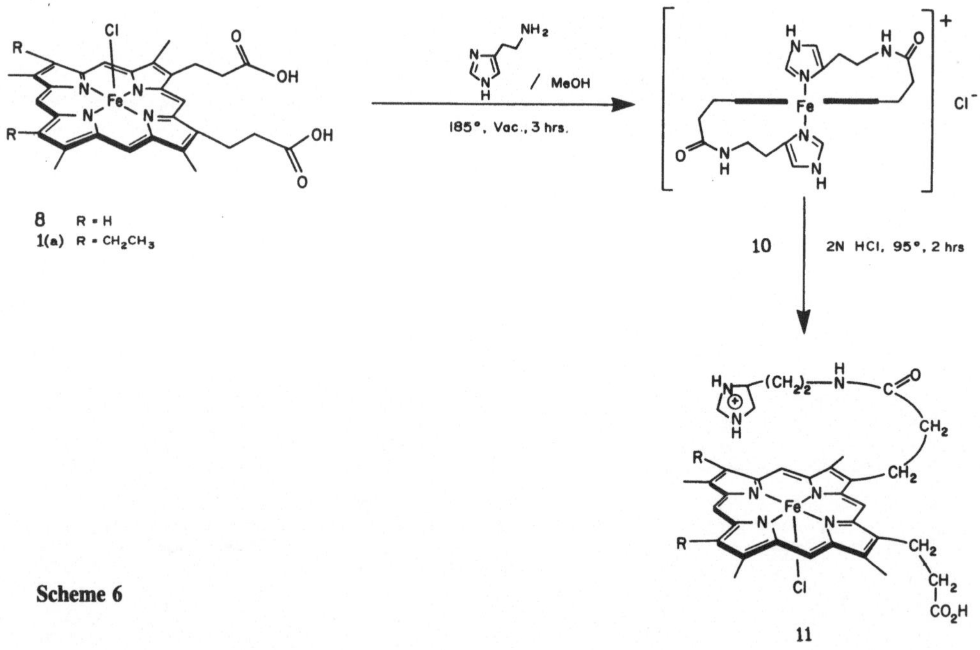

**Scheme 6**

More recently, Molokoedov et al.[44], have used protohemin monobenzyl ester **13**, (obtained from protohemin dibenzyl ester **12** in 61% yield by partial hydrolysis), to prepare a series of histidine-containing peptide derivatives **14a–e.** Coupling of peptide and hemin was completed by the mixed anhydride method using ethyl chloroformate and triethylamine. The yields of product decreased from 47% to 25% as the length of the peptide chain increased, the products being obtained as a mixture of the 6- and 7-isomers

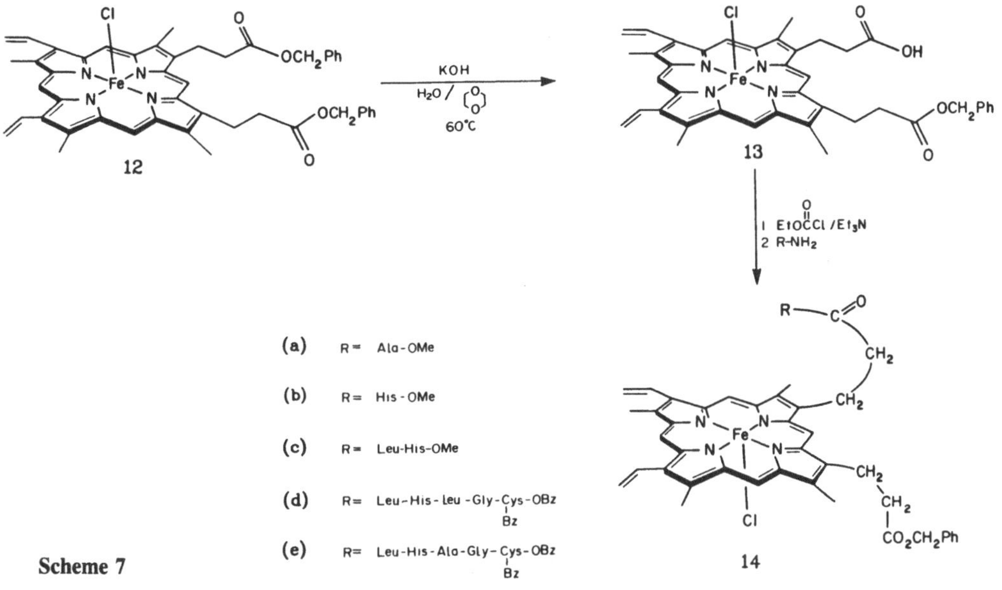

Scheme 7

(a)    R = Ala-OMe

(b)    R = His-OMe

(c)    R = Leu-His-OMe

(d)    R = Leu-His-Leu-Gly-Cys-OBz
                                    |
                                    Bz

(e)    R = Leu-His-Ala-Gly-Cys-OBz
                                    |
                                    Bz

(Scheme 7; only the 6-isomers are shown). The reduced pentapeptide **14d** and hexapeptide **14e** heme derivatives were reported to be stable for 35–40 minutes at room temperature in chloroform solution in the presence of air.

This work was extended to prepare a series of protohemin IX derivatives **15** with two appended peptide fragments[45]. The previously prepared protohemin IX monoaminoacyl derivatives (a mixture of 6- and 7-isomers) were condensed with various peptide fragments, using the pentafluoroester or mixed anhydride methods, to give the unsymmetric bisaminoacyl protohemin IX derivatives **15a–i** as a mixture of the 6- and 7-isomers (30–60%) (Scheme 8). Absorption spectra indicated that complexes having unsubstituted histidine residues in the peptide chain were capable of pentacoordination. Furthermore the hydrophobic environment of the bulky peptide chain was believed to enhance the stability of the 5-coordinate iron(II) derivatives.

There have been two attempts to model the active site of cytochrome c using heme-peptide compounds. Sequencing of various cytochromes c reveals the segment 14–18 to be invariant, with structure 14–Cys–X–X–Cys–His-18. The heme active site is bound to the polypeptide via sulfide bonds to the two cysteine residues 14 and 17, while the imidazole of His-18 provides one of the ligands for the iron atom of the heme. Momenteau and Loock[46], with a strategy similar to that of Lautsch et al.[38], attached a protected pentapeptide (Cys–Gly–Gly–Cys–His) to the ethyl side chains of 2,4-α,α′-dibromomesoporphyrin IX (**16**) to obtain the appended porphyrin **17** (Scheme 9). After iron insertion, optical spectra indicated that the imidazole of the histidine could bind to the metal centre of **17**. Binding of a methyl thioether e.g. methionine, to the iron would then mimic the coordination sphere observed in the natural cytochrome c. However, thioethers are poor ligands for iron. An attempt was made to overcome this poor binding by covalently attaching the thioether to the porphyrin. In this case a derivative of the synthetic porphyrin, tetraphenylporphyrin (TPP) was used. Various cysteine dipeptides were condensed with the TPP derivative **18** bearing a propenoate ester side chain at a β-

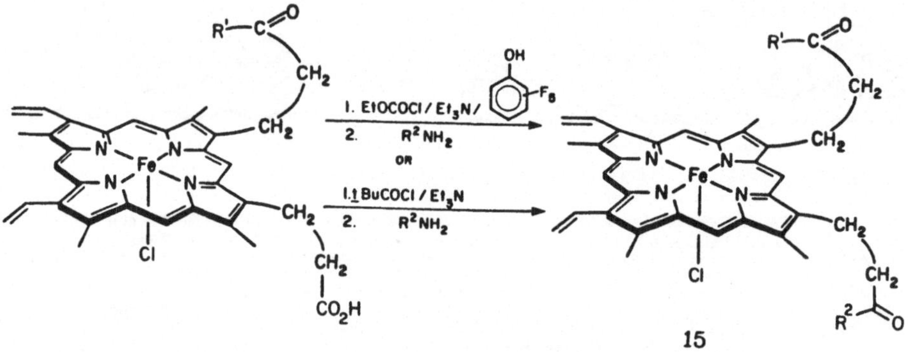

**15**

(a) $R^1$ = Leu−His(Bzl)−Lys(Z)−OMe
   $R^2$ = Ala−Leu−Ala−Phe−Ala−Cys(Bzl)−OMe

(b) $R^1$ = Leu−His(Bzl)−Lys(Z)−OMe
   $R^2$ = Ala−Leu−Ala−Phe−Ala−Cys(Bzl)−OBzl

(c) $R^1$ = Leu−His(Bzl)−Lys(Z)−OMe
   $R^2$ = Leu−His−Ala−Lys(Z)−Gly−Cys(Bzl)−OBzl

(d) $R^1$ = Leu−Lys(COCF₃)−Ala−OMe
   $R^2$ = Leu−His−OMe

(e) $R^1$ = Leu−Lys(COCF₃)−Ala−OMe
   $R^2$ = Leu−His−Leu−Lys(COCF₃)−Ala−OMe

(f) $R^1$ = Val−Phe−OMe
   $R^2$ = Leu−His−OMe

(g) $R^1$ = Leu−Leu−Val−Phe−OMe
   $R^2$ = Leu−His−OMe

(h) $R^1$ = Leu−Lys(COCF₃)−Ala−Ala−OMe
   $R^2$ = Ala−His−Lys(Z)−Leu−OMe

(i) $R^1$ = Lys(COCF₃)−Ala−Ala−OMe
   $R^2$ = Ala−His−Lys(Z)−Leu−OMe

**Scheme 8**

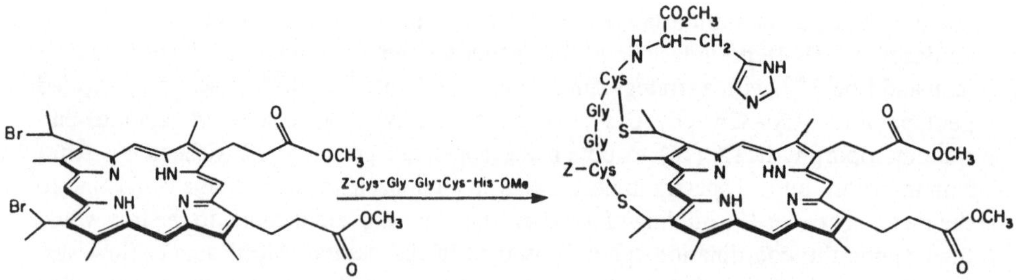

**16**

**17**

Z = benzyloxycarbonyl

**Scheme 9**

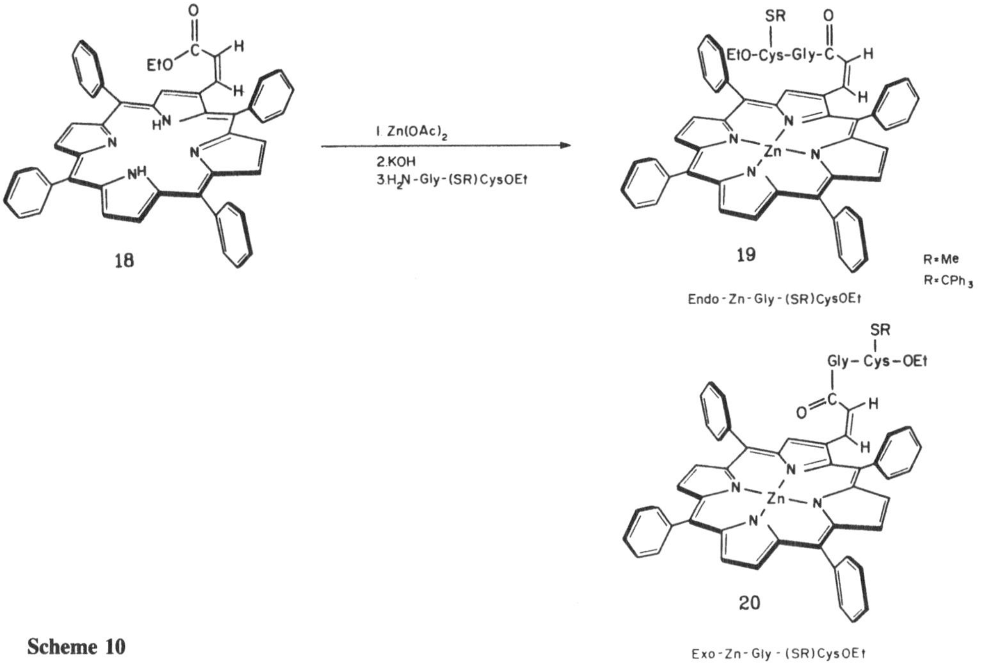

**Scheme 10**

pyrrole position[47–50]. Reaction of a dipeptide containing a terminal S-alkyl cysteine residue with cis-meso-tetraphenylporphyrin-3-propenoic acid, gave a mixture of two atropisomers, cis-endo **19** and cis-exo **20** (Scheme 10). In the cis-endo case, substituents on the peptide chain are disposed in a favourable conformation for binding to a metal in the porphyrin. When the dipeptide chain was Gly–(SR)Cys–OEt (R = Me, trityl), $^1$H–NMR and magnetic circular dichroism suggested that a metal sulfur bond involving the cysteine residue might be occurring. However, recent EXAFS data indicated that for the substituted Zn porphyrin, the Zn-sulfur interaction could only be weak and long range, if it occurred at all[50].

# 3. Chelated Hemes

## 3.A. Porphyrins Having Covalently Attached Imidazole or Pyridine Ligands

Chang and Traylor argued that in heme-peptide models the side chains containing the imidazole had either too few or too many atoms to achieve a strain-free iron-imidazole bond[51]. On the other hand it was argued that condensation of 1-(3-aminopropyl) imidazole **21** with the acid chloride of pyrroporphyrin XV **22** followed by insertion of iron would give a strain-free five-coordinate-system **23** (Scheme 11). This "chelated" heme was capable of binding dioxygen in a reversible manner in the solid state or when

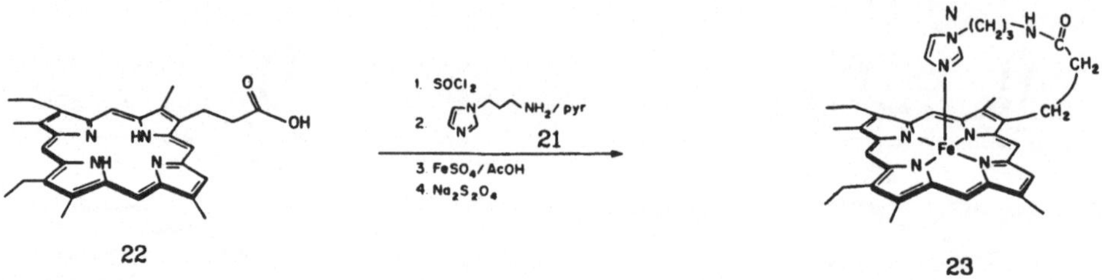

**Scheme 11**

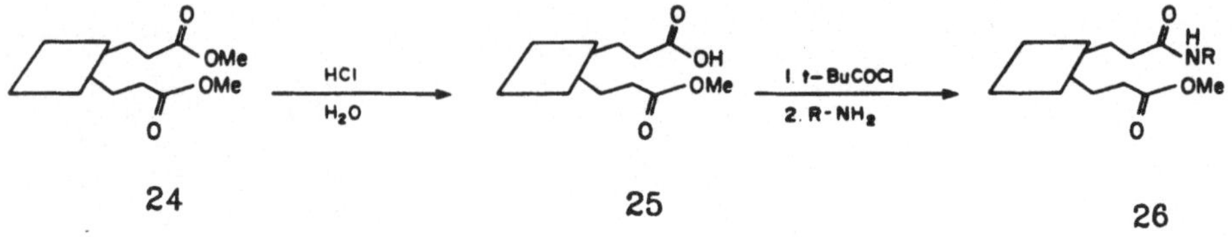

**Scheme 12**

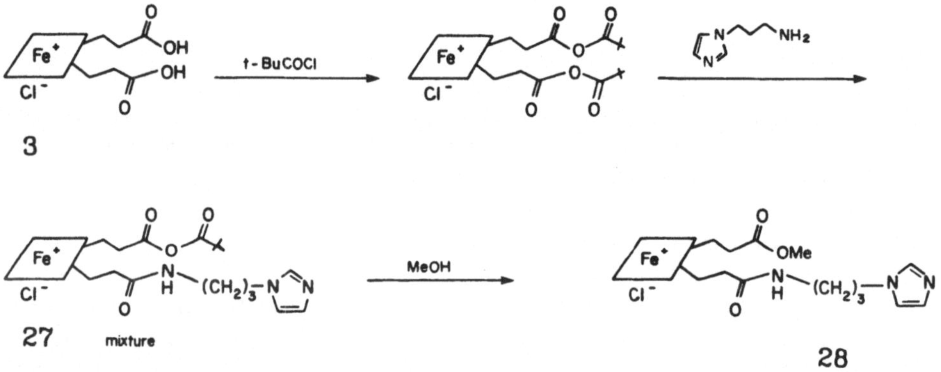

**Scheme 13**

dissolved in a polystyrene film. While capable of binding oxygen reversibly in solution at
$-45\,°C$, only irreversible oxidation occurred at room temperature[52]. A series of derivatives of pyrro-, proto- and mesoheme having pyridine or imidazole covalently bound to
the porphyrin ring through ester or amide linkages was investigated[26, 53-55]. For proto-
and mesoheme, the monochelated hemes were easily prepared as shown in Schemes 12
and 13[56]. In one approach the porphyrin dimethyl ester **24** was partially hydrolyzed. The
purified monoacid **25** was then coupled to a primary amine or alcohol containing an
imidazole or pyridine base using pivaloyl chloride. Alternatively commercially available
protohemin **3** was treated with excess pivaloyl chloride followed by one equivalent of the
base-containing primary amine. The reaction mixture was then quenched with water or
methanol. The monochelated products **26, 28** were isolated by chromatography as a
mixture of isomers in up to 30% yield. The dichelated compounds were similarly prepared.

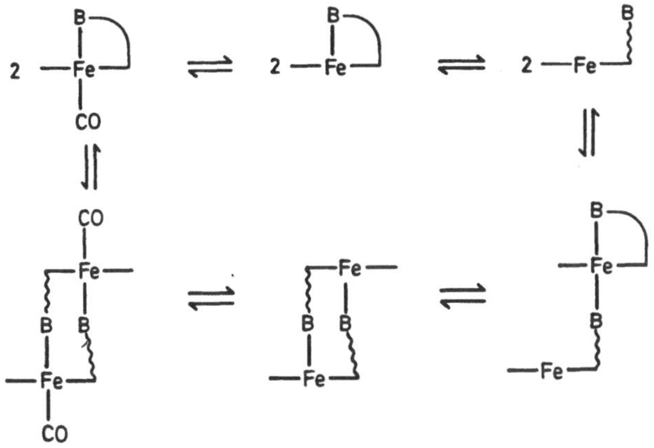

**Scheme 14**

The versatility of the chelated heme approach has allowed the systematic study of the kinetics of the $O_2$ and CO binding to these compounds. Changes in solvent, porphyrin side chains, the chelated base, and the length and nature of the chelation arm have been correlated with changes in the association and dissociation rates of $O_2$ and CO[57, 58].

Unlike the "chelated-histidine" system of Momenteau[42] which underwent dimerization at low temperatures, Traylor argued against the presence of any polymeric forms (Scheme 14) at the temperatures and concentrations used in his studies[26]. Indeed, Traylor exploited such dimerization to design a system exhibiting cooperativity[59]. For the iron(III) protoporphyrin IX derivative **29**, the side chain is too short to allow intramolecular binding of the pyridine (Scheme 15). While pyridine binds very poorly to hemin, addition of cyanide greatly increases pyridine affinity and vice versa. Therefore titration of the side-chain pyridine protohemin **29** with cyanide leads to clean conversion to the pyridine-hemin-CN$^-$ dimer **30**. A Hill plot (log Y/1 − Y vs. log [CN]) for this titration gave a straight line of slope n = 2.1, indicating cooperativity between the metal centres.

Axial base chelation was similarly used to prepare a symmetric diheme[60]. Meso-1,2-di-(3-pyridyl)ethylenediamine **(31)** was coupled with mesoporphyrin monomethyl ester **32** through the pivaloyl anhydride, followed by iron insertion to give the diheme **33** (Scheme 16). The reaction with CO exhibits two rate constants, indicating either two environments or a sequential change of environment due to cooperativity.

The paramagnetic $^1$H-NMR spectra of the imidazole-cyanide complexes of dichelated protohemins **34** and **35** were studied[61]. Chelation of the imidazole maintains a fixed orientation of the base with respect to the porphyrin ring, causing different chemical shifts for the methyl and vinyl protons. Comparison of these shifts with the values published for various heme proteins provided confirmation for the heme-imidazole orientations proposed in the natural systems.

Tabushi et al., have used the chelated heme approach to mimic the orientation of the heme groups in cytochrome $c_3$[62-64]. The "gable-porphyrin" **37** was prepared by the condensation of (*m*-formylphenyl)triphenylporphyrin **36** with pyrrole and benzaldehyde (22%) (Scheme 17). After metal insertion, the use of dimeric bridging ligands such as N,N'-diimidazolylmethane **38** and γ,γ'-dipyridylmethane **39** resulted in the formation of stable gable porphyrins with bridging ligands which displayed cooperative behaviour to the binding of 1-MeIm and CO.

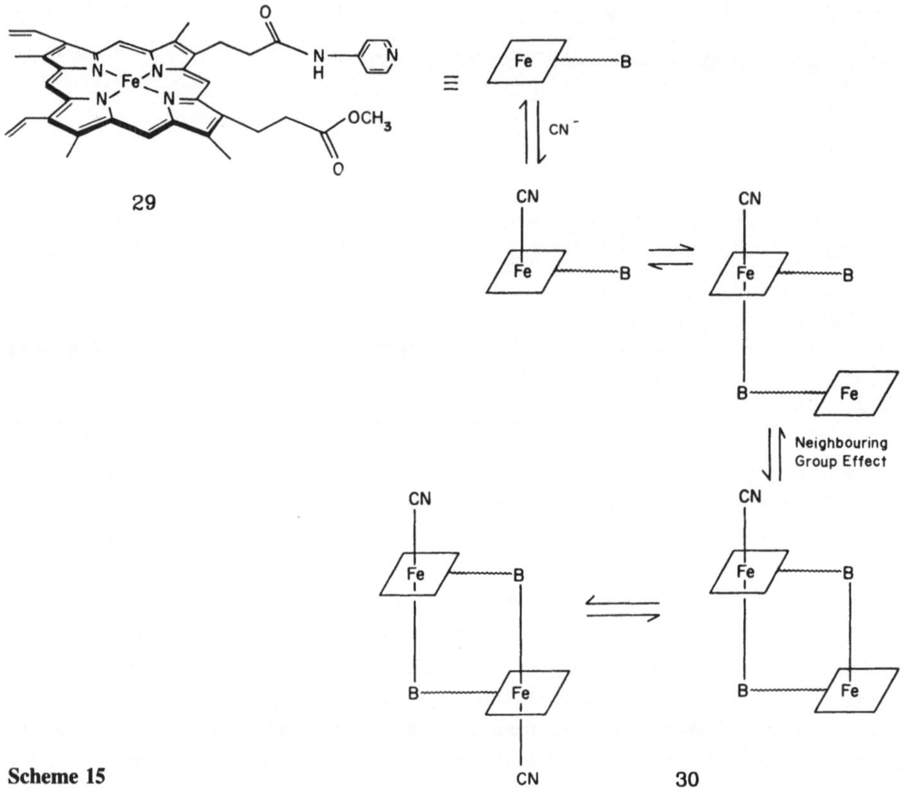

29

CN⁻

Neighbouring Group Effect

**Scheme 15**

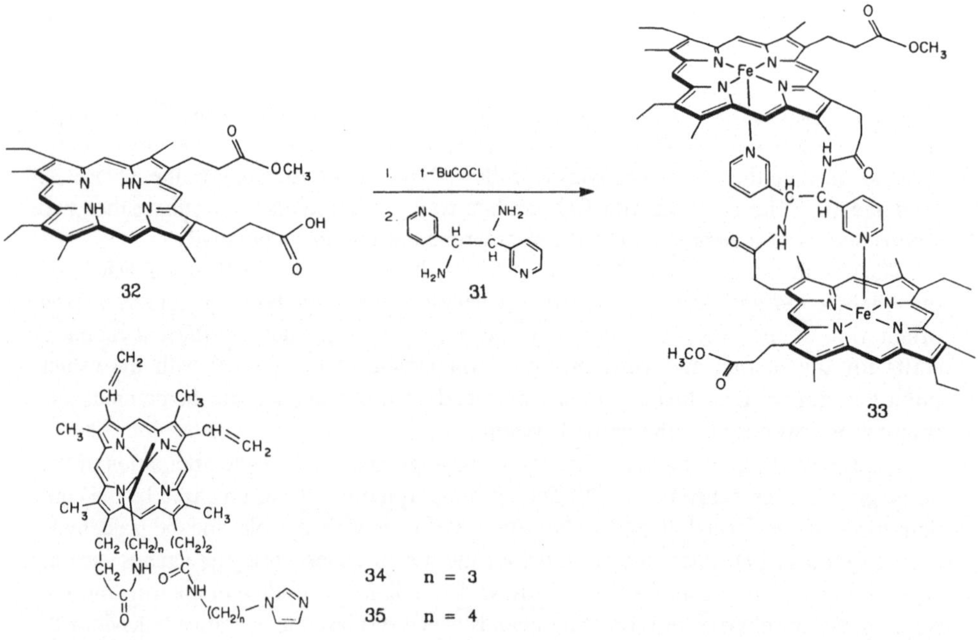

30

32

1. t-BuCOCl

2. 

31

33

34   n = 3

35   n = 4

**Scheme 16**

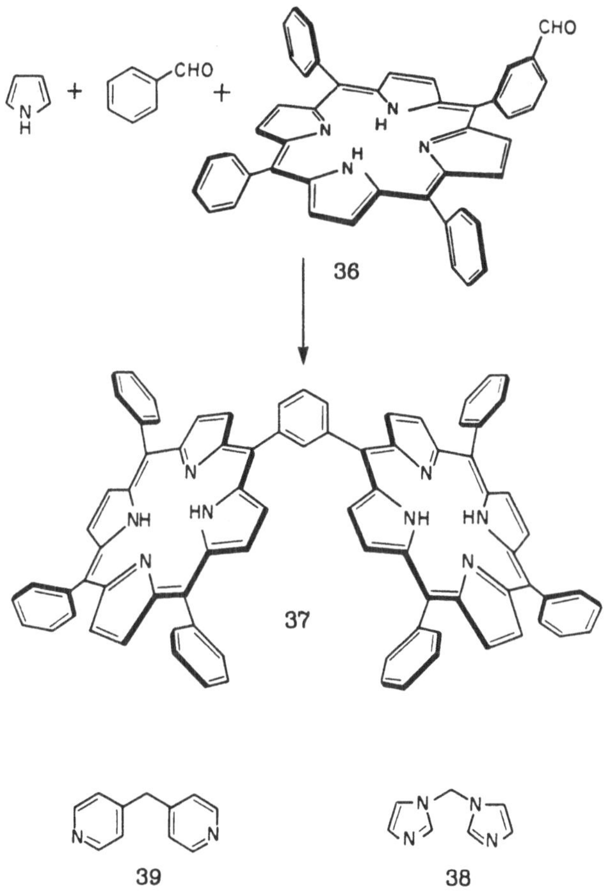

**Scheme 17**

The $^1$H-NMR analysis of chlorophyll is often complicated by its tendency to form dimers or higher aggregates. Denniss and Sanders[65] removed the phytyl group of chlorophyll-a and converted it to the mixed anhydride with triethylamine/ethyl chloroformate. Reaction with 1-(3-hydroxypropyl)imidazole **40** resulted in the chelated chlorophyll derivative **41** (Scheme 18). Intramolecular binding of the imidazole to the magnesium prevented aggregation and gave well resolved $^1$H-NMR spectra. A similar "chelated chlorophyll" **42** was used by Boxer and Wright to model the complex formed when apomyoglobin is reconstituted with chlorophyll derivatives[66].

Momenteau et al. have synthesized a similar series of chelated heme compounds[67]. In this series the base is attached to the β-pyrrole position of a tetraphenylporphyrin (TPP) ring, via amide or ester linkages (Scheme 19). Vilsmeier formylation of Cu(TPP) (**43**)[68] afforded the monoformyl derivative **44,** which was elaborated by a Wittig condensation to yield the acrylate **45** as a mixture of cis and trans isomers. Demetallation, hydrogenation and saponification gave the propionic acid derivative **46.** Treatment of the corresponding acid chloride **47** with 1-(3-aminopropyl)imidazole **48** or 3-(3-hydroxypropyl) pyridine **49** resulted in the appropriate chelated porphyrin **50** or **51** (70% and 60% respectively) (Scheme 19). These models were used to study the kinetics of base binding

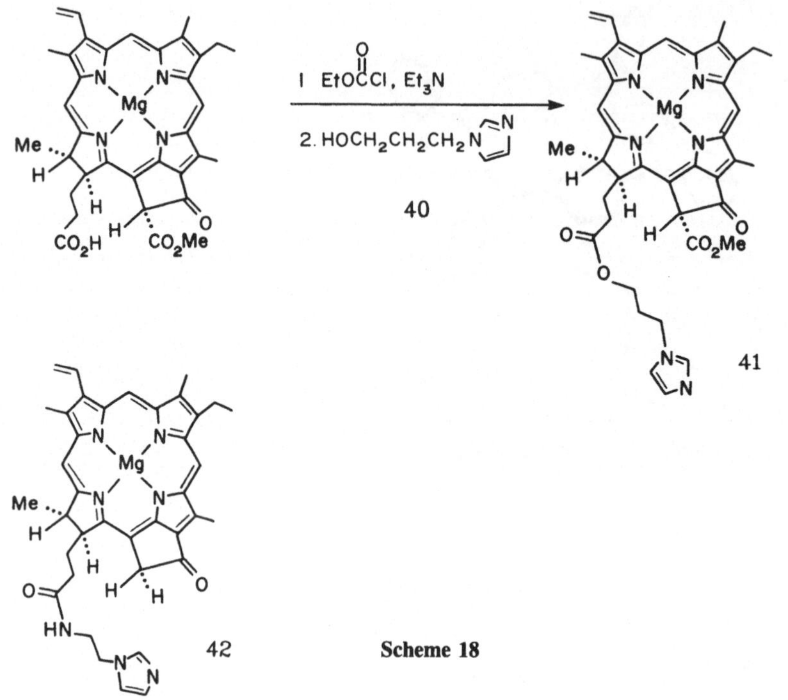

**Scheme 18**

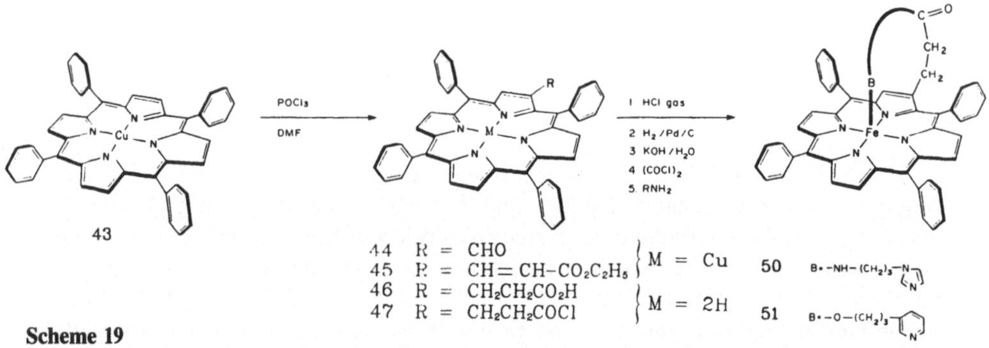

| |  |  | | | |
|---|---|---|---|---|---|
| 44 | R | = | CHO | | |
| 45 | R | = | CH=CH–CO₂C₂H₅ | } M = Cu | 50 |
| 46 | R | = | CH₂CH₂CO₂H | | |
| 47 | R | = | CH₂CH₂COCl | } M = 2H | 51 |

**Scheme 19**

to 4- and 5-coordinated iron(II) TPP[69], and also to study the transient oxygenation of iron(II) carbonmonoxy TPP after photolytic displacement of CO[70].

The same TPP-acrylic acid system was used by Eaton et al.[71, 72]. Treatment of the acid chloride of Cu–TPP acrylic acid **52a** with a nitroxyl resulted in the appropriate spin-labelled Cu–TPP derivatives **52b–f** (Scheme 20). The extent of metal-nitroxyl interaction was investigated using EPR by varying the nature of the nitroxyl, the linkage (amide or ester), and the geometry of the complex (cis or trans). A similar study was carried out on a vanadyl porphyrin[73].

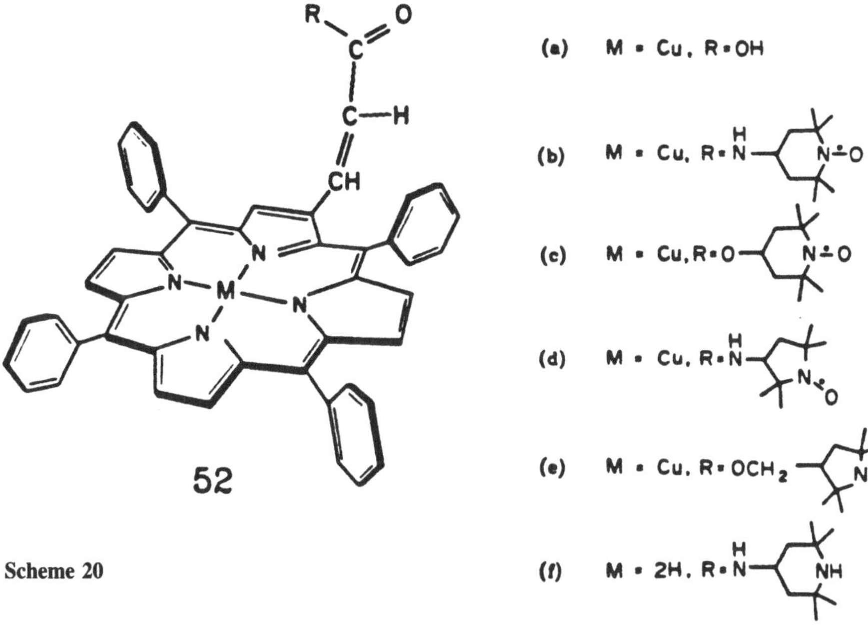

**52**

**Scheme 20**

(a)  M • Cu, R • OH

(b)  M • Cu, R • N—⟨peptide NH⟩N±O

(c)  M • Cu, R • O—⟨⟩N±O

(d)  M • Cu, R • N—⟨⟩N≗O

(e)  M • Cu, R • OCH₂—⟨⟩N≗O

(f)  M • 2H, R • N—⟨⟩NH

Collman et al.[74], have prepared chelated TPP compounds where the chain is attached to the *ortho* position of a *meso*-phenyl ring. Condensation of *o*-nitrobenzaldehyde (**53**), benzaldehyde (**54**) and pyrrole (**55**) in hot glacial acetic acid gave a 2% yield after chromatography, of the *meso*-mono-(*o*-nitrophenyl)triphenylporphyrin **56**. Reduction with stannous chloride produced the mono-*o*-amino-TPP (**57**) which was coupled with various imidazole chains (Scheme 21). Collman and his colleages merely used the compounds **58a–c** as substitutes for the less readily accessible "tailed-picket fence" porphyrins which are discussed in Sect. 4B. Mashiko et al.[75], used the same chelated TPP

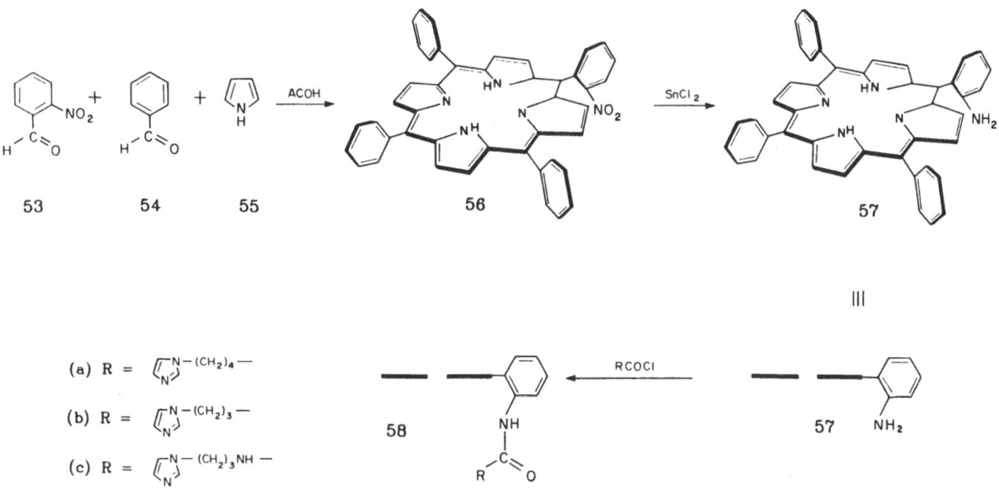

53        54        55        ACOH →        56        SnCl₂ →        57

III

(a) R = ⟨imidazole⟩N—(CH₂)₄—

(b) R = ⟨imidazole⟩N—(CH₂)₃—

(c) R = ⟨imidazole⟩N—(CH₂)₃NH—

**58**        ← RCOCl ←        **57** NH₂

**Scheme 21**

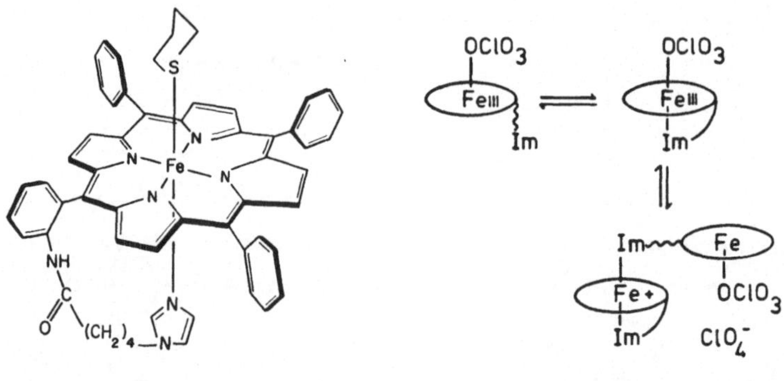

**Scheme 22**

compounds to control coordination in a mixed ligand system. Attempts to prepare models for cytochrome c, which contain histidine and methionine as the axial ligands, are frustrated by the greater affinity of heme iron for imidazole rather than thioether. By covalently attaching the imidazole to the porphyrin ring a stoichiometric amount of ligand was provided; addition of thioether then furnished the mixed six-coordinate system. These authors prepared several complexes with different tail lengths and various thioethers. For the $C_5$ tail with tetrahydrothiophene as the thioether, a crystalline iron(II) complex, $Fe^{II}(C_5Im)(TPP)(THT)$ (59) (Scheme 22) was obtained and its crystal structure determined (Fig. 1). Efforts to obtain the corresponding iron(III) complex were defeated by "head-to-tail" dimerization.

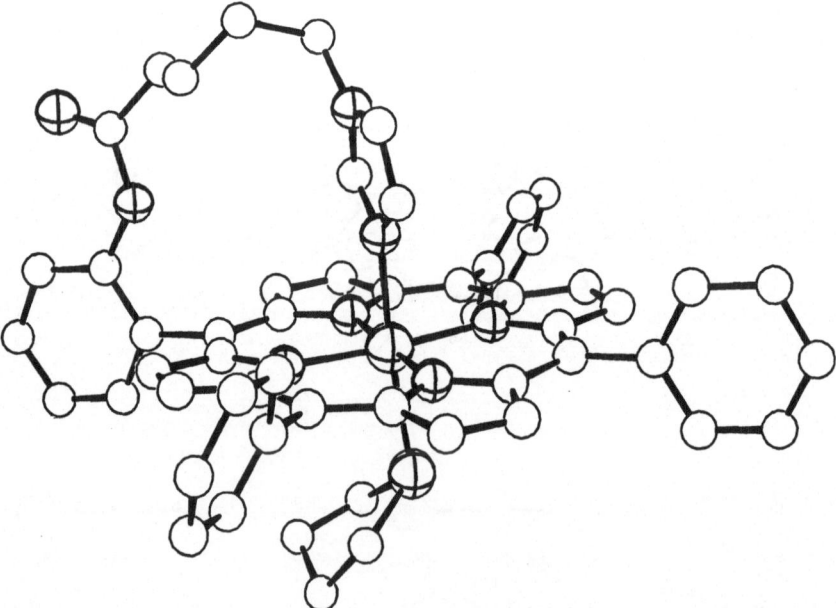

**Fig. 1.** Computer produced perspective of $Fe^{II}(C_5Im)(TPP)(THT)$ **59.** Adapted from Ref. 75

A similar chelated TPP compound **60** was used by Walker to study the effect of axial ligand plane orientation on the ${}^1$H-NMR shifts of the pyrrole protons in iron(III) (TPP)bis(imidazole) complexes[76]. In this case the tail was made shorter to study the effect of axial ligand bond strain. In addition, Walker and Benson have used mono-(*o*-aminophenyl)triphenylporphyrin (**57**) to prepare a series of derivatives **61a–e** containing a pyridine ligand bound to a zinc TPP[77]. ${}^1$H-NMR and visible spectroscopy were used to study the displacement of the 3-pyridyl ligand by free 3-picoline (Scheme 23).

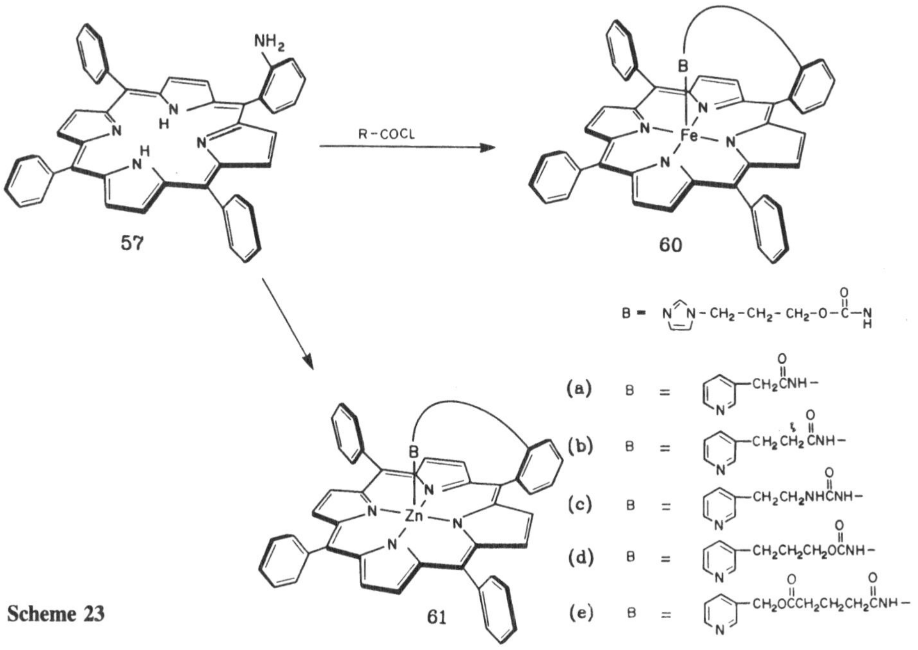

**Scheme 23**

Condensation of **57** with trans-urocanic acid chloride (**62**) furnished the chelated porphyrin **63**[78]. In this case the imidazole was attached at the C-4 rather than the more common N-1 position to allow deprotonation to the imidazolate. After iron insertion and deprotonation addition of Cu(acac)$_2$ yielded the μ-imidazolato binuclear complex **64** (Scheme 24), a potential model for the [Cu$_u^{2+}$/Cyta$_3^{3+}$] center of cytochrome oxidase.

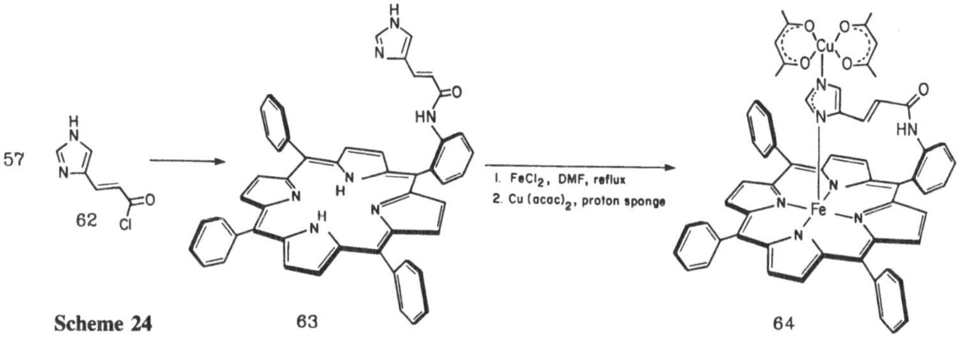

**Scheme 24**

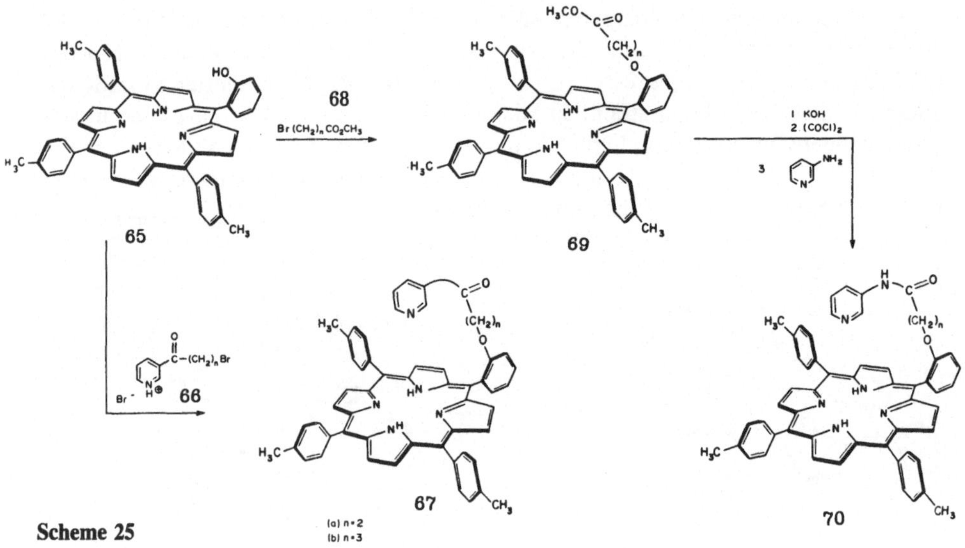

**Scheme 25**

(a) n = 2
(b) n = 3

However magnetic and EPR studies of **64** indicated that the Fe$^{III}$ and Cu$^{II}$ centers were essentially non-interacting, unlike the strongly coupled [Cu$_u^{2+}$/Cyta$_3^{3+}$] pair in the natural system.

Molinaro et al.[79] used mono-(o-hydroxyphenyl)tritolylporphyrin (**65**) to covalently attach a series of pyridine ligands to the porphyrin ring via an ether linkage. Reaction of o-hydroxyphenyl-TTP (**65**) with 3-(bromoalkyl)pyridine hydrobromide (**66**) furnished the chelated porphyrin **67** in 40% yield. The condensation of **65** and methyl-4-bromobutyrate (**68**) gave the porphyrin butoxy ester (**69**), which was elaborated to **70** in 27% overall yield (Scheme 25). The cobalt derivatives reacted reversibly with oxygen at low temperatures (− 50 °C to − 80 °C) but the presence of the axial base did not enhance oxygen affinity. Goff used the same synthetic strategy to prepare porphyrins with appended imidazoles[80]. Reaction of o-hydroxyphenyl-TTP (**65**) with a dibromoalkane (**71**) forms the corresponding ether **72**, which, when allowed to react with imidazole in DMF solvent (using K$_2$CO$_3$ or Et$_3$N), gives the required chelated TTP derivative **73** (Scheme 26). By varying the length of the alkane chain, "tension" may be introduced into the molecule in the form of tilting or elongation of the iron-imidazole bond. This was found to have an effect on the splitting and shift of the pyrrole resonances in the $^1$H-NMR spectrum.

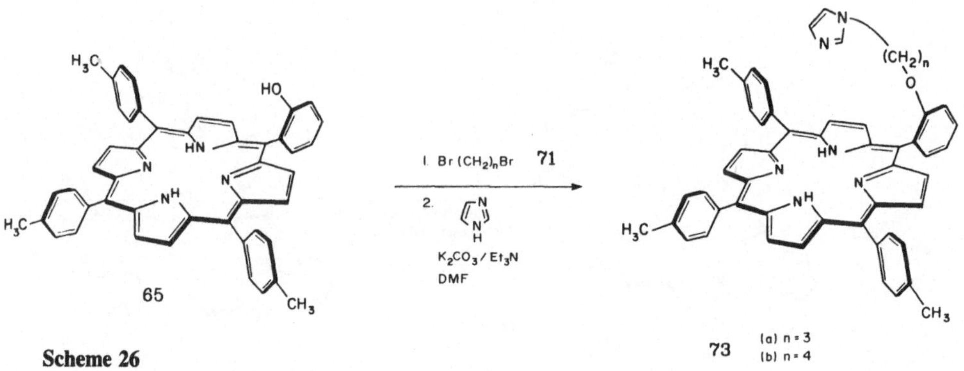

**Scheme 26**

1. Br (CH$_2$)$_n$Br   **71**

2. imidazole
K$_2$CO$_3$ / Et$_3$N
DMF

**73**   (a) n = 3
(b) n = 4

## 3.B. Porphyrins Having Covalently Attached Sulfur Ligands

Because of the poor affinity of iron(II) porphyrins for mercaptide anion, models of cytochrome P450 have usually consisted of solutions of porphyrins in the presence of large concentrations of excess mercaptide ion. However, Traylor[81] has used the chelated heme approach to covalently attach mercaptide to the porphyrin periphery, making it available for binding to the metal without the necessity of excess external ligand (Scheme 27). Protohemin chloride monodimethylamide monoacid (74) was coupled to 1-

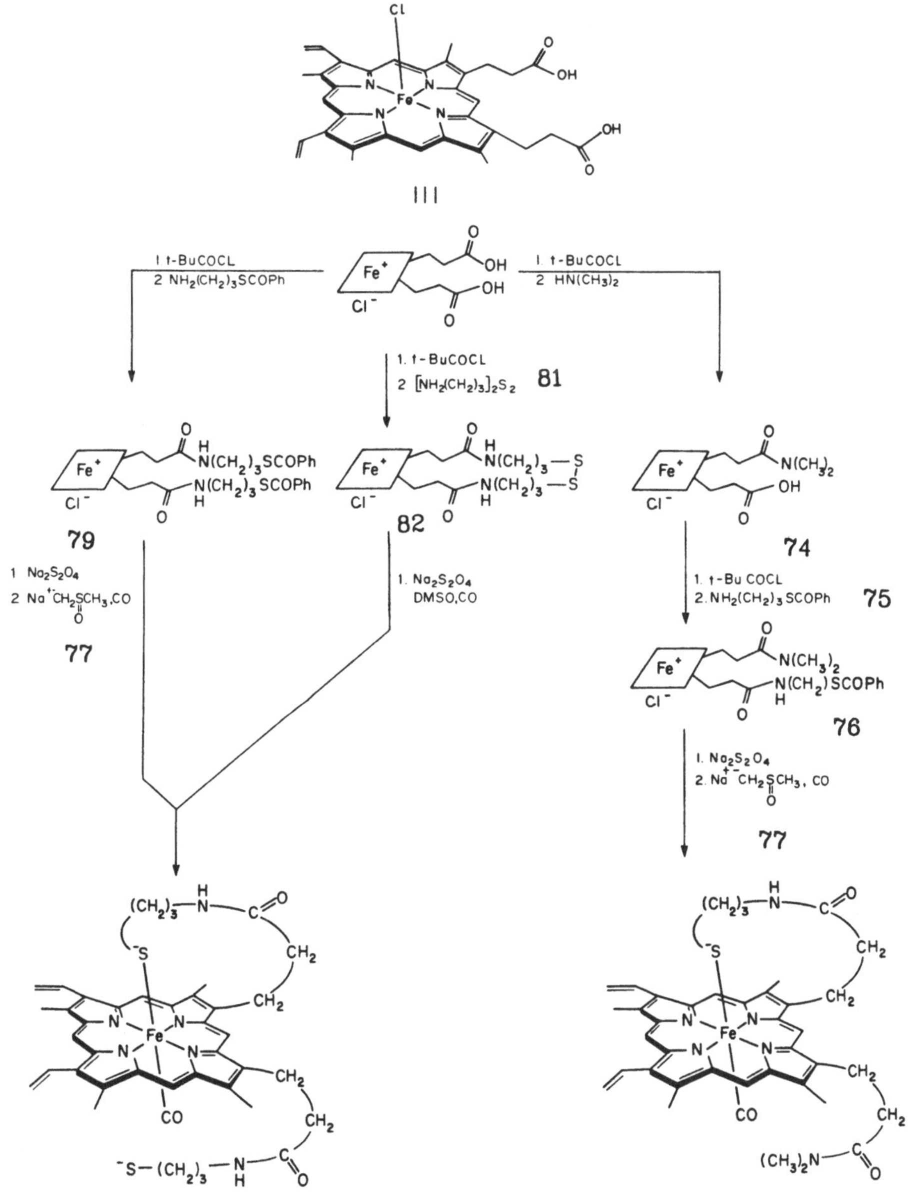

**Scheme 27**          80                                                                      78

amino-3-mercaptopropane benzoyl ester (75). After reduction with sodium dithionite, addition of dimsyl anion (77) and warming removed the benzoyl group, and addition of CO resulted in formation of the carbonmonoxy-mercaptide complex 78. A similar compound 79, containing two masked mercaptides was also prepared and deprotected to give the analogous CO complex 80. Protection of the mercaptide as the benzoylthio derivative before reduction of $Fe^{III}$ was necessary because of the reducing ability of mercaptans.

$$2\,Fe^{III} + 2\,RSH \rightleftharpoons 2\,Fe^{II} + RSSR + 2\,H^+ \tag{18}$$

Alternatively protohemin was coupled with bis(3-aminopropyl)-disulfide (81). The resultant disulfide 82 was treated with sodium dithionite, the iron being reduced faster than the disulfide. Addition of CO then furnished the carbonmonoxy complex 80.

$^1$H-NMR of the CO complexes indicated that the sulfide underwent intramolecular binding without appreciable dimer formation.

$$(19)$$

UV/visible spectroscopy in DMSO solution or aqueous suspension showed a split Soret band (384/460 or 363/446 nm) which was similar to the hyper spectrum of carbonylated cytochrome P450.

A series of alkyl and aryl "mercaptan-tail" porphyrins has been prepared by Collman and Groh (Scheme 28)[82]. The $C_6$ alkyl chain may be prepared directly by treating mono-(o-aminophenyl)triphenylporphyrin (57) with S-acetyl or S-tritylthiohexanoyl chloride (83). Since the S-protected pentanoic acid derivatives were more difficult to obtain, the bromoalkyl chain was first attached to the porphyrin to give 85 and the thio group introduced by treatment with either $TrS^-K^+$ or $AcS^-K^+$. Deacetylation ($MeOH/NH_3$) or detritylation ($Hg^{II}/H_2S$) gave the free thiol. However, after iron insertion, visible spectra indicated that the alkyl chain was too flexible to hold the mercaptan at the metal site; the spectra, in toluene, were similar to those of square planar four-coordinate iron(II) species. The introduction of CO does lead to the formation of six-coordinate low-spin $Fe^{II}$–CO complexes.

$$(20)$$

To ensure greater rigidity, tails derived from o-mercaptobenzoic acid (86; X = H or $CH_3$) or (m-mercaptophenyl)acetic acid 87 were attached to the aminoporphyrin. In this case the potential thiol was introduced as the disulfide (see 86 or 87) which was subsequently cleaved with sodium borohydride to give the free aryl "mercaptan-tail" porphyrins 90 and 91. As in the alkyl case, the aryl iron(II) species did not show five-coordina-

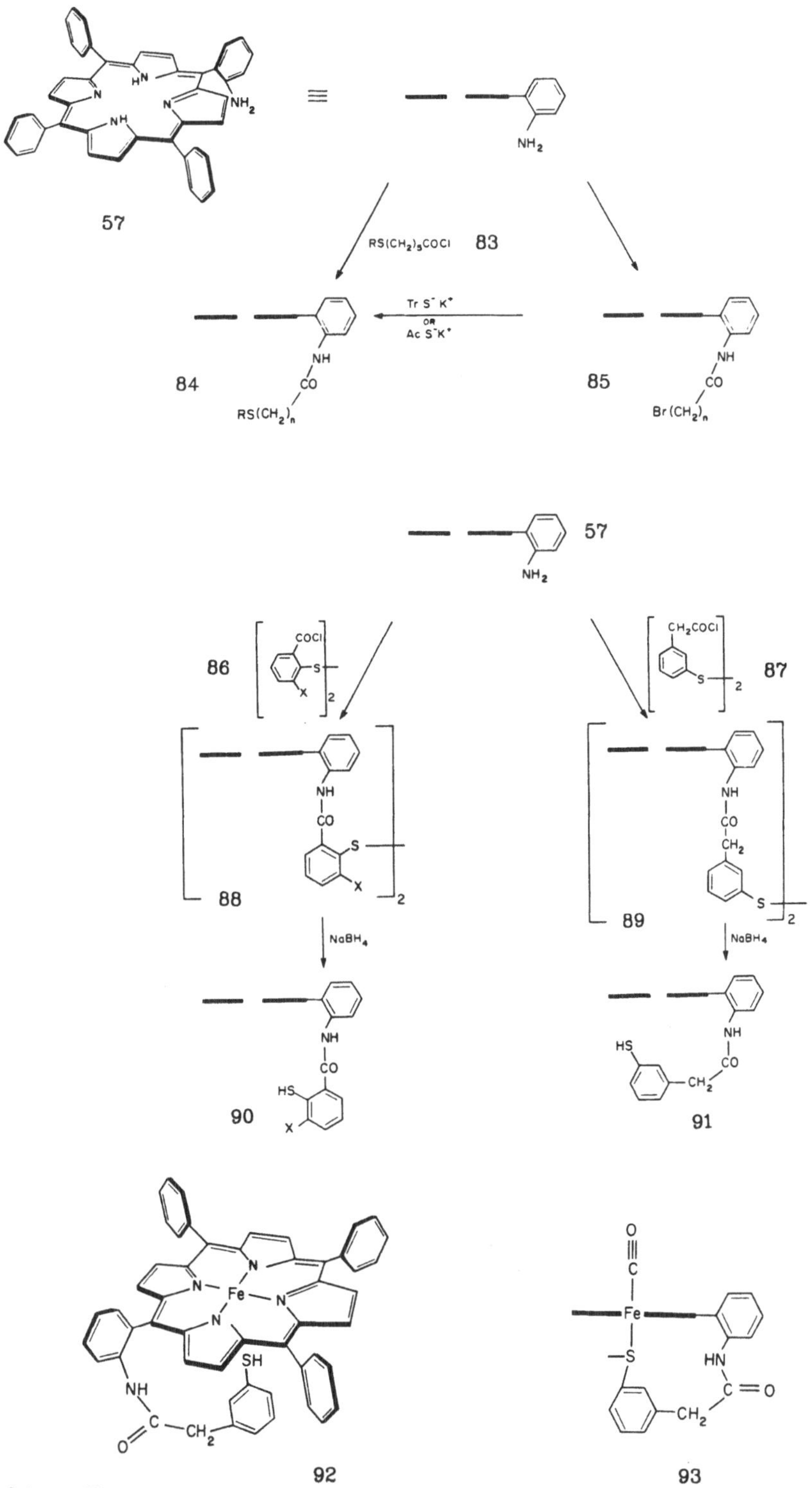

**Scheme 28**

tion. Furthermore, addition of CO gave mixtures of five- and six-coordinate species, depending on the nature of the mercaptan and the temperature, suggesting a tail-off/tail-on equilibrium.

Deprotonation of the mercaptan to give the mercaptide was attempted. The extent of mercaptide formation depended both on the nature of the base

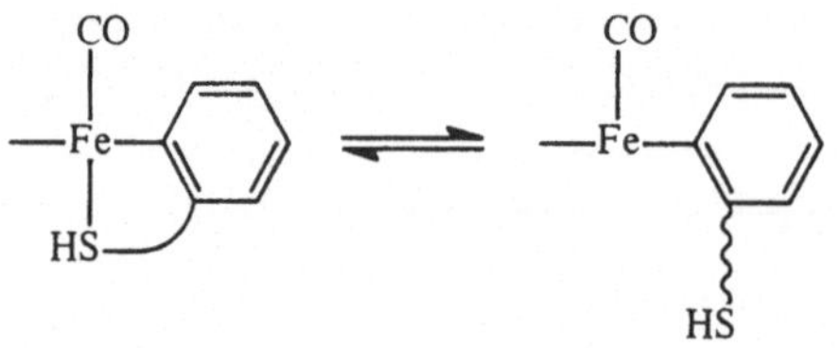

$$(21)$$

and of the tail. Indeed, for most tails only incomplete deprotonation occurred. However for the (*m*-mercaptophenyl)-acetamide tail system **92**, deprotonation to the mercaptide was clean and complete using acetanilide anion as base. In the presence of CO this system gave a six-coordinate iron(II)-mercaptide-CO complex **93** whose visible spectrum exhibited a split Soret absorption at 450 and 380 nm, typical of cytochrome P450.

In Sect. 3.A. we referred to the synthesis of $Fe^{II}(C_5Im)(TPP)(THT)$ (**59**) as a model for cytochrome c[75]. Buckingham and Rauchfuss adopted the alternative strategy of attaching the thioether ligand to the porphyrin periphery[83]. Reaction of (*o*-amino-phenyl)triphenylporphyrin **57** with the corresponding anhydrides **94** afforded the tailed porphyrins **95a–c**, containing thioether, sulfoxide, or sulfone groups in 60–90% (Scheme 29). After iron insertion and reduction, spectrophotometric titration with base allowed the following ordering of ligand affinities: $R_2SO > R_2S > R_2SO_2$. The easy displacement of sulfide by pyridine or imidazole precluded this system as an effective model for cytochrome c.

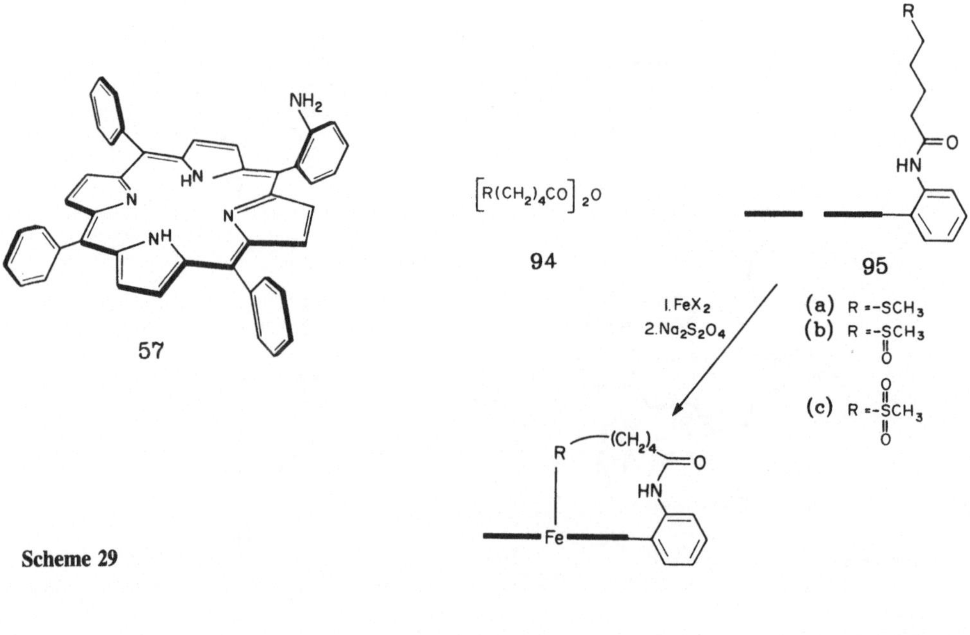

**Scheme 29**

Smith and Bisset have also used the chelated heme approach to synthesize a potential P450 model[84], but in this case the substituents were attached to the meso position of an octalkylporphyrin. A *meso*-acetoxymethyl substituent is susceptible to nucleophilic displacement at the "benzylic" carbon atom leading to the ready introduction of suitably functionalized chains. Heating the acetoxymethylporphyrin **96** in a melt with 1,6-hexanediol gave the ether **97** with no sign of dimer formation (Scheme 30). Conversion to the bromide followed by refluxing with thiourea afforded the thiouronium salt **98,** the hydrolysis of which was accompanied by oxidation to give the disulfide **99.** Unfortunately, attempts to generate the characteristic $RS^- - Fe^{II} - CO$ spectrum of **100** from the disulfide complex **99** were unsuccessful. Alternatively, treatment of acetoxymethylporphyrins with dithiols yielded only the *meso*-methylporphyrins. Treatment of *meso*-acetoxymethyl octaethylporphyrin with excess 1-(3-aminopropyl)imidazole **21** in refluxing tetrahydrofuran containing a suspension of sodium hydride provided the *meso*-chelated imidazole porphyrin **101** (Scheme 30), a potential model for T-state hemoglobin.

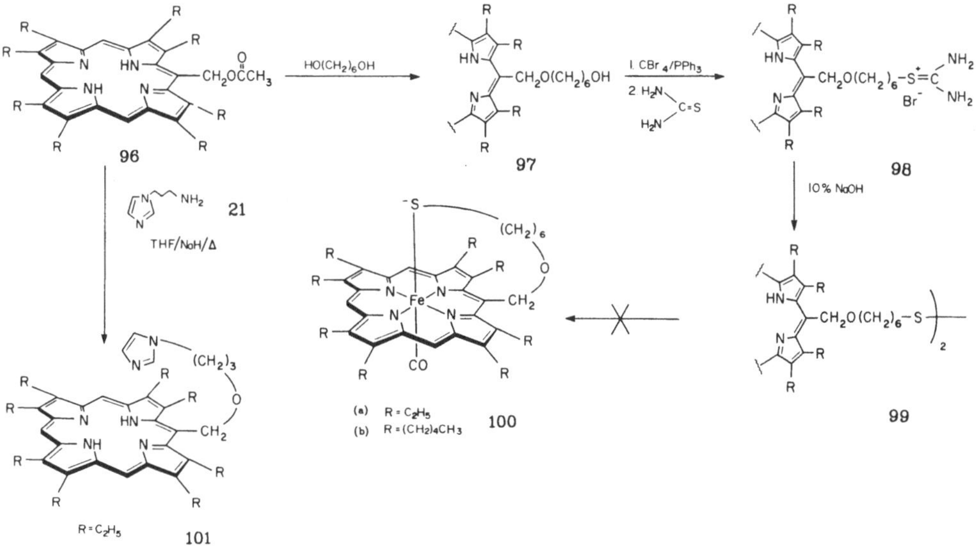

**Scheme 30**

The success of the chelated heme approach to model heme proteins is due to its ability to control the coordination of a metalloporphyrin. For poorly binding ligands e.g. mercaptide, thioether, covalent attachment to the porphyrin increases the local concentration and enhances binding without the need of excess external ligand. In the case of strongly binding ligands e.g. imidazole, pyridine, chelation can be used to lower their binding ability. Addition of one equivalent of base to an iron porphyrin results in a mixture of four- and six-coordinate species, since $K_2 > K_1$ in Eq. 22.

$$-Fe- + B \underset{K_1}{\rightleftharpoons} \overset{\overset{B}{|}}{-Fe-} + B \underset{K_2}{\rightleftharpoons} \overset{\overset{B}{|}}{\underset{\underset{B}{|}}{-Fe-}} \tag{22}$$

However covalent attachment of the base to the porphyrin provides a stoichiometric equivalent of base which can bind intramolecularly to give the desired five-coordinate species, provided dimerization is not significant.

$$
\begin{array}{c}
\text{B—} \\
| \\
\text{—Fe—}
\end{array}
\rightleftharpoons
\text{—Fe——}\wedge\wedge\text{B}
\rightleftharpoons
\begin{array}{c}
\text{B}\wedge\wedge\text{——Fe—} \\
| \\
\begin{array}{c}\text{—Fe—}\\ | \\ \text{—B}\end{array}
\end{array}
\tag{23}
$$

Addition of a second ligand then gives the mixed ligand complex. As models for hemoglobin or myoglobin the chelated hemes of Traylor were found to bind oxygen reversibly in solution at low temperature ($-40\,^\circ$C) and the kinetics of oxygen binding at room temperature could be measured. However the oxygen complexes were not stable at room temperature, irreversible oxidation to the μ-oxo complex occurring.

In an attempt to prepare stable oxygen-binding complexes it was realized that steric hindrance about one face of the porphyrin might prevent irreversible oxidation. If a base bound to the open face and oxygen bound at the sterically hindered face, close approach of two porphyrin rings would be discouraged, preventing μ-oxo bridge formation.

$$
\text{B—Fe}
\rightleftharpoons
\text{B—Fe——O}_2
\xrightarrow{\times}
\text{B—Fe——O}_2 \quad \text{Fe——B}
\tag{24}
$$

Such an approach has been followed by many groups to produce an array of different model porphyrins e.g. picket-fence, capped, cyclophane and crowned, strapped and basket-handle systems.

## 3.C. Porphyrins with Covalently Attached Quinone Groups

An approach, similar to that of the chelated heme model systems, has been adopted by many researchers to prepare porphyrins with appended quinone groups. Such systems have stimulated interest as possible models for the primary electron transfer event of photosynthesis, where photoinduced charge transfer occurs from excited singlet state chlorophyll donors to nearby quinone acceptors.

One of the earliest such models was that of Kong and Loach who prepared a meso-(p-carboxyphenyl)tritolylporphyrin (102)[85]. Condensation with a suitably substituted dimethoxybenzene 103 followed by demethylation and oxidation furnished the desired porphyrin-quinone pair 104 (Scheme 31).

A similar series of substituted tritolylporphyrins 105 have been prepared by McIntosh et al.[86, 87], who varied both the length of the chain (n = 2, 3, 4) and the nature of the linkage (ester or amide). EPR and laser flash photolysis studies indicated that these compounds, especially 105 (a, b; n = 3), could adopt a conformation in which electron transfer occurred from porphyrin to quinone; subsequent flipping of the chain to an extended conformation could then prevent recombination and lead to a long-lived radical

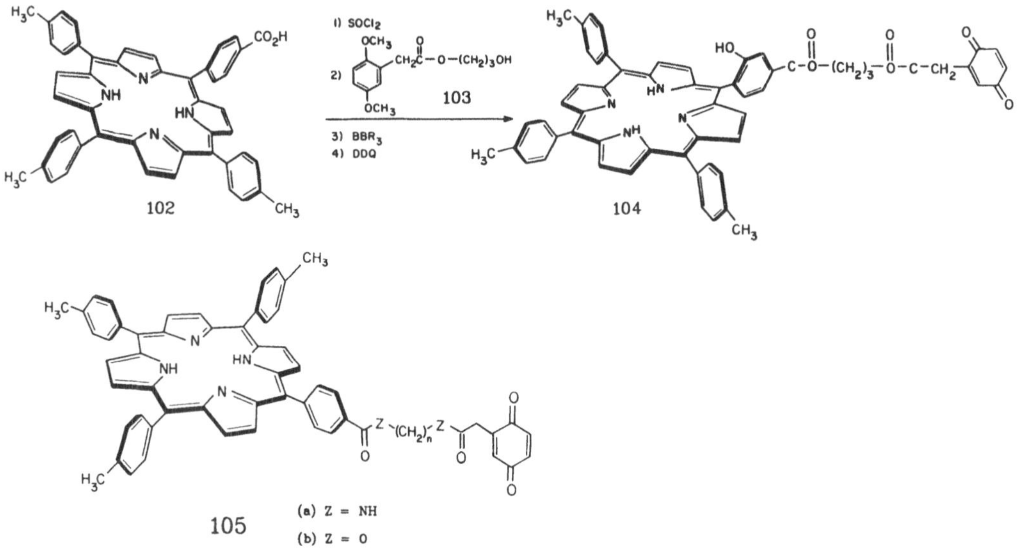

**Scheme 31**

ion pair. Wang et al.[88], prepared a series of porphyrins **106, 107** having either one or two quinone rings attached to the porphyrin periphery by amide linkages (Scheme 32). Similar porphyrins with an appended carotenoid group were also prepared, and both classes of compound have been incorporated into bilayer lipid membranes. Photoconductivity in such membranes was enhanced relative to simple membranes[89].

The use of amide or ester linkages to join the porphyrin and quinone moieties can lead to complications as both the separation and orientation of the two centers can vary. Possible solutions to the problems of porphyrin-quinone orientation and of prolonging the change-separated species have included:

(i)   the use of more rigid spacers to separate the porphyrin and quinone (models employing capped, strapped and doubly strapped quinones are described later in the appropriate sections);

(ii)  attempts to stabilize the transient radical ion pair by introducing other groups onto the porphyrin.

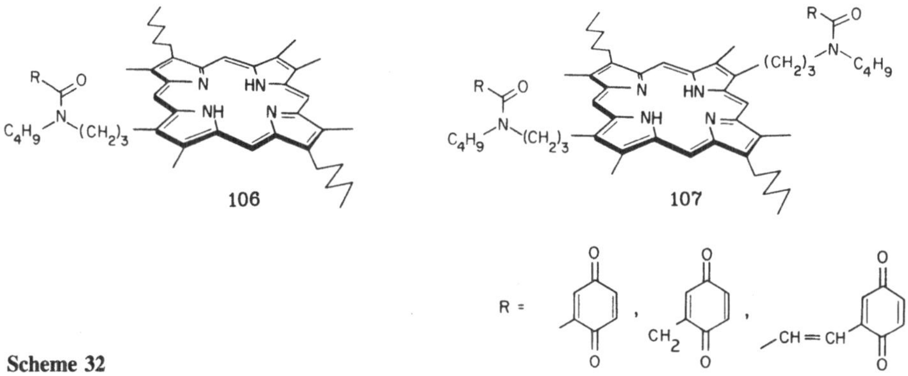

**Scheme 32**

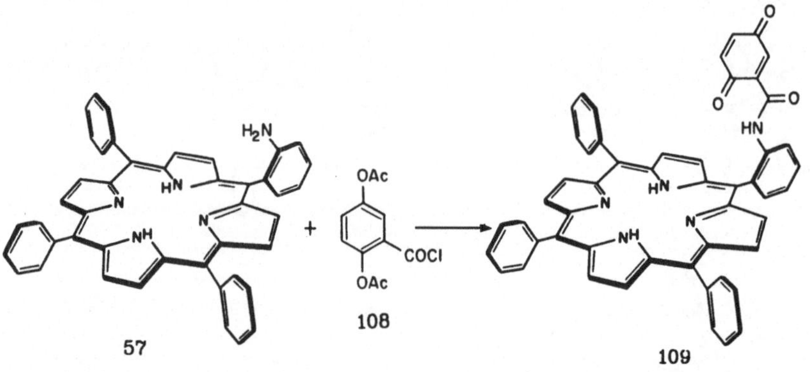

**Scheme 33**

Tabushi et al.[90], have reacted mono-*o*-amino TPP **57** with 2,5-diacetoxybenzoyl chloride (**108**), which, after selective hydrolysis and oxidation, furnished **109** (Scheme 33) in which the porphyrin-quinone distance is fixed. Efficient fluorescence quenching was observed but no mechanism was proposed. The porphyrin-quinone distance has been extended and the relative orientations fixed in compounds **111a–c** (Scheme 34)[91]. Reaction of 2-anthraldehyde, benzaldehyde and pyrrole yielded *meso*-(2-anthracenyl) triphenylporphyrin **110** (25%). Diels-Alder reaction with benzoquinone gave the product TPPBQ (40%) **111a.** TPPNQ **111b** and TPPAQ **111c** were prepared by condensation of the corresponding 2-aldehyde with pyrrole and benzaldehyde (15% and 18% respectively). The porphyrin-quinone distances were estimated at 10, 10.5 and 11 Å. The rate constants for photoinduced charge separation and recombination for these compounds have been reported[92]. The condensation of a linear tetrapyrrole **112** and an aldehyde to

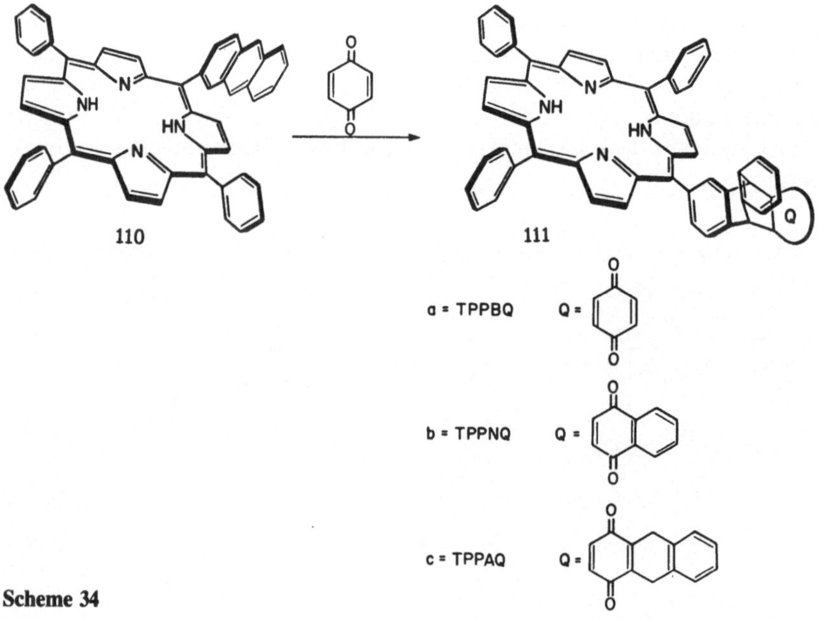

**Scheme 34**

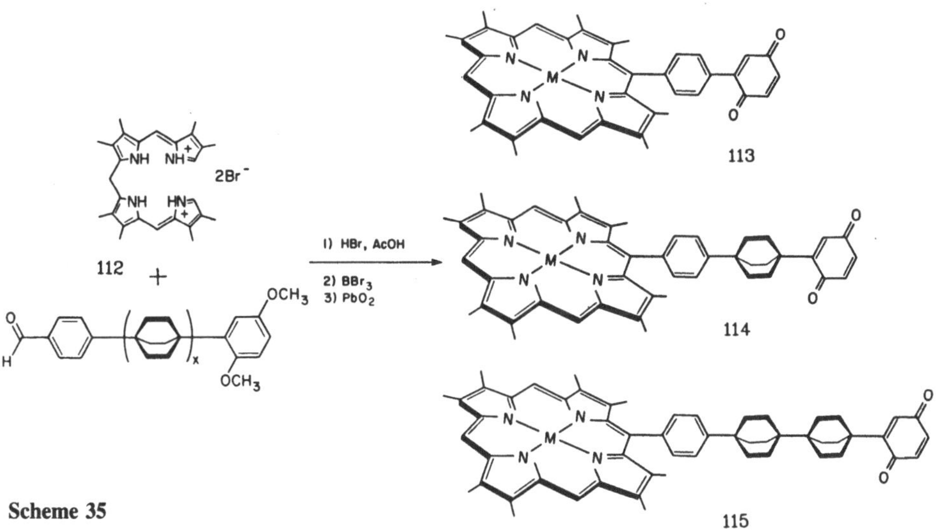

**Scheme 35**

form *meso*-substituted porphyrins was used to prepare another series of compounds **113–115** (Scheme 35) with increasing porphyrin-quinone distance (6, 10 and 14 Å)[93]. The bicyclooctane units eliminate flexibility and only rotational freedom is allowed. Preliminary fluorescence yields of the free base and zinc porphyrins indicate an incremental effect of distance on photochemical electron transfer[93].

A similar incremental effect of porphyrin-quinone separation was observed with the systems shown in Scheme 36 which were prepared by Wittig condensation of the meso-substituted porphyrin **116** (as the nickel complex) with the phosphorus ylide **117**[94]. Demethylation, reduction of the double bonds and then oxidation furnished the free base porphyrins **118** and **119a, b**. The rate of photoinduced electron transfer in such systems showed an inverse exponential dependence on the length of the chain[95]. In order to demonstrate a multistep electron transfer the bis-quinone porphyrin **120** was prepared[96] in which the pair of quinone rings provide a redox potential gradient and may thus stabilize charge separation. Comparison with the mono-quinone etioporphyrin **119a**

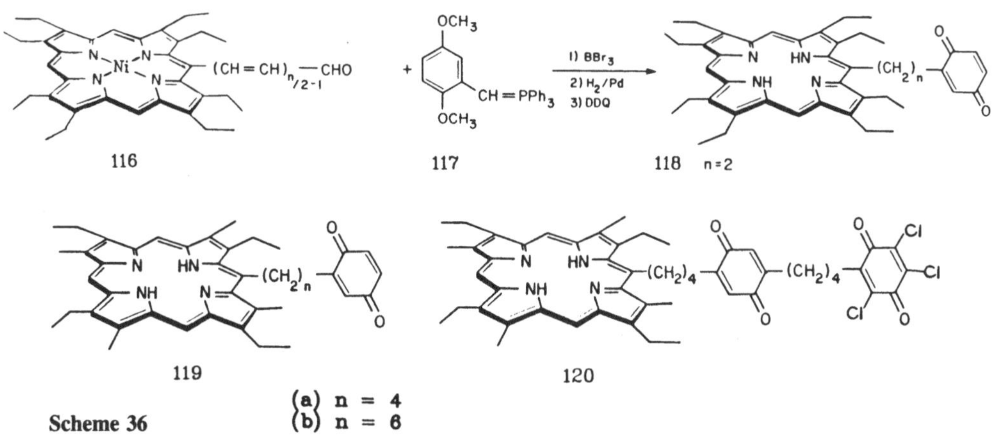

**Scheme 36**    (a) n = 4
                 (b) n = 6

showed approximately exponential decay of the charge transfer state in both cases, but the decay time was much longer for **120**. Similarly photoinduced charge transfer in **119b** is much longer than for **120** demonstrating the importance of the second quinone in **120** for the generation of long-lived charge-separated states. Moore et al.[97], used 5,15-bis(4-aminophenyl)-bistolylporphyrin to prepare a system incorporating both a quinone and a carotenoid group **121**. A photodriven charge separated complex was observed since charge recombination was inhibited by electron transfer from the carotenoid to the porphyrin cation radical, to give a long-lived (μs scale) $C^{+\cdot}$–P–$Q^{\cdot-}$ species.

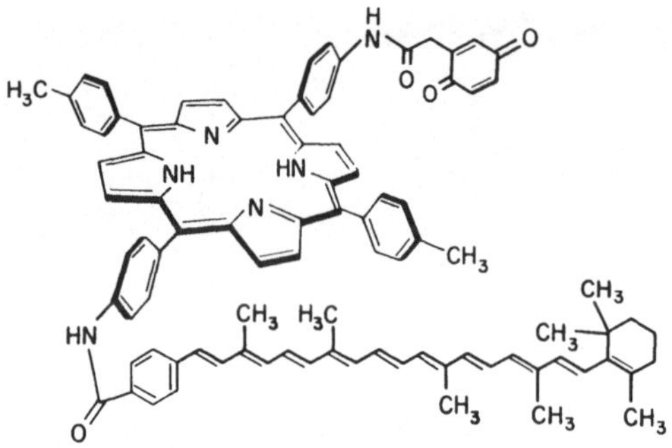

121

## 3.D. Porphyrins with Covalently Attached Interactive Groups

A number of other "tailed" porphyrins have been prepared which contain groups other than potential ligands. Using mono-(m- or p-hydroxyphenyl)triphenylporphyrin, Maiya and Krishnan[98] have prepared a series of derivatives containing a phenoxyglycol chain of variable length **122**. Both fluorescence and EPR data indicated intramolecular interaction between the porphyrin ring and the terminal phenoxy group, suggesting that the tail existed in a folded-over conformation.

Similarly mono-(o-hydroxyphenyl)triphenylporphyrin (**65**) was covalently attached to a cyclodextrin unit, in the hope of preparing a water-soluble oxygen carrying model[99]. While **123** did not display the desired properties, guest inclusion in the cyclodextrin moiety appears to induce novel conformational changes in aqueous solutions.

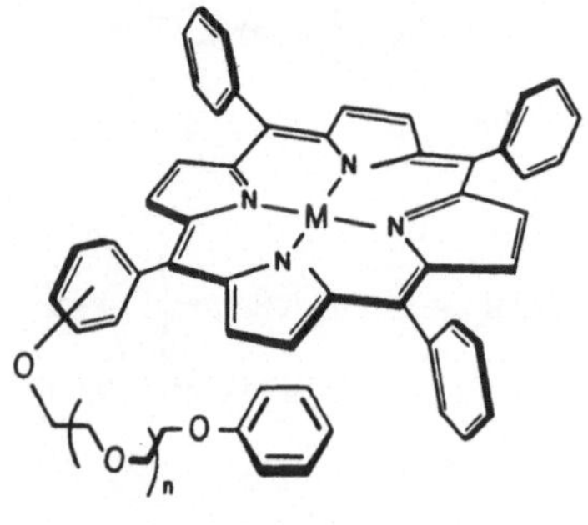

122

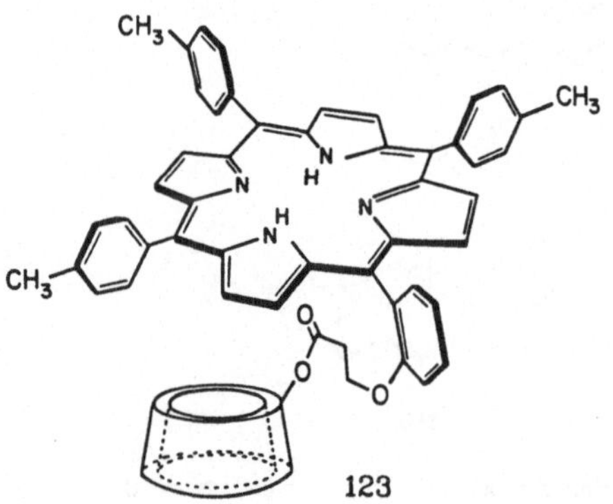

123

Other cyclodextrin capped hemes have been prepared[100] using 1-substituted 2-methylimidazoles included in an α-cyclodextrin which formed a pentacoordinated complex with protoheme.

The ability of Fe$^{II}$ porphyrins to generate reactive oxygen radicals has been exploited by several groups to prepare systems with potential antitumour properties. Covalent attachment of a suitable DNA intercalator to an iron porphyrin might deliver the heme to the DNA helical surface where reactive oxygen species could give rise to DNA scission. Lown and Joshua[101] have prepared a series of deuterohemins with an attached acridine **124a–c** which mimic the properties of the glycopeptide antibiotic bleomycin. A similar series of compounds **125, 126**, derived from mono-(p-aminophenyl)-tritotylporphyrin and protohemin, in which a suitable intercalator is bound to the porphyrin via a spermine chain, has also been prepared which exhibits oxygen-dependent DNA cleaving ability[102].

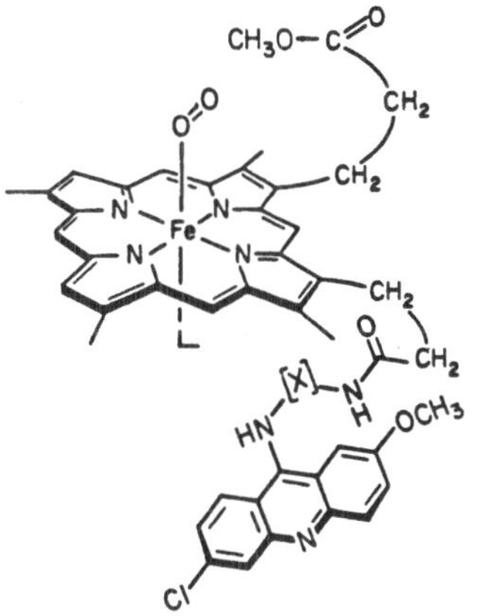

(a)  X = (CH$_2$)$_2$
(b)  X = (CH$_2$)$_3$
(c)  X = (CH$_2$)$_4$

124

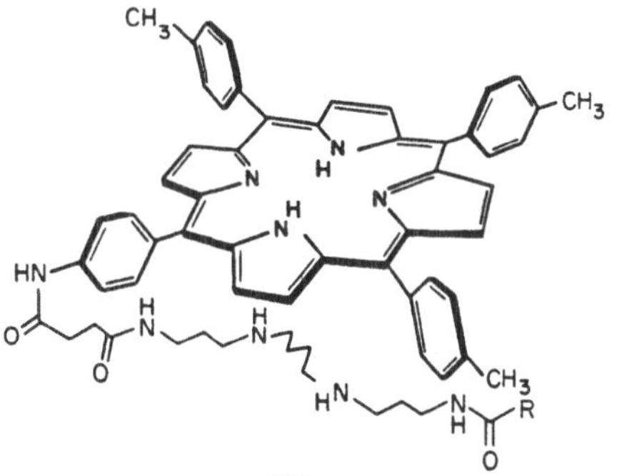

125

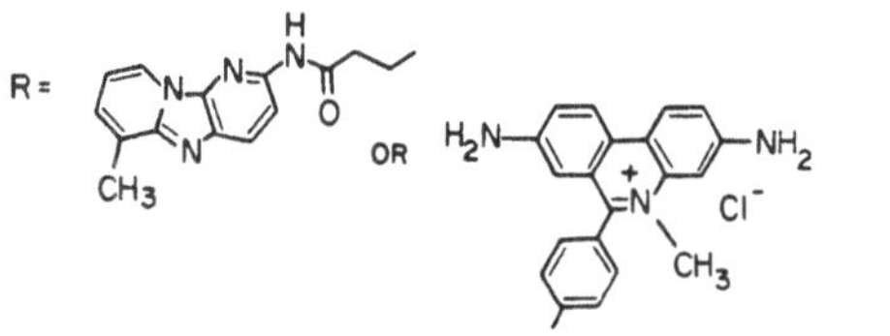

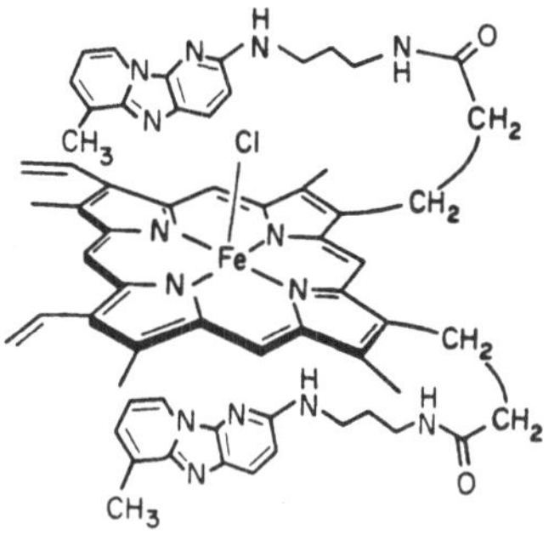

126

# 4. "Picket-Fence" Porphyrins and Related Species

## 4.A. "Picket-Fence" Porphyrins

Perhaps the most successful of the heme protein active site models is the "picket-fence" porphyrin of Collman. Steric encumbrance about the metal site of these substituted TPP molecules depends on two factors:

(i)   due to steric repulsion, the TPP will adopt a conformation in which the four *meso*-phenyl rings are essentially perpendicular to the porphyrin ring; substituents at the *ortho*-positions of the phenyl rings will lie above and below the porphyrin plane, and

(ii)  for TPP molecules containing mono-*ortho*-substituted phenyl rings, separation and, depending on the bulk of the substituent, interconversion of the four possible atropisomers may be achieved.

Collman reasoned that synthesis of a substituted iron(II) tetraphenylporphyrin having four pivalamido groups located on the same side of the porphyrin ring would give a "protected pocket". Ligands e.g. imidazole, could bind to the metal on the open face but could not penetrate the pocket, thereby ensuring five-coordination even in the presence of excess ligand. The much smaller dioxygen molecule would not be sterically encumbered and a six-coordinate complex could form. This oxygenated complex should be stable since the bulky pivalamido groups should prevent irreversible oxidation through close approach of two porphyrins and formation of a μ-peroxo complex.

Condensation of pyrrole (**55**) and four equivalents of *o*-nitrobenzaldehyde (**53**) in acetic acid gave *meso*-tetra(*o*-nitrophenyl)porphyrin **128** which was reduced by stannous chloride to the *meso*-tetra(*o*-aminophenyl)porphyrin (Scheme 37)[103, 104]. The four atropisomers (4 α, 3 αβ, 2 α2 β) were separated by chromatography, the slowest moving of which was the desired αααα isomer **129**. Interconversion of the atropisomers was sufficiently slow at room temperature to afford clean separation. Refluxing the unwanted

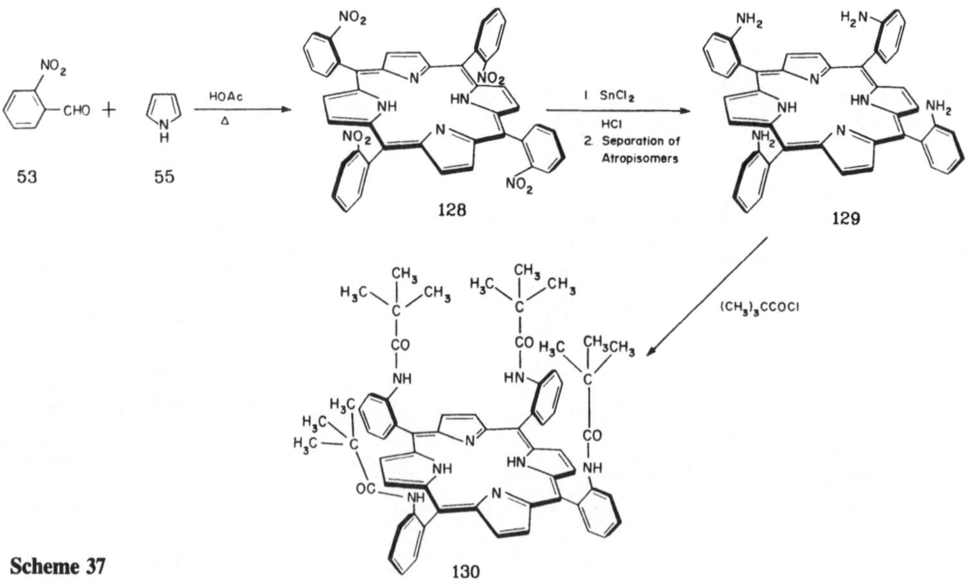

**Scheme 37**

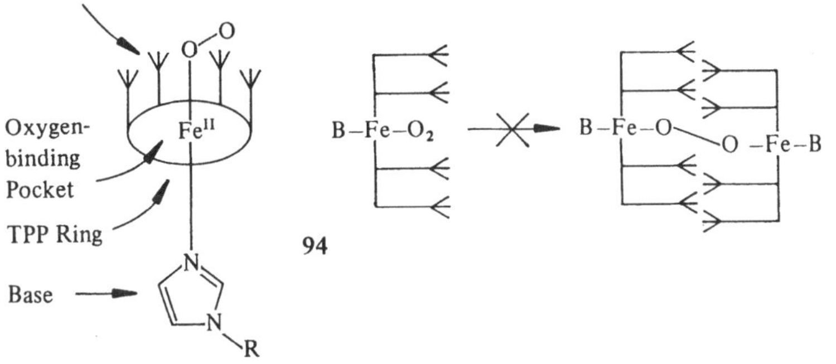

"Picket-fence"

Oxygen-
binding
Pocket

TPP Ring

Base

$$B-Fe-O_2 \longrightarrow\!\!\!\times\!\!\!\longleftarrow B-Fe-O{\diagdown}_{O}-Fe-B \qquad (25)$$

94

products in toluene for 20 min effected reequilibration to the statistical mixture allowing further isolation of the $\alpha\alpha\alpha\alpha$ isomer. Reaction of the amino groups with pivaloyl chloride gave the "picket-fence" porphyrin $\alpha,\alpha,\alpha,\alpha$-$H_2$(TpivPP) **130,** in which the configuration is frozen by the bulky substituents. Several studies on the physical properties and isomerization of these atropisomers have been made[105–107]. Treatment with $FeBr_2$, followed by reduction with $Cr(acac)_2$ gave $Fe^{II}(\alpha,\alpha,\alpha,\alpha,$-TpivPP) **131.**

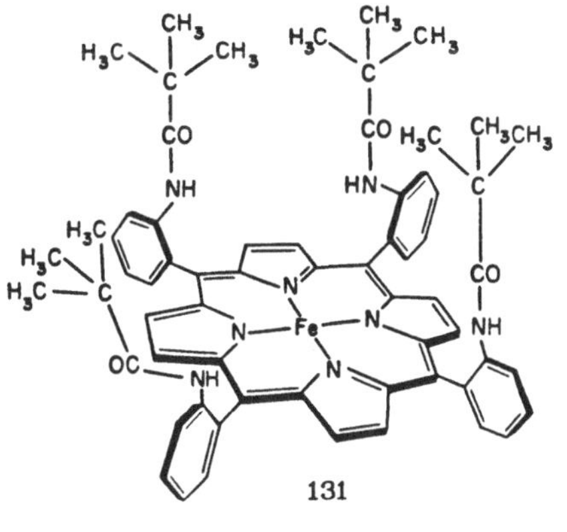

131

Although addition of strong field ligands gave low spin six-coordinate complexes, $FeB_2(\alpha,\alpha,\alpha,\alpha$-TpivPP), it was suspected that the binding constant of the base on the "picket-fence" side was less than that on the "open" side. A series of six-coordinate compounds was prepared (B = Im, 1-MeIm, 1-n-BuIm, 1-tritylIm, 4-t-BuIm, 1,2-$Me_2$Im, pyridine, piperidine, tetrahydrothiophene, tetrahydrofuran)[104]. All of these showed reversible oxygen binding behaviour in benzene solution at 25 °C, without appreciable amounts of decomposition. Indeed, the oxygen complexes, Fe(TpivPP) (N-RIm)($O_2$) were stable for long periods ($t_{1/2}$ 2–3 months) in solution provided 2–4 equivalents of axial

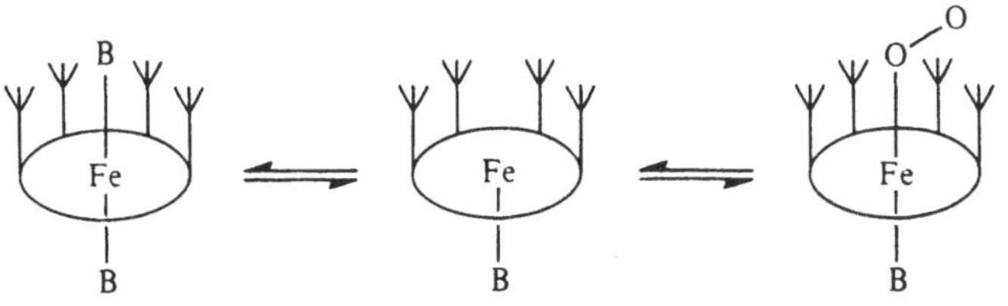

$$\qquad\qquad \rightleftharpoons \qquad\qquad \rightleftharpoons \qquad\qquad \qquad (26)$$

base were present to protect the unshielded face. Furthermore, analytically pure, crystalline dioxygen complexes could be obtained[104]. The crystal structures of $Fe^{II}(TpivPP)(1\text{-}MeIm)(O_2)$ **132a** (Fig. 2)[104, 108], $Fe^{II}(TpivPP)(2\text{-}MeIm) \cdot EtOH$ (Fig. 3) and its dioxygen adduct[109] (Fig. 4) have been determined. Further structural information

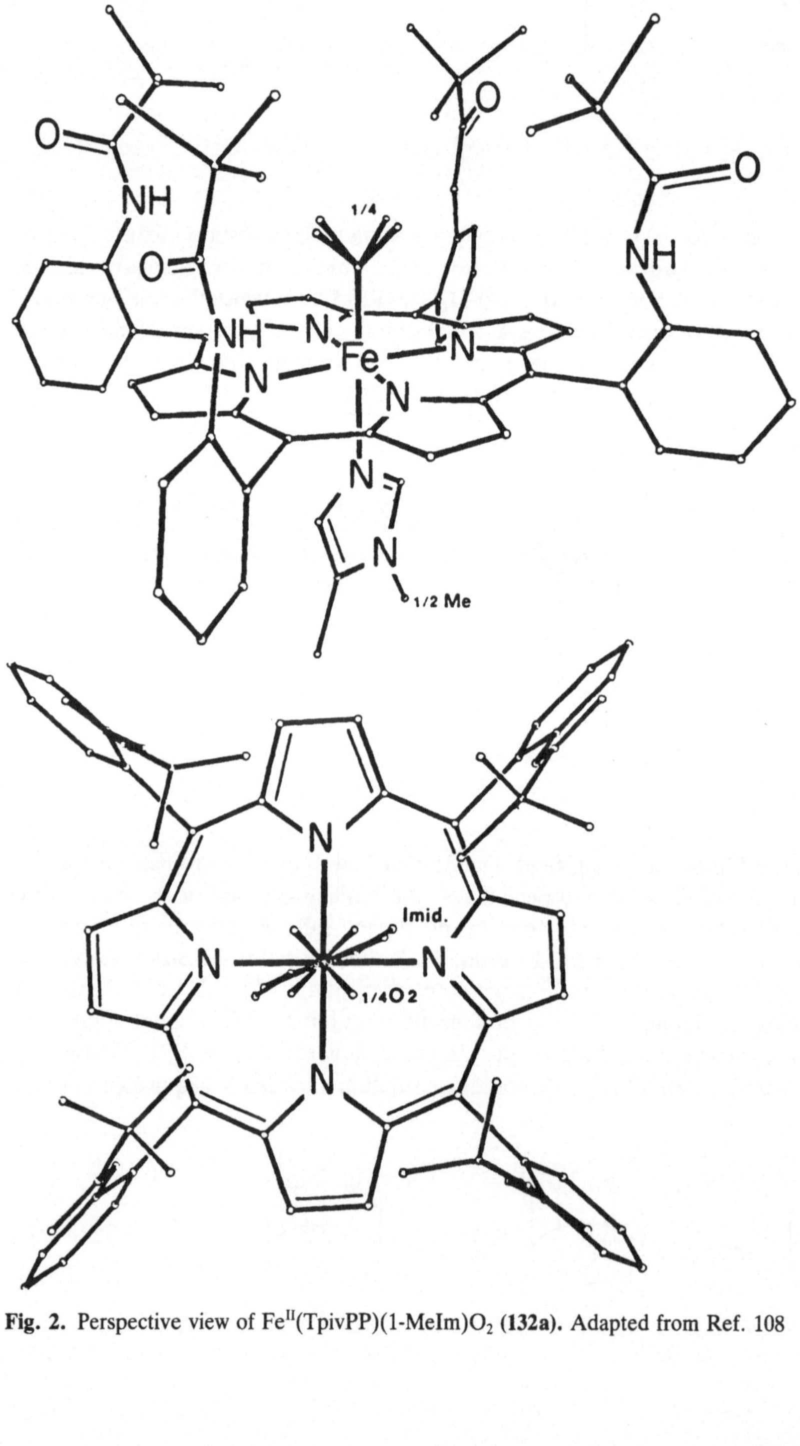

**Fig. 2.** Perspective view of $Fe^{II}(TpivPP)(1\text{-}MeIm)O_2$ (**132a**). Adapted from Ref. 108

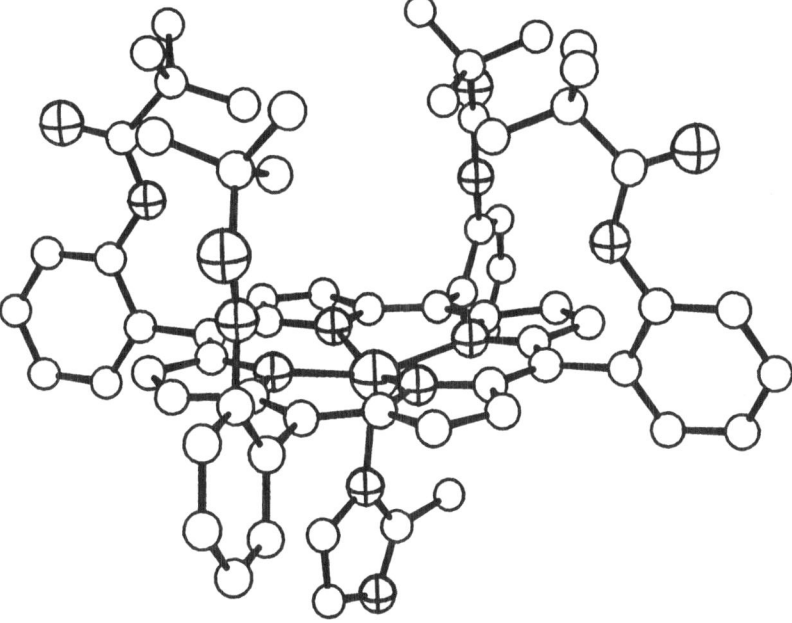

**Fig. 3.** Perspective view of Fe(TpivPP)(2-MeIm) adapted from Ref. 109

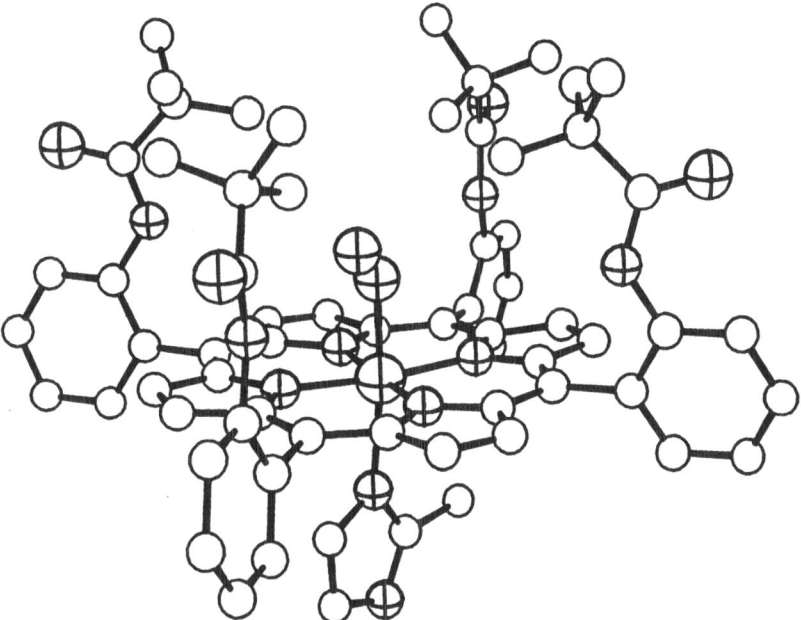

**Fig. 4.** Perspective view of Fe(O₂)(TpivPP)(2-MeIm). The dioxygen and imidazole are disordered. The disorder has been idealized and only one concentration is shown in this figure, which is adapted from Ref. 109

a  B = 1-MeIm
b  B = 1-BuIm
c  B = THF
d  B = THT

**132**

was obtained by I.R.[108, 110, 111], and Mössbauer spectroscopy and magnetic susceptibility measurements[82, 108]. The reversible binding of oxygen to cobalt complexes of TpivPP has also been studied[112]. Reviews on oxygen binding to picket fence, and related systems have also appeared[32, 37, 113–115].

Binding of a second base on the "picket-fence" side of the porphyrin prevented determination of oxygen binding to the five-coordinate heme in solution. However, crystals of $Fe^{II}(TpivPP)(1\text{-}MeIm)(O_2)$ could be deoxygenated under vacuum to give the five-coordinate species which could be re-oxygenated. This cycling between $O_2$ and vacuum produced no observable irreversible oxidation over many cycles[116]. Similar reversible oxygenation was demonstrated by solid samples of Fe(TpivPP)B (B = 2-MeIm, 1,2-Me$_2$Im). From the Hill plot (log Y/1 – Y vs. log $P_{O_2}$) it was concluded that solid state oxygen binding for these two compounds showed two regions of non-cooperative binding (at high and low $O_2$ pressures) and an intermediate region of cooperative binding. Collman rationalized this in terms of a shrinking of the molecules' dimensions on oxygenation as the iron and its bound imidazole move towards the porphyrin ring. As increasing numbers of molecules in the solid are oxygenated this change in molecular dimensions presumably induces sufficient strain in the crystallite to induce a conformational change in the solid which enhances the oxygen affinity of the remaining deoxy sites. This behaviour is reminiscent of the cooperative oxygen binding of hemoglobin.

Various other sterically hindered TPP derivatives have been prepared by the condensation of meso-α,α,α,α-tetra(o-aminophenyl)porphyrin 129 with bulky acid chlorides. Collman et al. have also prepared the compounds, $H_2(T_{Phth}PP)$ (135) and $H_2(T_{Tos}PP)$ (136), by reaction of $H_2(T_{Am}PP)$ (129) with iso-phthaloyl dichloride 133 and p-toluenesulfonyl chloride 134 respectively (Scheme 38). Despite the bulky "picket-fence", the ferrous complexes of both compounds exhibited only irreversible oxidation on exposure to oxygen at 25 °C. This was attributed to the acidic amide protons which were presumably directed into the cavity, allowing protonation of the coordinated dioxygen and consequent heme oxidation[104]. A similar series of TPP derivatives 137a–g have been synthesized under identical conditions by Bogatskii et al.[117, 118].

The reaction of o-$H_2(T_{Am}PP)$ (129) with excess isonicotinic chloride hydrochloride yielded the tetra-isonicotinamide TPP as a statistical mixture of atropisomers which could be separated by chromatography. The isonicotinamide groups were then methylated with CH$_3$I, followed by anion exchange, to yield the water soluble tetrakis-(N-methylisonicotinamidophenyl)porphyrin-tetracation (138) (Scheme 39). The corresponding para and meta derivatives were also prepared and reduction potentials, basicity and reactivity with metal ions of the isomers were compared[119].

Ortho-substituted TPP derivatives have been used as binuclear ligand systems by choosing suitable substituents on the phenyl rings. For example, treatment of α,α,α,α-$H_2(T_{Am}PP)$ 129 with maleic anhydride gave the tetrakis(o-maleamoylphenyl)porphyrin

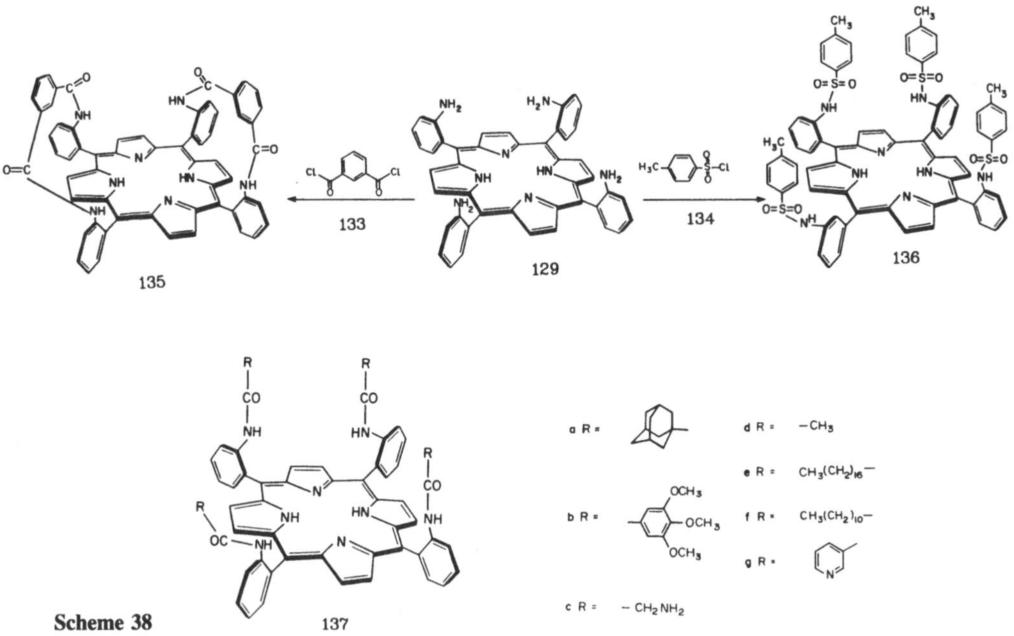

**Scheme 38**  137

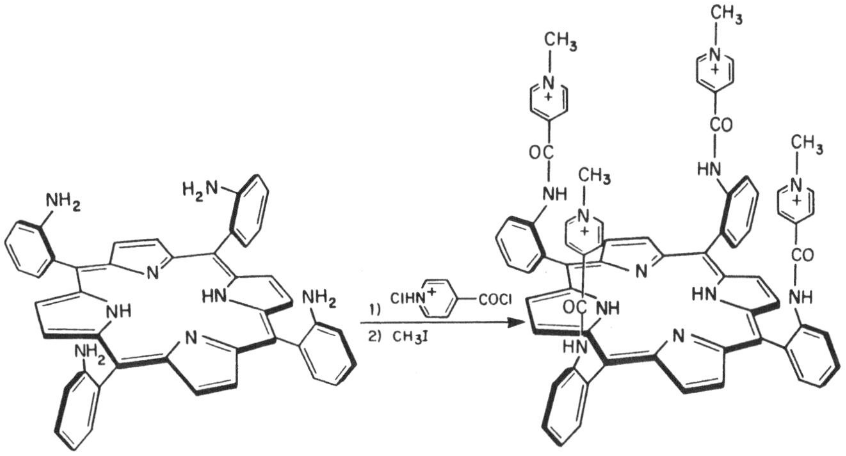

**Scheme 39**  129  138

(139) in 90% yield (Scheme 40). In aqueous DMF this porphyrin undergoes copper insertion at a rate much faster than for unsubstituted porphyrins. Rapid complexation of the copper by the carboxylates holds the metal ion in a position favourable for rapid intramolecular transfer to the porphyrin nucleus[120].

Binuclear porphyrins capable of binding iron and copper have been investigated as synthetic models for the iron/copper site of cytochrome oxidase. One such model **141** consists of a tetraphenylporphyrin ring with a covalently attached tetrapyridine ligand

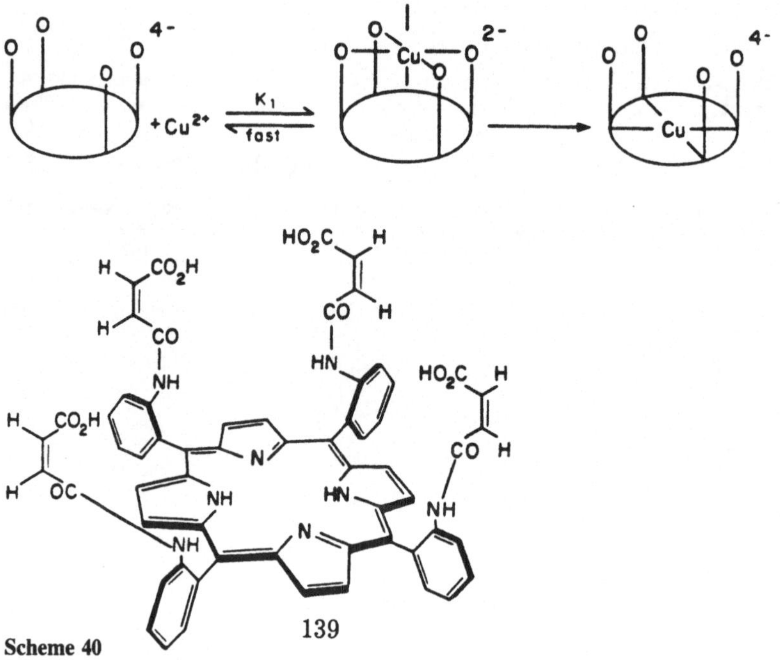

**Scheme 40**

139

system, obtained by treating $\alpha,\alpha,\alpha,\alpha$-H$_2$(T$_{Am}$PP) **(129)** with excess nicotinic anhydride **(140)** (Scheme 41)[121]. The mixed metal compound with iron inserted into the porphyrin ring and copper coordinated to the four nicotinamide groups was prepared, and the EPR and magnetic susceptibility of the complex examined[122]. In contrast, Elliott and Krebs claim that the conditions necessary for metal insertion into this complex cause isomerization of the nicotinamide groups[123]. Instead, these authors have coordinated Ru$^{II}$ to the nicotinamide groups, which locks the "pickets" into place allowing more forcing conditions to be used for the introduction of divalent and trivalent first-row transition metals into the porphyrin ring, without fear of isomerization.

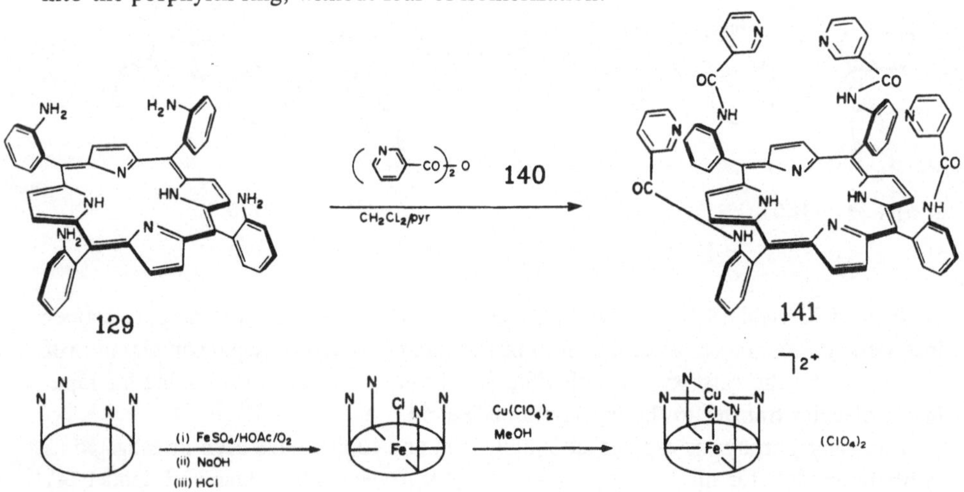

**Scheme 41**

Iron porphyrin catalysis of epoxidation and hydroxylation using iodosylarenes as oxidants is believed to proceed via a reactive iron-oxo intermediate. Groves and Meyens have attempted the catalytic asymmetric epoxidation of olefins using suitably substituted chiral "picket-fence" porphyrins to control the stereochemistry of approach of the substrate olefin to the iron-oxo species[124]. The chiral porphyrins were prepared by reacting $\alpha,\beta,\alpha,\beta$-H$_2$(T$_{Am}$PP) (142) with an optically active acid chloride (Scheme 42). With an (R)-2-phenylpropanamido group as the chiral appendage the porphyrin 143a was formed in high yield (95%). However, in the epoxidation of various olefins with iodosylbenzene very little selectivity was obtained (%ee, 9–31%). Instead the binaphthyl group was chosen as an appendage which could form a large and relatively rigid chiral cavity about the porphyrin core. The diacid chloride of 1,1'-binaphthyl-2,2'-dicarboxylic acid (144) was reacted with $\alpha,\beta,\alpha,\beta$-H$_2$(T$_{Am}$PP) (142), followed by methanolysis and iron insertion. This catalyst 143b was more efficient and enantiomeric excesses of 20–50% were observed for variously substituted styrenes and aliphatic olefins.

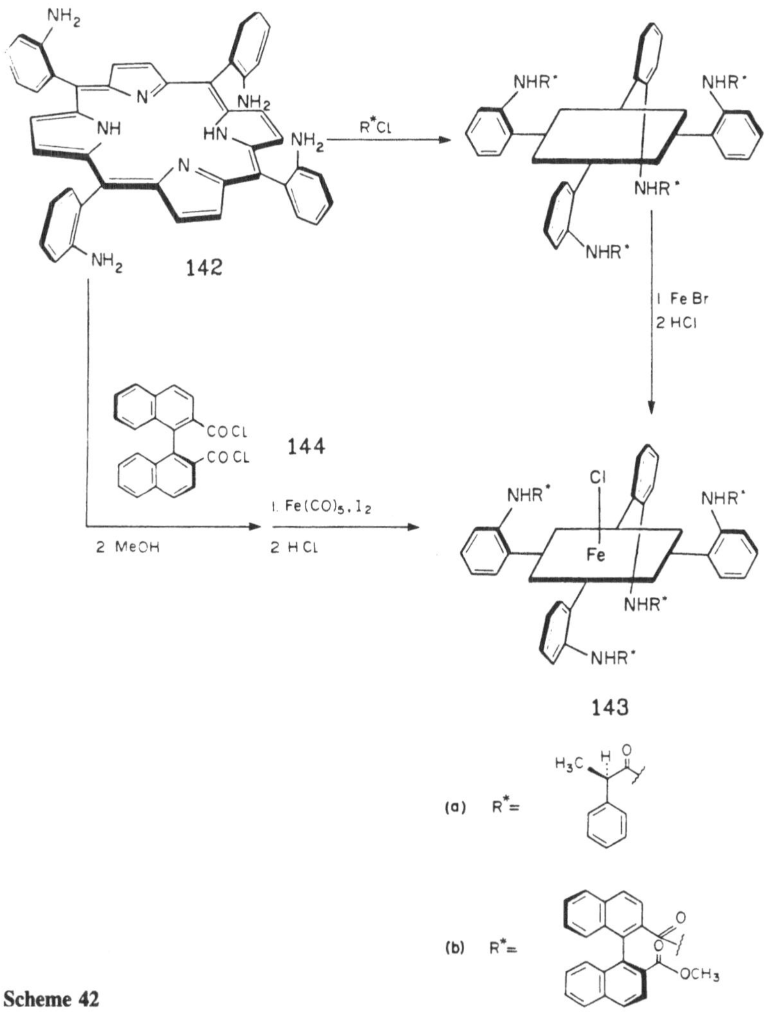

**Scheme 42**

More recently Tabushi et al.[125], have used oxygenated $Fe^{III}$(TpivPP)Cl in the presence of HCl and $H_2$-colloidal platinum supported on poly(vinylpyrrolidone) in the presence of acid chlorides as a cytochrome P450 mimic. Under such conditions cyclohexene is converted to its epoxide and slow reduction of the ferric complex followed by formation of the dioxygen ferrous complex completes the catalytic cycle.

A porphyrin **145** bearing pickets on both faces of the ring has been prepared (Scheme 43; R' = $CH_3$)[126]. However, only two opposite *meso*-positions are substituted with the bis-*ortho* substituted phenyl rings so it is unlikely that this compound will be a successful oxygen binding model.

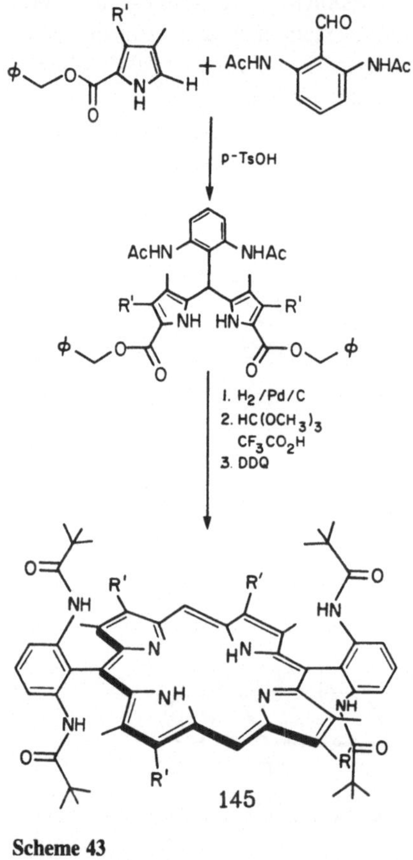

**Scheme 43**

## 4.B. *"Tailed Picket-Fence" Porphyrins*

While the direct oxygenation of five-coordinate $Fe^{II}$ "picket-fence" was observable in the *solid* state, such studies in solution were not possible since the "picket-fence" could not prevent six-coordination in the presence of excess sterically unhindered base. (Excess base is necessary to ensure complete coordination on the "open" face and prevent μ-oxo complex formation.) To control coordination, Collman et al., adopted the "chelated heme" approach. Dispensing with external ligand, the base was covalently attached to

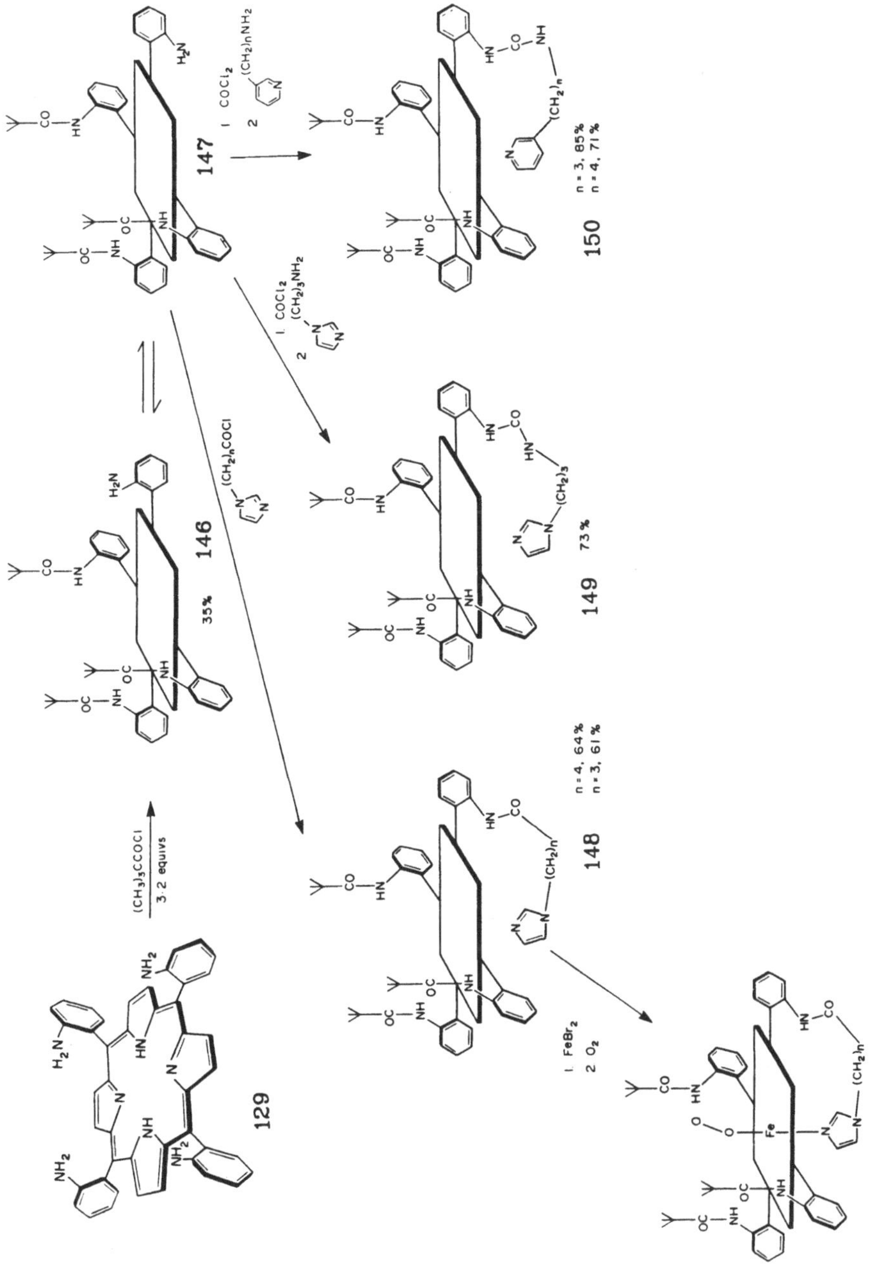

**Scheme 44**

the *ortho*-phenyl position of TPP, and so constrained into a position promoting intramolecular binding to the porphyrin metal. The other three *meso*-phenyl rings carry the "pickets" necessary to prevent irreversible oxidation.

Treating $\alpha,\alpha,\alpha,\alpha$-H$_2$(T$_{Am}$PP) (129) with 3.2 equivalent of pivaloyl chloride gave the "3-picket" $\alpha$-aminophenylporphyrin 146 (35%) (Scheme 44). Refluxing in benzene solution for 2 h equilibrated the free aminophenyl group to a 1:1 mixture of the $\alpha$ and $\beta$ atropisomers 146, 147 which were separated by chromatography. (The unwanted $\alpha$-isomer could be re-equilibrated to increase the yield of the $\beta$-product.) Using amide or urea linkages an imidazole was attached to the $\beta$-aminoporphyrin 147 by chains of varying length to yield the porphyrins 148 and 149. Direct metal insertion using anhydrous FeBr$_2$ gave nearly quantitative yields of the five-coordinate iron(II) "tail picket-fence" porphyrins, e.g. 151[74]. $^1$H-NMR spectra confirmed the proposed five-coordinate high spin (S = 2) iron(II) formulation, but on decreasing the temperature ($-25\,°C$) peaks due to a diamagnetic (S = 0) complex were observed, presumably due to a dimerization process. (Momenteau, but not Traylor, had observed such dimerization in the chelated heme systems.)

$$2 \quad \underset{S=2}{\overset{B}{\underset{|}{\overset{|}{\underset{Fe}{\Big|}}}}} \quad \rightleftharpoons \quad \underset{S=0}{\overset{B}{\underset{|}{\overset{|}{\underset{Fe}{\Big|}}}}} \quad \underset{S=1}{\overset{}{\underset{B\sim\!\!\sim\!\!\sim Fe}{}}} \tag{27}$$

Addition of oxygen to solutions of the high spin five-coordinate iron(II) compounds gave the expected spectra of a diamagnetic compound for the oxygenated species 151. The peak pattern for the "pickets" suggests that the oxygen may be ordered in these complexes, presumably towards the open side of the pocket, a suggestion which is awaiting confirmation by X-ray crystallography.

A similar series of "tailed picket-fence" porphyrins 150 has been synthesized with a pyridine covalently attached to the porphyrin periphery via urea linkages[127]. O$_2$ and CO binding to both series of porphyrins has been carried out[127–129].

The use of *meso*-(*o*-aminophenyl)triphenylporphyrin 57 to prepare a series of alkyl and aryl mercaptan-tail porphyrins as cytochrome P450 models has been described in Sect. 3.B. A similar series of compounds has been prepared by acylation of the tripivalamide-$\beta$-aminophenylporphyrin 147 to give the alkyl mercaptan "tailed picket-fence" porphyrins 154 (Scheme 45)[82]. A similar compound 155 with an appended thioether chain has also been prepared and is reportedly capable of reversibly binding dioxygen[74].

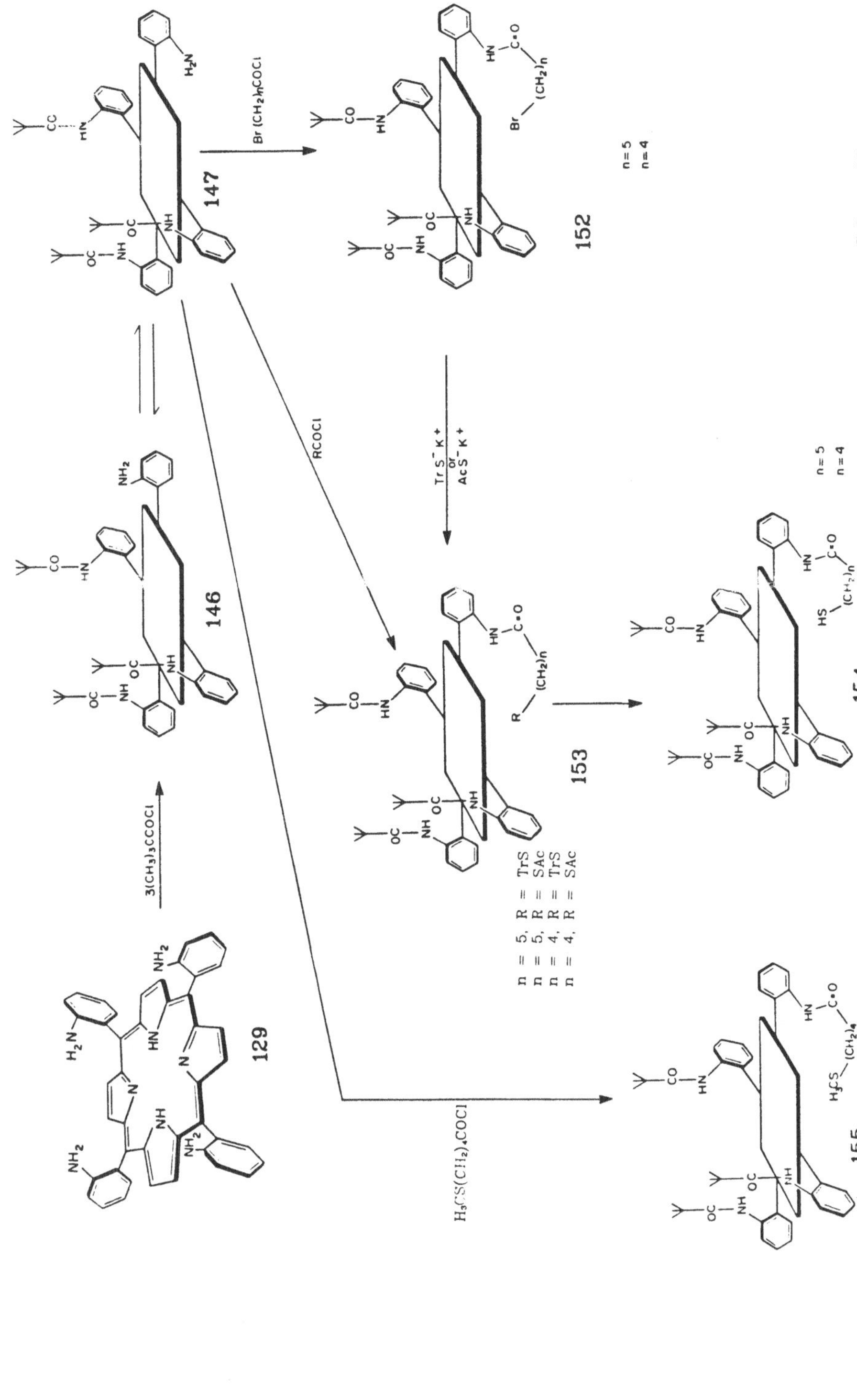

**Scheme 45**

# 5. "Capped" Porphyrins and Related Species

## 5.A. "Capped" Porphyrins

The direct condensation of aromatic aldehydes and pyrrole to form tetraphenylporphyrins was exploited by Baldwin and co-workers to prepare "capped" porphyrins. In these
molecules a benzene ring was covalently attached to all four *ortho*-positions of the *meso*-
phenyl rings, enclosing a volume of space above one face of the porphyrin ring. If the cap
was sufficiently tight the binding of bases (e.g. alkylimidazoles, pyridine) should be
prevented on the enclosed face; binding on the open face would result in a five-coordinate species. On the other hand, the smaller dioxygen molecule would be able to fit
under the cap, which should provide a physical barrier to μ-oxo complex formation. It
was recognized that attempts to condense a benzene ring with a porphyrin by four ester
linkages would probably result in very low yields. Instead the necessary units were
attached to the "cap" to give a tetraaldehyde which was condensed with pyrrole to give
the "capped" porphyrin. Unlike the "picket-fence" porphyrin, cyclization of the porphyrin ring is the last step of the synthesis, so chromatographic separation of atropisomers is
not required.

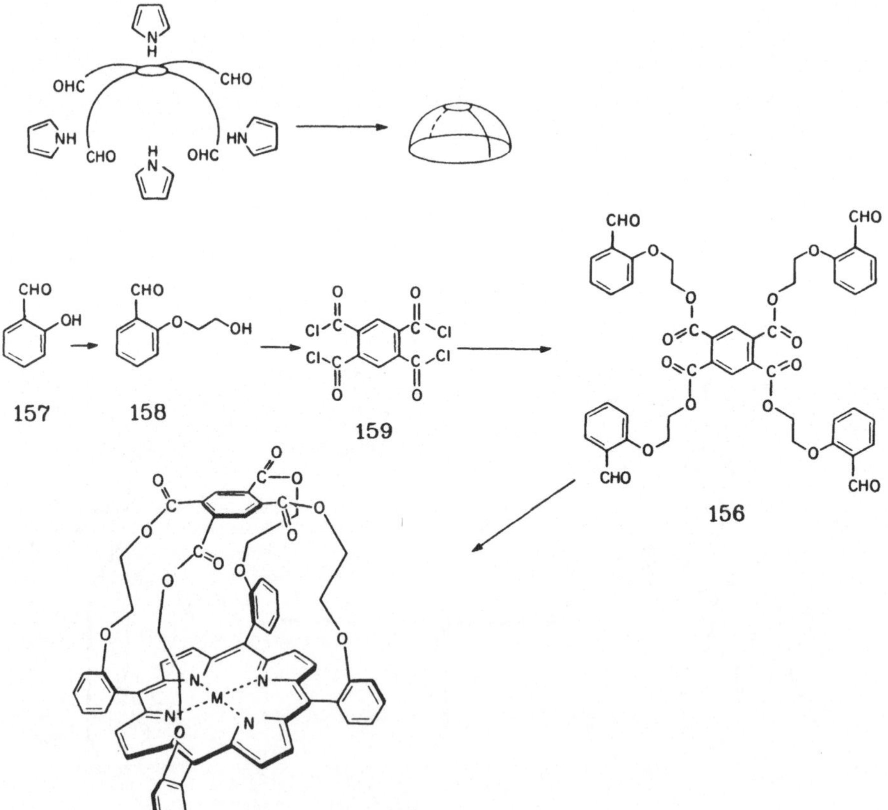

**Scheme 46**

The required tetraaldehyde **156** was prepared by alkylation of salicylaldehyde **(157)** with bromoethanol to yield **158**, followed by condensation with pyromellitoyl chloride **159**. Reaction of the tetraaldehyde with pyrrole in refluxing propionic acid yielded the "capped" porphyrin **160** after chromatographic purification (Scheme 46) (Fig. 5)[130, 131].

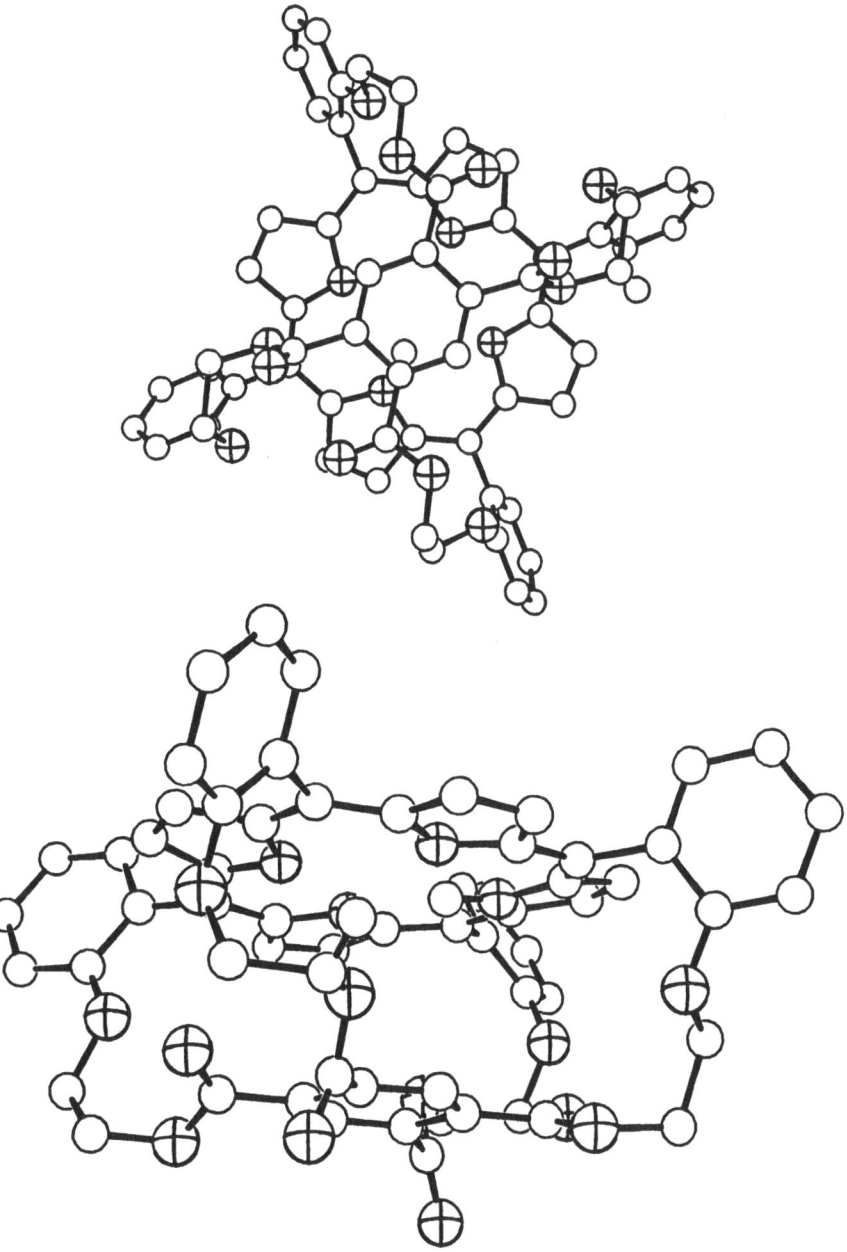

**Fig. 5.** Two perspective views of the "capped" porphyrin **160** as the free base. Adapted from Ref. 140

The same reaction sequence using 2-(3-hydroxypropoxy)benzaldehyde yielded the corresponding "homologous" or "C₃-capped" porphyrin 161 in which there is an extra methylene group in each link of the cap[131]. In the latter case the yield of the cyclization reaction was much lower (5%), probably reflecting the extra entropy factors required to form the larger cap. To provide steric hindrance on the uncapped face a "naphthyl-C₂-capped" porphyrin 165 was similarly prepared (Scheme 47) from 159 and 162 via 163 and 164[131].

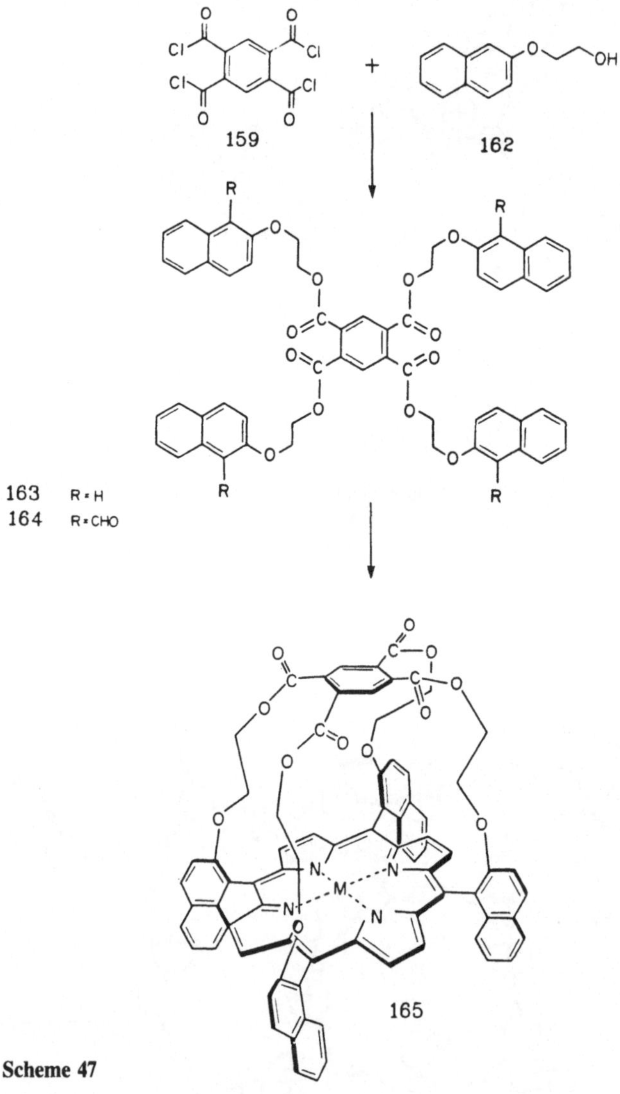

163    R•H
164    R•CHO

165

**Scheme 47**

Insertion of iron into the "C₂-capped" porphyrin, followed by reduction, gave a crystalline four-coordinate high spin iron(II) porphyrin 166 (Scheme 46). In solutions containing excess axial base the five-coordinate heme was formed which was capable of reversible dioxygen binding at 25 °C. The stability of the dioxygen adduct depended on

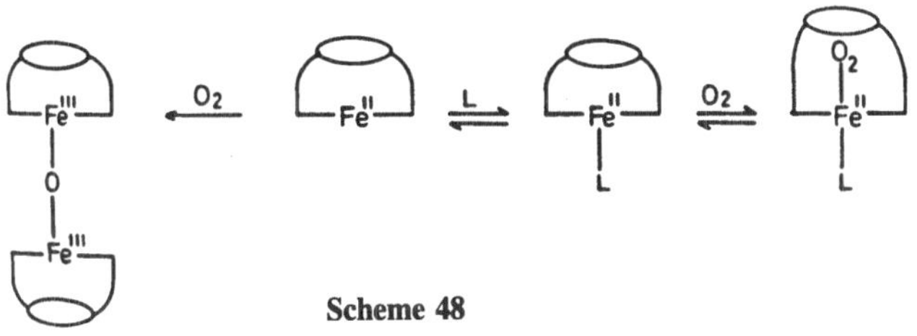

**Scheme 48**

the nature and concentration of the axial base and the position of the equilibria in Scheme 48[130]. Unlike the "$C_2$-capped" porphyrin **166** which could only bind a single axial base, the larger size of the "$C_3$-Cap" permitted binding of two axial bases, provided that they were small, e.g., propylamine. For intermediate size bases such as 1-MeIm, it appeared that a second base could weakly coordinate to the iron, probably through the side of the cap. Oxygen binding was still reported to occur, giving a pseudo-seven-coordinate complex[132–134].

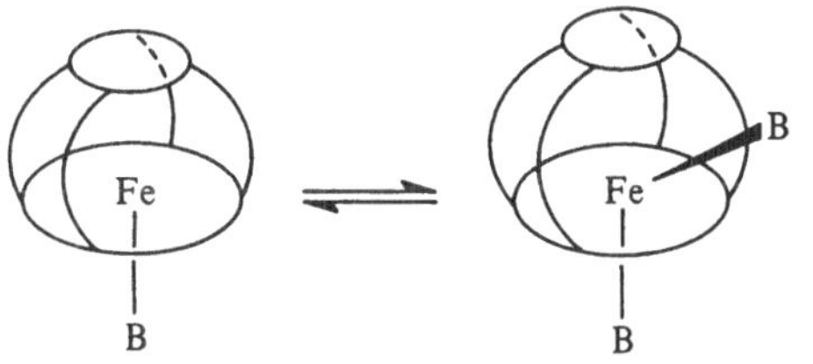

$$(28)$$

The $O_2$, CO and NO affinities of a series of $Fe^{II}$ and $Co^{II}$ capped porphyrins have been studied[131, 137, 138]. It was found that $O_2$ affinities were much lower than those of natural porphyrins and other synthetic models e.g., the $Fe^{II}$ complexes of $H_2(TAP)$ > $H_2(NapC_2–Cap)$ **(165)** > $H_2(C_2–Cap)$ **(160)** > $H_2(C_2–CapNO_2)$ **(167)** > $H_2(C_3–Cap)$ **(161)**. In contrast CO bound more quickly than $O_2$, and the rate of binding was indepen-

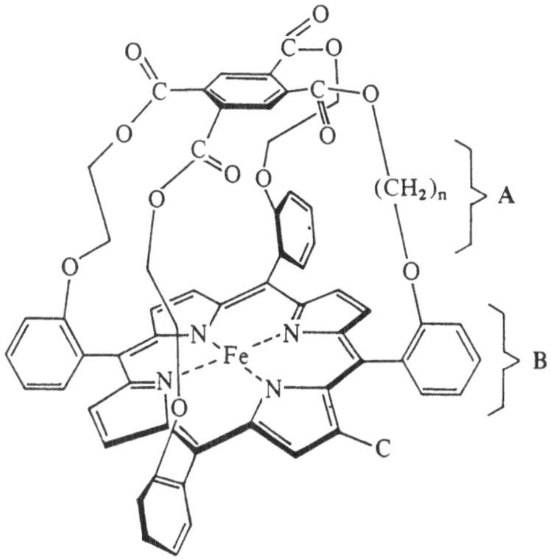

| Porphyrin | | A | B | C |
|---|---|---|---|---|
| $C_2$–Cap | **160** | n = 2 | Phenyl | H |
| $C_3$–Cap | **161** | n = 3 | Phenyl | H |
| $C_2$–CapNO$_2$ | **167** | n = 2 | Phenyl | NO$_2$ |
| NapC$_2$–Cap | **165** | n = 2 | Naphthyl | H |

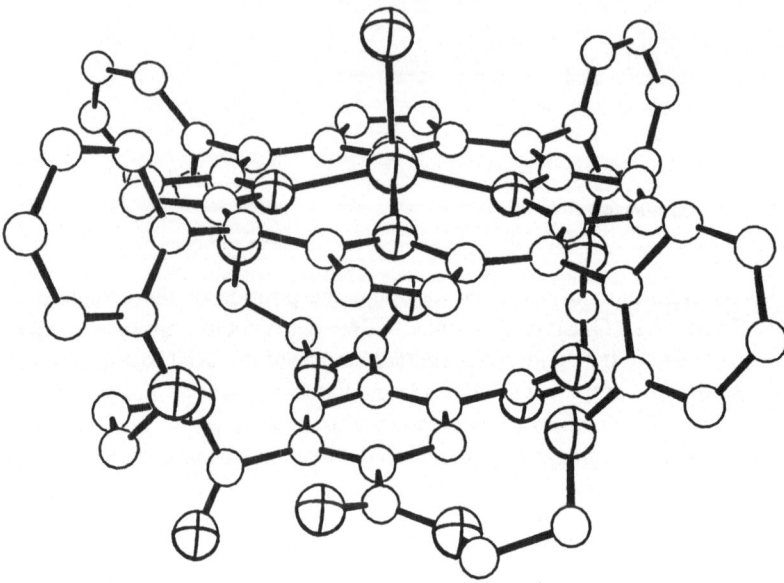

**Fig. 6.** Perspective view of Fe$^{III}$Cl(C$_2$–Cap). Adapted from Ref. 141

dent of the cap size and comparable to unhindered model porphyrins[139]. This was rationalized in terms of a steric effect of the cap. Although the crystal structures of both H$_2$(C$_2$–Cap) (**160**) (Fig. 5)[140] and Fe$^{III}$Cl(C$_2$–Cap) (Fig. 6)[141] indicated that the phenyl cap-porphyrin separation was too small to accommodate either CO or O$_2$, a considerably more expanded version must exist in solution. That the linear Fe–C–O system could be accommodated under the cap argued against a "central" steric effect. However, the bent Fe–O–O system might be destabilized by a "peripheral" steric effect of the methylene chain linkages.

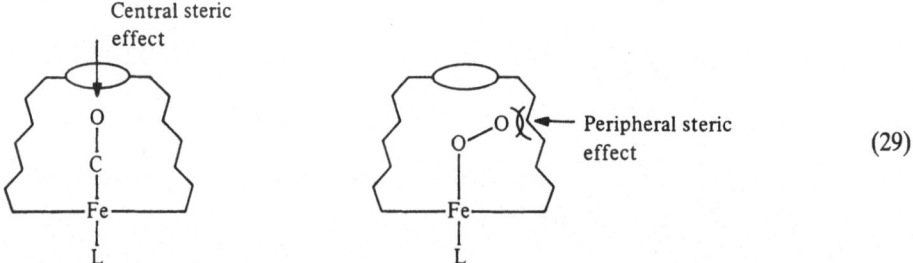

(29)

Studies[142, 143] of the paramagnetic shifts in the $^1$H-NMR of Co$^{II}$Cap porphyrins were used to deduce the cap-porphyrin separation. The relative cavity size was in the order NapC$_2$–Cap > C$_2$–Cap > C$_3$–Cap, which correlates with the relative oxygen affinity of the iron(II) "capped" porphyrins.

Recently[144] the "capped porphyrin" approach has been extended to prepare a modified "C$_2$-capped" porphyrin in which a pyridine is covalently bound to two opposite *meso*-aromatic rings of the parent "C$_2$-capped" porphyrin, forming some kind of "strap" (see Chap. 6) over the porphyrin face opposite to the "cap". Bis-amino "$C_2$-capped" porphyrins were prepared by condensation of a dinitro-tetraaldehyde with pyrrole followed by reduction. For solubility reasons it was necessary to benzylate the amino functions before condensation with the appropriate 3,5 pyridine diacid compound. Both of the corresponding iron(II) $C_2$-capped strapped porphyrins **168a, b** exhibit reversible

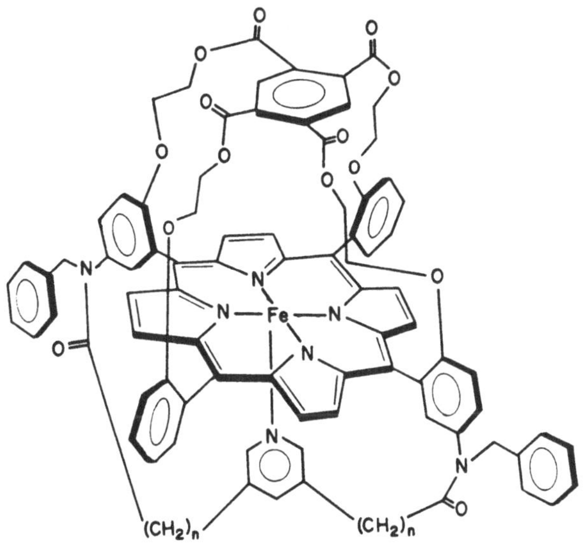

(a) n = 4

**168**

(b) n = 5

oxygen binding in toluene solution at room temperature and show good stability to autoxidation reactions (the $C_5$-strapped complex **168b** has a $t_{1/2}$ of several days while that of the $C_4$-strapped complex **168a** is several months). Furthermore, decomposition does not result in a μ-oxo complex, the strap and cap preventing dimer formation. Although a kinetic study has not been made, it appears that the stability to autoxidation is due to the low affinity of the iron(II) complexes for dioxygen. Comparison of the $P_{1/2}$ values with that of iron(II) $C_2$-capped complex (**166**) (which also has a low $O_2$ affinity) shows that introduction of the strap decreases the $O_2$ affinity by a factor of 4 for the $C_5$-strapped complex **168b,** and by a factor of 40 for the $C_4$-strapped complex **168a**. This is due to a combination of steric interactions. The presence of the cap locks the unoxygenated complex in a domed configuration which resists movement of the metal towards the porphyrin upon oxygenation. The presence of the strap and the binding of pyridine to the metal offers further resistance to the movement of the metal on oxygenation. Furthermore steric interaction may occur between the pyridine side-chains and the *meso*-aromatic rings preventing motion of the pyridine towards the porphyrin center. This would be more severe for the shorter $C_4$-strap, resulting in lower dioxygen affinity.

## 5.B. "Pocket" and "Tailed Pocket" Porphyrins

To prepare a system which discriminates against the binding of CO relative to that of $O_2$, Collman et al. have used a combination of the "picket-fence" and the "capped" porphyrin approach to prepare a series of "pocket" porphyrins[145, 146]. As above, a benzene ring is used to provide steric encumbrance at one face of the porphyrin, but in this case it is linked to only three *meso*-phenyl groups, leaving an open side. Oxygen may bind within the pocket by orientation of the bent Fe–O–O unit toward the open side. Carbon monoxide can only be accommodated by bending and/or tilting of the linear Fe–C–O unit leading to decreased CO affinity.

Treatment of $\alpha,\alpha,\alpha,\alpha$-$H_2(T_{Am}PP)$ (129) with slightly more than one equivalent of pivaloyl chloride formed the "mono-picket" porphyrin 169 (Scheme 49). Condensation with a benzene tris-acid chloride 170 under high dilution conditions afforded the pocket porphyrins 171 in good yield (> 60%). The volume of the pocket was dictated by the choice of acid chloride and the presence of the single picket provided protection against irreversible oxidation. A similar strategy was employed to prepare the "tailed-pocket" porphyrins 172. However, in these compounds, the remaining *ortho*-amino group is used

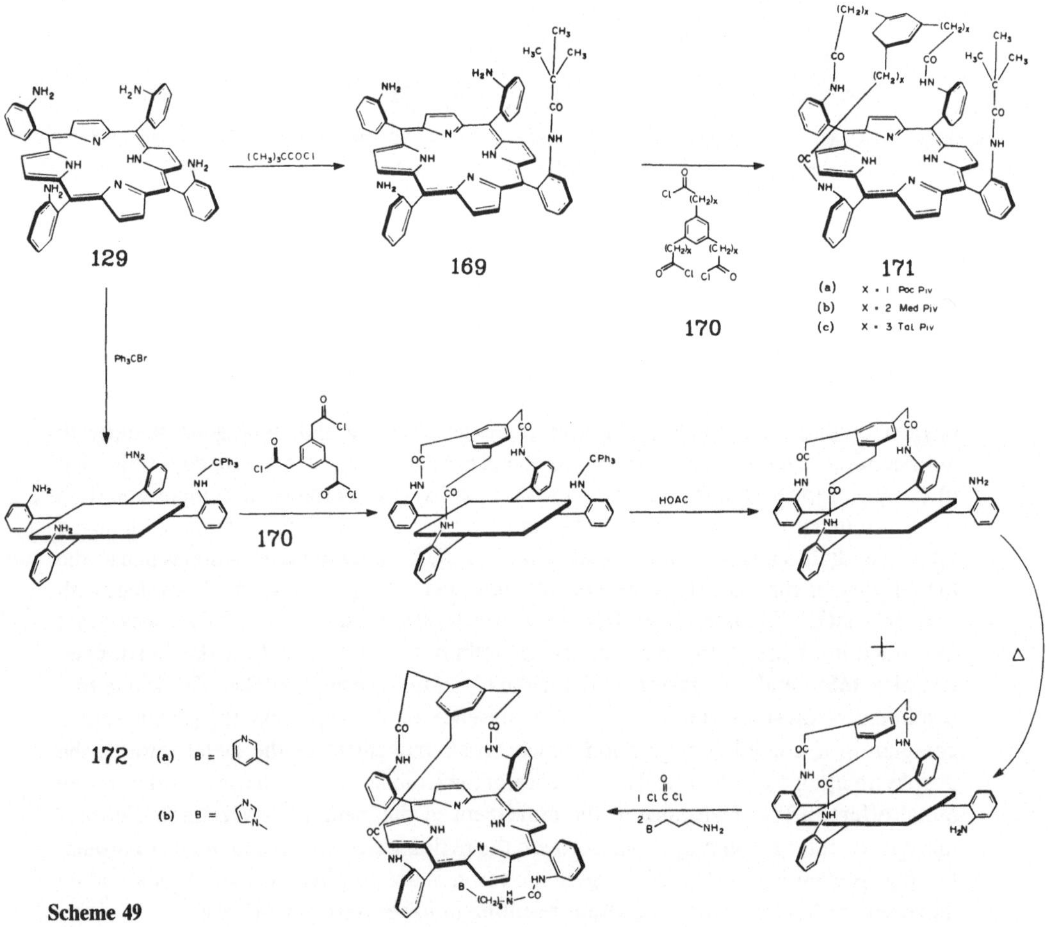

**Scheme 49**

to attach the base, leaving no protection on the open face of the pocket. In contrast to the "tailed picket-fence" porphyrins, these complexes undergo rapid oxidation to the μ-oxo complex[145].

For the iron(II) "pocket" porphyrins, the coordination state of the iron depended on the size of the pocket. Visible absorption and MCD spectral data showed that $Fe^{II}$(Poc-Piv) derived from **171a** remained five coordinate even in the presence of excess base. Although the medium and large size pockets showed increasing six-coordination, concentration ranges could be determined within which five-coordinate iron(II) was the dominant species. The $O_2$ and CO binding of the "pocket" and "picket-fence" porphyrins were compared[147]. While the $O_2$ affinities for both systems were similar, the "pocket" porphyrins showed a reduced CO affinity. Since electronic and solvent effects were similar in the two model systems, the reduction in CO affinity was attributed to the steric hindrance of the cap which distorted the Fe–C–O unit from linearity.

## 5.C. "Bis-Pocket" Porphyrins

Eventual irreversible oxidation of both the "picket-fence" and the "capped" porphyrins occurs because steric encumbrance is present only on one side of the molecule. To avoid this, octa-*ortho*-substituted TPP complexes have been prepared. By choosing the correct steric bulk for the ortho substituents a protected pocket may be formed on both sides of the porphyrin ring. The pockets could still be penetrated by axial bases and diatomic molecules but would form a barrier to the close approach of two metal centres, thus stabilizing the oxygenated porphyrin. Amundsen and Vaska have prepared the hemes $Fe^{II}$(P) (where ($P^{2-}$) is the dianion of tetrakis[2,4,6-tris(methoxy)phenyl]porphyrin or tetrakis[2,4,6-tris(ethoxy)phenyl]porphyrin) by condensation of the appropriate trisubstituted benzaldehyde with pyrrole[148]. Balch has shown by $^1$H-NMR that although the

173

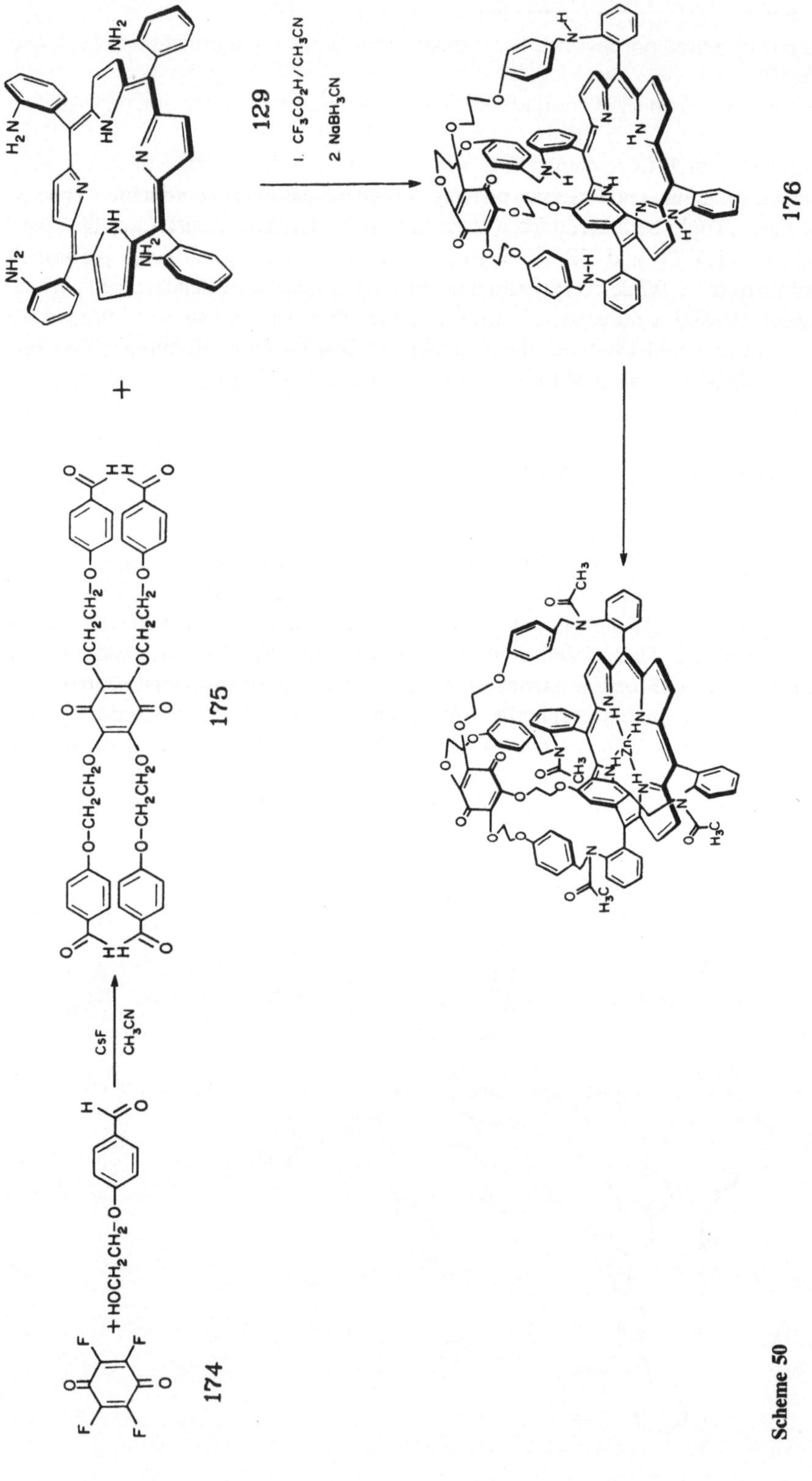

**Scheme 50**

*ortho*-methoxy substituents prevent μ-oxo complex formation, oxidation at room temperature proceeds to form $Fe^{III}(P)OH$ and $Fe^{III}(P)Cl^{11)}$. A more hindered complex was prepared by Suslick and Fox[149], who condensed 2,4,6-triphenylbenzaldehyde with pyrrole in refluxing propionic acid to obtain the porphyrin in 1% yield. Metallation and reduction gave the four-coordinate iron(II) species **173** in 80% yield. Addition of the sterically hindered base, 1,2-dimethylimidazole, gave a five-coordinate iron(II) complex which was capable of completely reversible oxygenation in non-polar solvents at 30 °C. However, the oxygen affinity is very low compared to other model complexes and the natural systems, an observation which was attributed to the non-polar nature of the binding site.

Covalent attachment of a group to the four *ortho*-positions of TPP may also be used to prepare porphyrin complexes having a rigid structure. Such an approach was adopted by Lindsey and Mauzerall[150] to prepare a cofacial porphyrin-quinone system where the separation between the two rings was estimated to be 10 Å. The quinone-tetraaldehyde **175** was prepared by alkoxylation of fluoranil **174**, which was then reacted with α,α,α,α-$H_2(T_{Am}PP)$ (**129**). The reversibility and the intramolecular nature of the Schiff base reaction were responsible for the high yield of the reaction; 85% yield of the "capped" porphyrin after stirring at room temperature for 24 h. Subsequent reduction of the Schiff bases with $NaBH_3CN$ yielded the desired porphyrin-quinone **176** in 80–95% yield (Scheme 50). Acetylation of the amino groups and metal insertion furnished the zinc porphyrin quinone **177**. The photochemical properties of this compound were investigated and the rate constant for electron transfer from porphyrin to quinone was estimated[151].

# 6. Strapped Porphyrins

The strapped porphyrin class of heme protein models embraces all those compounds in which some group is covalently linked to two corners (usually diagonally opposite) of a porphyrin macrocycle. The usual synthetic strategy has been to tie the strap to an already formed porphyrin, thus allowing great versatility in the types of structures made (e.g. cyclophane, crowned, pagoda, basket-handle, etc.). The porphyrins may be singly- or doubly-strapped and may be classified according to the nature of the chain:

(i)   Simple non-functionalized alkyl chains whose role is to span one face of the porphyrin, discouraging μ-oxo bridging and providing a more hydrophobic environment.
(ii)  Straps incorporating some bulky group which will provide more steric encumbrance than a simple alkyl chain.
(iii) Functionalized straps which incorporate some group capable of binding to or interacting with the metal at the porphyrin core. These may be used to maintain five-coordination or to form six-coordinate mixed ligand systems L–M–L', where one ligand binds poorly to the metal.

## 6.A. Non-Functionalized Alkyl Straps

A number of research groups have reported the synthesis of simple strapped porphyrins. Ogoshi et al.[152-154] condensed long-chain diamines with a difunctional etio-type porphy-

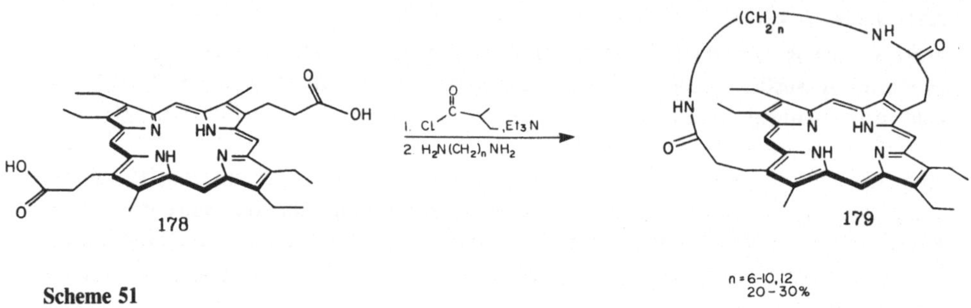

**Scheme 51**

$n = 6-10, 12$
$20 - 30\%$

rin **178** in the presence of isobutyl chloroformate and triethylamine under high dilution to obtain the strapped porphyrins **179** (n = 6–10, 12) in 20–30% yield after chromatography (Scheme 51). On the basis of visible absorption spectroscopy it was claimed that, for short straps, binding of a second bulky axial ligand under the strap was inhibited, giving five-coordinate iron(II) species. Attempts to observe oxygen binding to the ferrous porphyrins at 15 °C resulted in rapid formation of the μ-oxo complex. Battersby et al.[155] used the same strategy, reacting the bis-acid chloride of mesoporphyrin II (**180**) with 1,12-aminododecane (**181**) under high dilution to give the bridged porphyrin **182** in 25% yield (Scheme 52). Alternatively, elaboration of the porphyrin carboxyl side chains by ester or amide formation, followed by intramolecular oxidative coupling or Dieckmann condensation gave the ester-, or amide-linked strapped porphyrins **183, 184** in good yield. Because the strap can easily move sideways attempts to oxygenate the ferrous complexes in THF or aqueous acetone at 20 °C showed only rapid irreversible oxidation to the ferric state. Chang and Kuo[156] have also prepared an amide-linked strapped porphyrin **185a** (Scheme 53). In this instance the system was used as a cytochrome P450 model to investigate the porphyrin-catalyzed hydroxylation of unactivated alkanes by iodosyl benzene. In some cases the strap itself served as the substrate, the porphyrin ligand **185a** being transformed into **185b**. A similar series of amide-linked strapped porphyrins has also been prepared containing 5–7 methylene units in the strap[157]. Such models were prepared in an attempt to mimic the differentiation of $O_2$ and CO displayed by the natural systems. Using resonance Raman spectroscopy a correlation between increased steric strain (shorter straps) and the Fe–CO stretching and Fe–C–O bending vibrations has been observed[158].

A different synthetic strategy was adopted by Baldwin et al., to prepare strapped porphyrins where the *ortho*-positions of opposite *meso*-phenyl rings are linked[159, 160]. Unlike the previous syntheses, it is the strap which is prepared initially and the two halves of the porphyrin attached to both ends. In the final step the porphyrin ring is formed by intramolecular cyclization and the strap is stretched into position (Scheme 54). Condensation of the anion of salicylaldehyde with 1,12-dibromododecane gave the "strapped-dialdehyde" **186.** Acid catalyzed condensation with benzyl 3,4-dimethylpyrrole-2-carboxylate **187** afforded the chain-linked bis-dipyrromethane, which after hydrogenolysis gave the unstable tetraacid **189.** This was immediately condensed with trimethyl orthoformate to give the "alkyl-strapped" porphyrin **190** in 23% overall yield. As in the previous examples the alkyl strap was not able to enforce five-coordination of the respective iron(II) complex. In the presence of excess base the six-coordinate species **191** was formed which did not bind oxygen. Reducing the concentration of base led to an increase

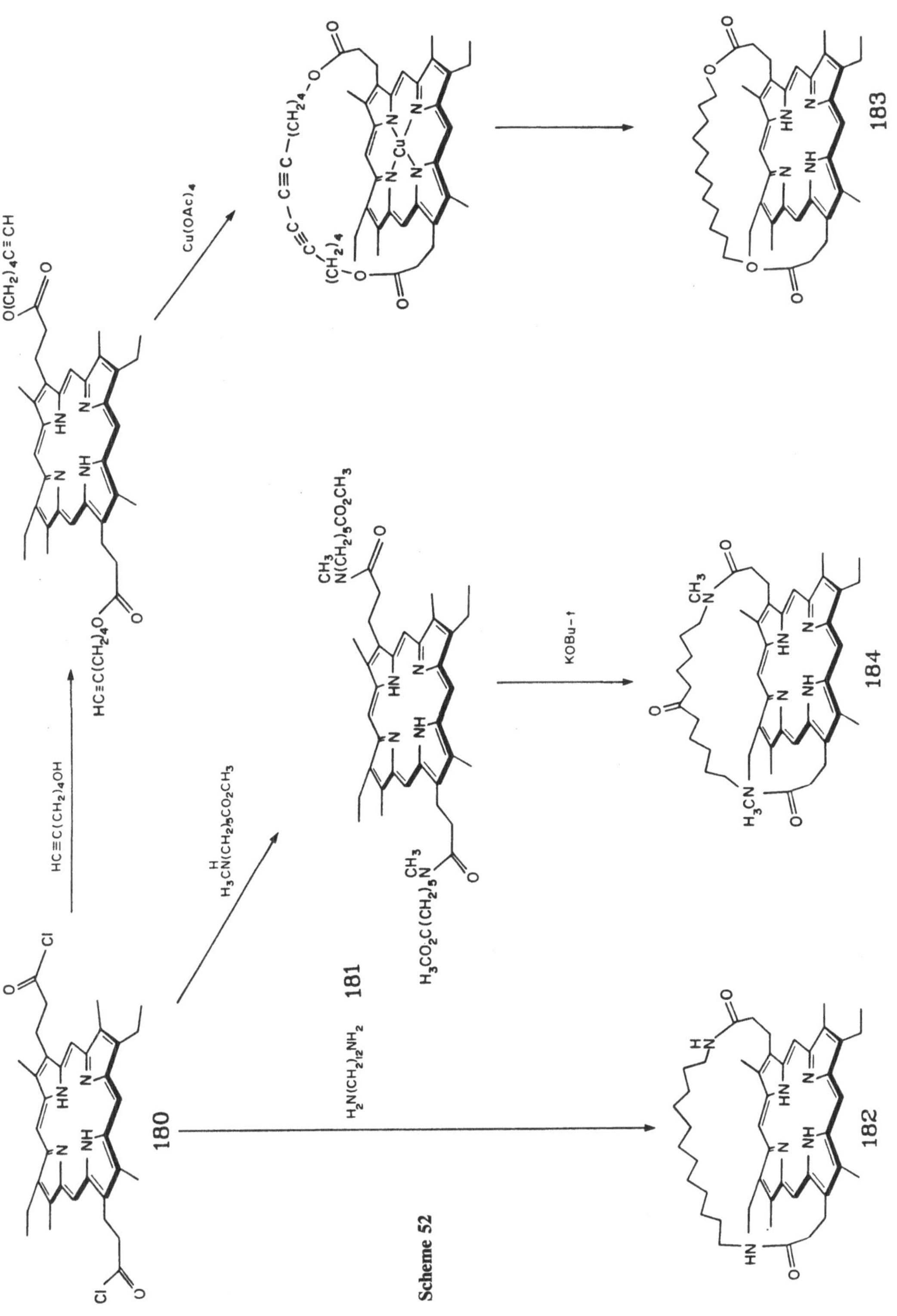

**Scheme 52**

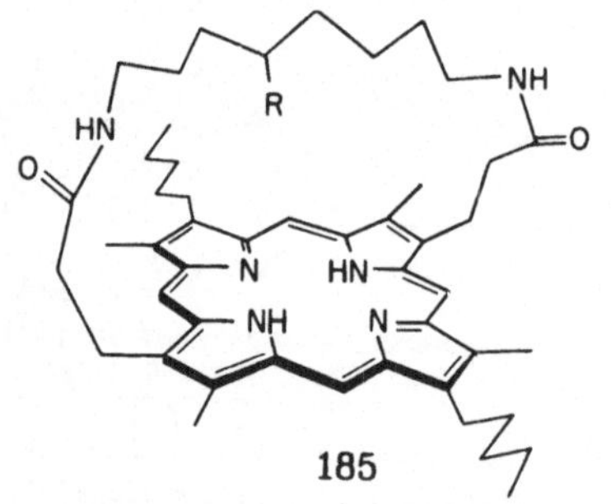

(a) R = H
(b) R = OH

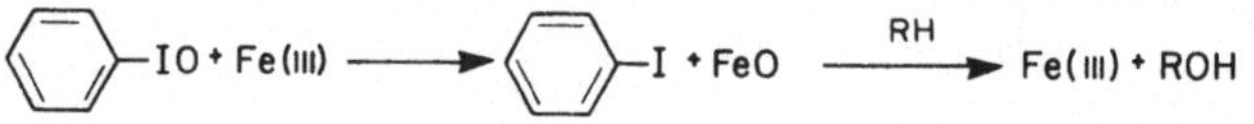

**185**

**Scheme 53**

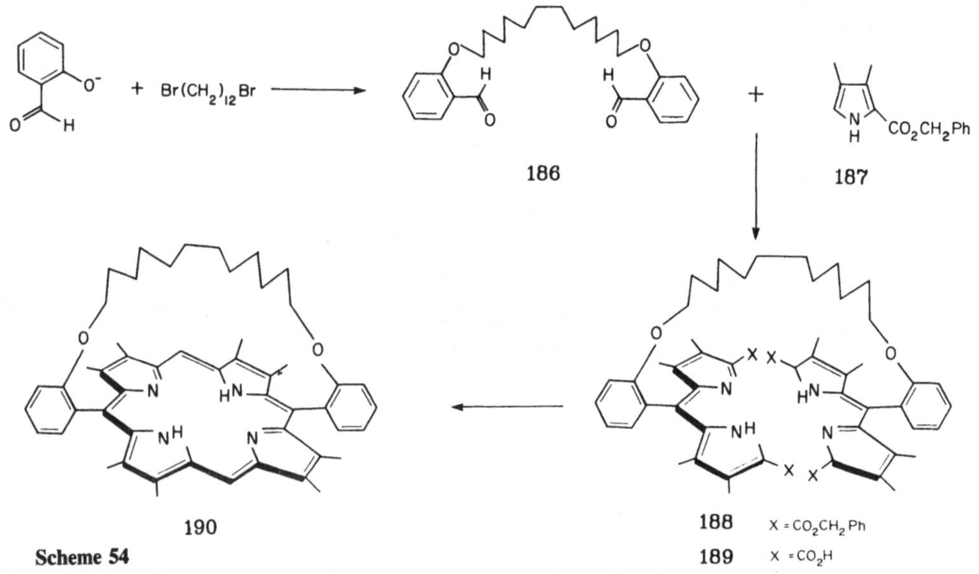

**186**

**187**

**190**

**Scheme 54**

**188**  X = CO₂CH₂Ph
**189**  X = CO₂H

**191**

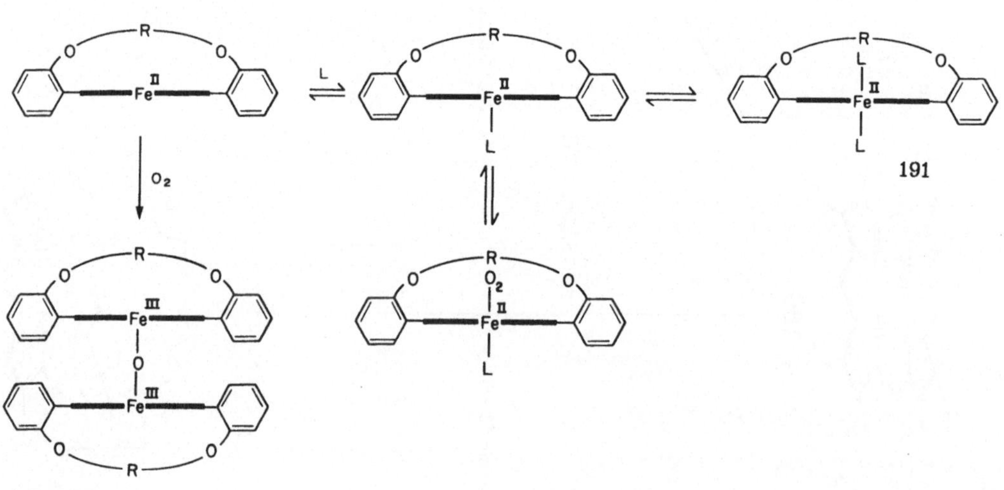

**Scheme 55**

of the four-coordinate species which underwent irreversible oxidation. While reversible oxygenation was observed at $-55\,°C$ similar to unhindered porphyrins, warming to room temperature caused μ-oxo bridge formation (Scheme 55).

A similar approach was used by Wijesekera et al.[161, 162], who were attempting to strap a porphyrin with very short alkyl chains – short enough to cause deformation of the porphyrin. Obviously in this case, linking opposite corners of a preformed porphyrin would, at best, give very poor yields. Instead the two halves of the porphyrin were assembled at each end of the strap, and only at the last step was the porphyrin **192** formed by acid-catalyzed intramolecular cyclization under high dilution conditions (Scheme 56).

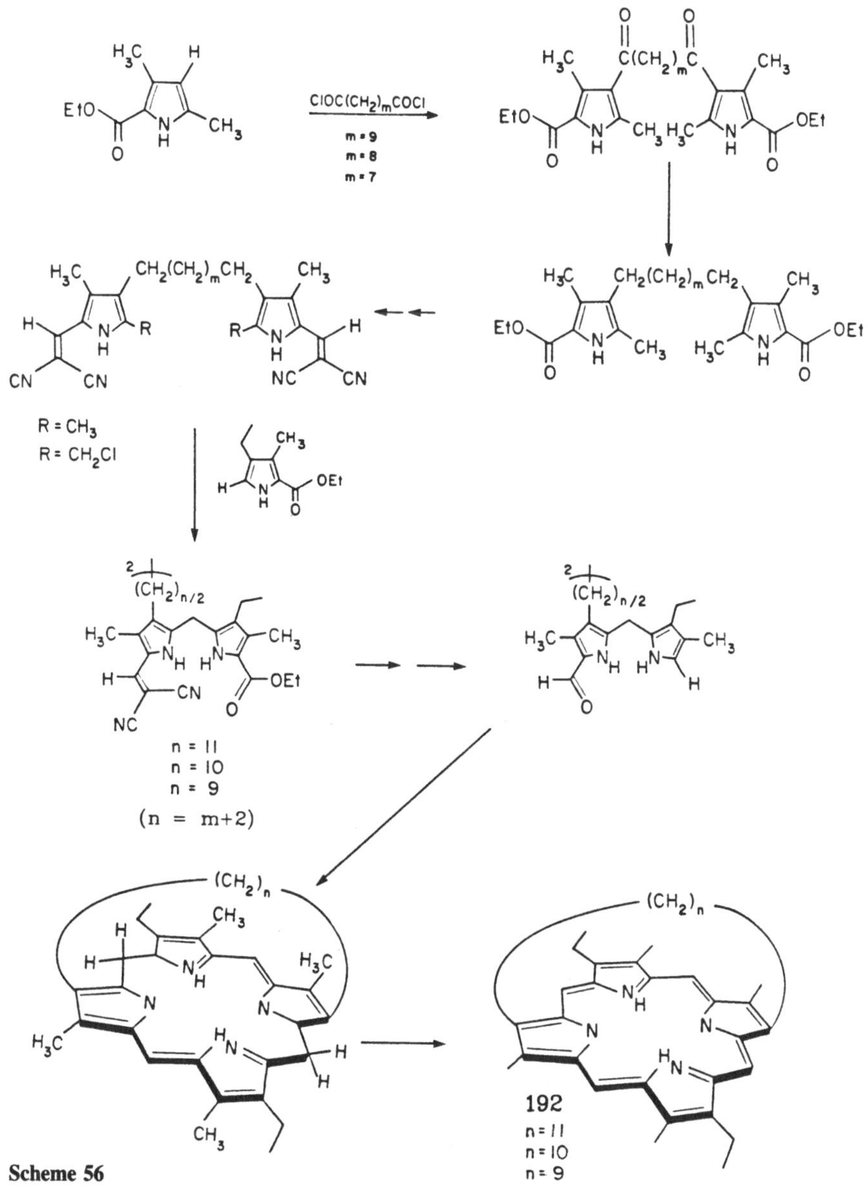

**Scheme 56**

Visible and $^1$H-NMR spectroscopy and X-ray crystallography (Fig. 7) all point to increasing distortion of the ring as the length of the strap is decreased (**192,** n = 11, 10, 9). A chain length of nine methylene units appears to be the lower limit; attempts to prepare even more strained porphyrins with shorter straps were unsuccessful. Ligand binding to the metal complexes showed that the straps provided no steric protection.

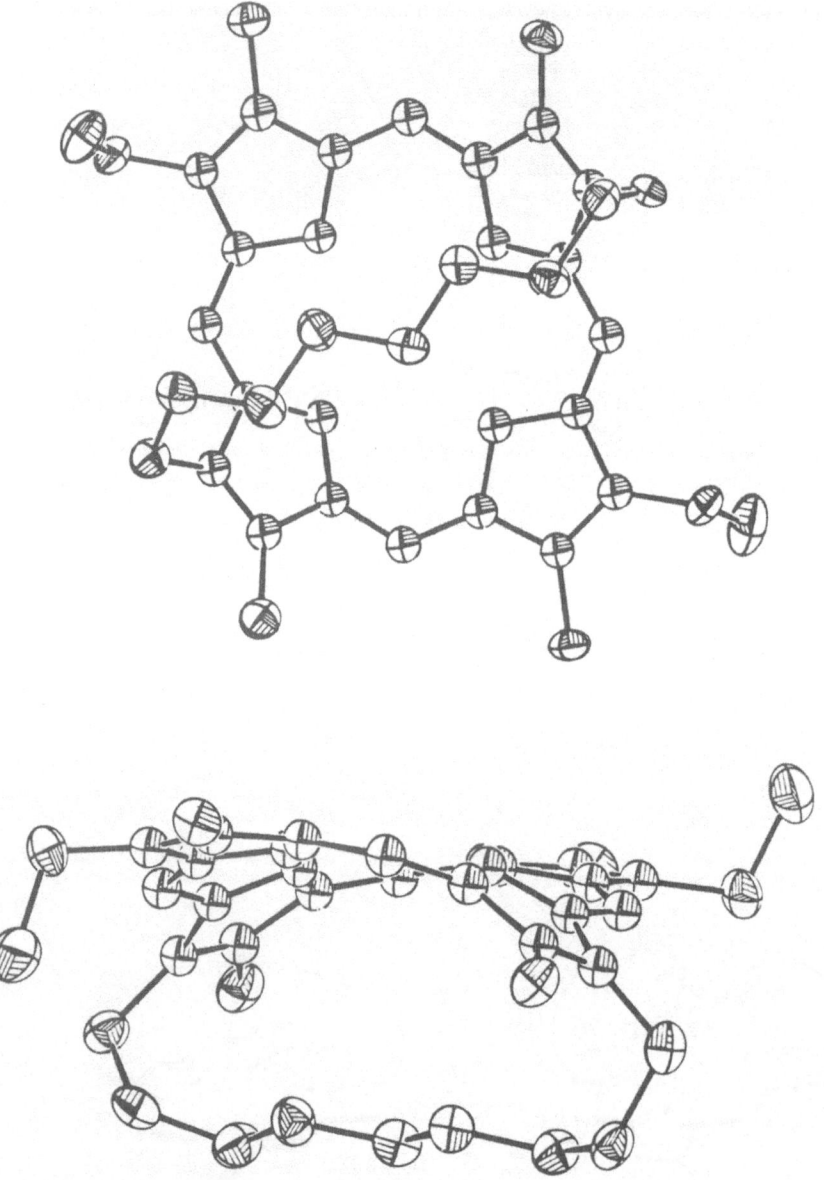

**Fig. 7.** Ortep drawings of **192** (n = 9). Adapted from Ref. 162

## 6.B. Straps Containing Bulky Blocking Groups

Porphyrins strapped with simple alkyl chains are poor models for oxygen binding heme proteins. In most cases the strap is too "floppy" and can be pushed to one side allowing μ-oxo bridge formation. In addition, base is not prevented from binding under the strap, leading to oxygen binding on the open face and, consequently, irreversible oxidation. The logical extension is to incorporate some bulky group into the strap to increase the steric encumbrance about one face.

One of the initial examples of the strapped porphyrin approach to heme protein models was the cyclophane porphyrin **193** of Diekmann et al.[163]. In this example steric

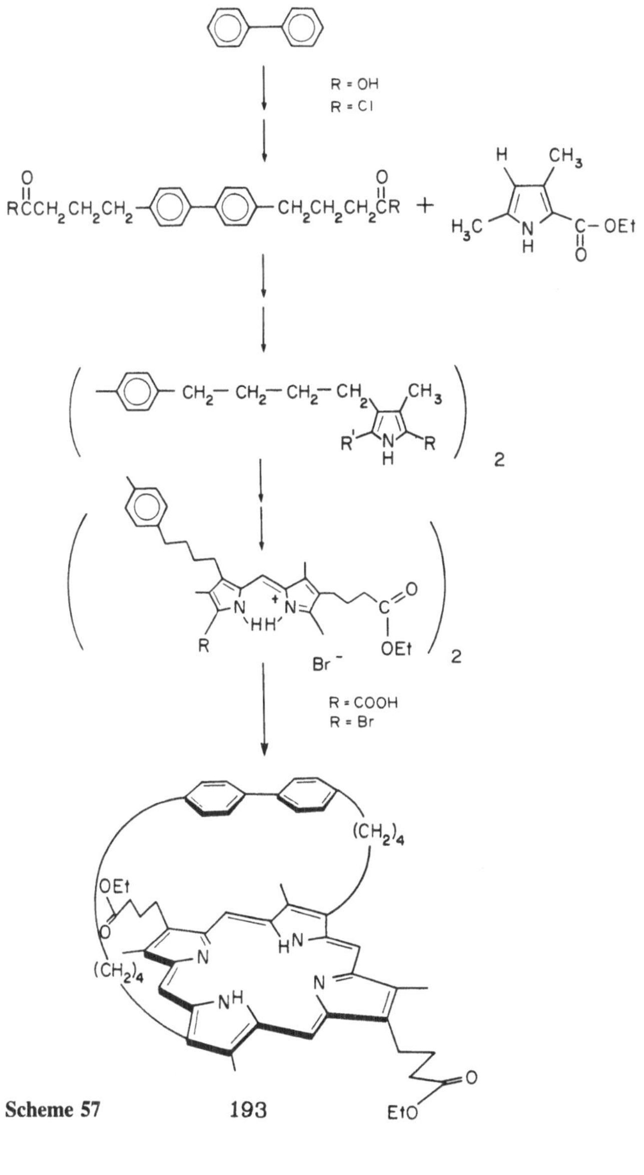

**Scheme 57**          **193**

encumbrance was provided by a biphenyl group in the strap. To ensure a tightly fitting strap, porphyrin cyclization was delayed until the final step (Scheme 57). Because of poor yield (5% for the cyclization step after repeated chromatographic purification) this porphyrin was not used as a heme protein model. Instead an anthracene was strapped across a preformed porphyrin (194) by means of amide linkages (Scheme 58). For the

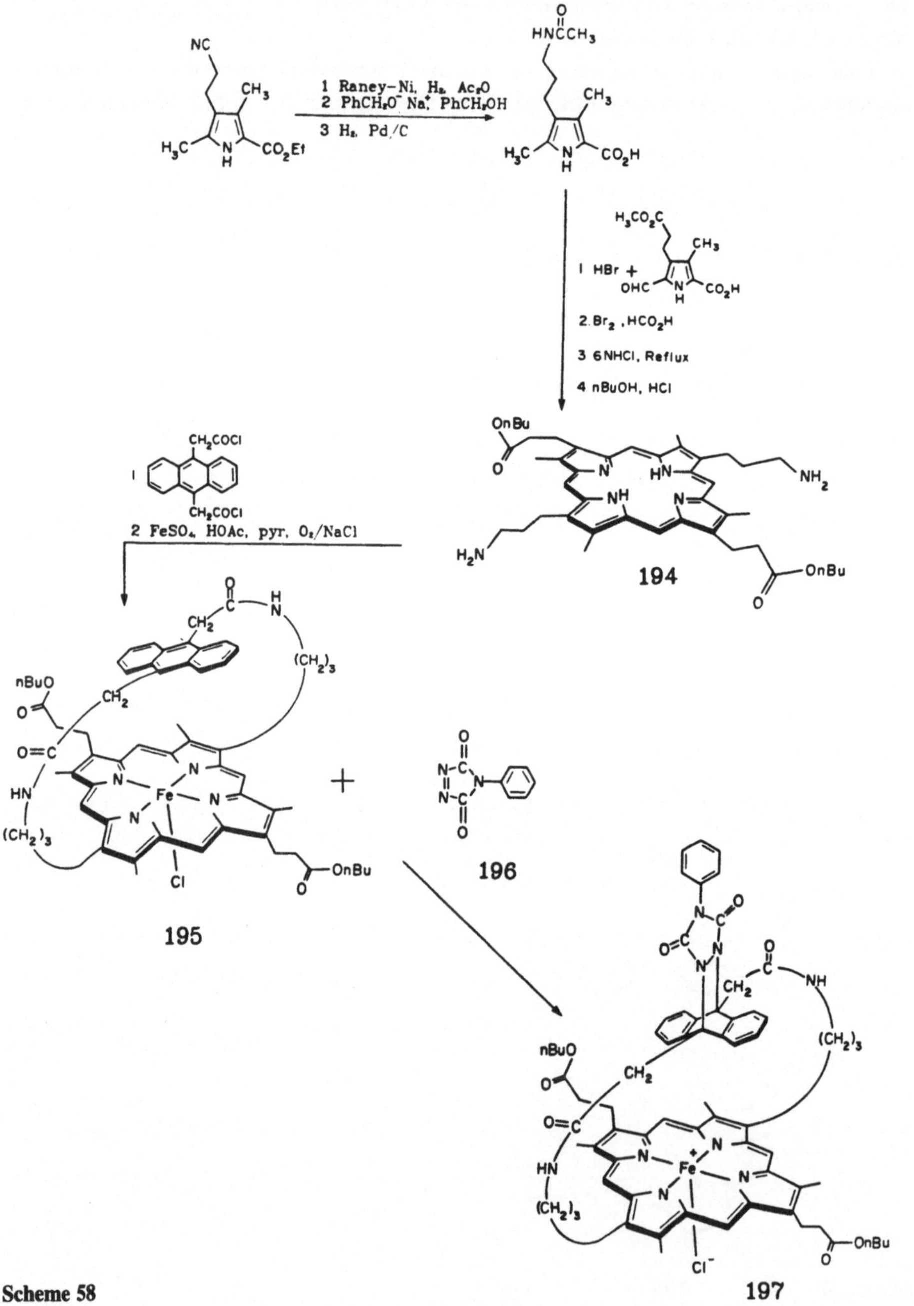

Scheme 58

anthracene-heme[6,6]cyclophane **195** the two aromatic rings were estimated to be ~ 4.5 Å apart. A Diels-Alder addition on the anthracene with 1-phenyl-triazine-2,5-dione (**196**) gave the "pagoda porphyrin" **197** possessing an even tighter pocket. Binding of a second axial base underneath the anthracene ring was not observed, the iron(II) complexes **198** being five-coordinate even in 1 M 1-MeIm. The anthracene-heme[6,6]cyclophane **198a** and the homologous[ 7,7] compound **198b** were used to study the binding of isonitriles, CO and $O_2$ within the pocket as models for the distal side steric effects in heme proteins[164–166].

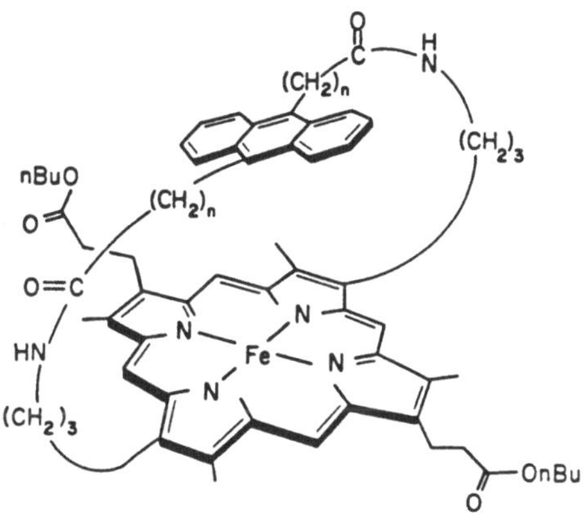

198

(a) n = 1

(b) n = 2

The Fe[6,6] cyclophane **198a** showed a large reduction in affinity for CO and $O_2$ compared to flat unhindered hemes (the Fe[7,7] cyclophane **198b** showed only a small effect), and this steric effect was manifested primarily in the ligand association rate[167]. This suggested that distal side steric effects are not due to repulsion in the bound state, but due to limited access to the heme face. While distal steric effects in the model compounds could differentiate bulky isocyanide ligands, they could not differentiate between linear and bent diatomic molecules. These results led Traylor and his colleagues to postulate that reported bending or tilting of bound CO in heme proteins may be of minor chemical significance.

Baldwin adapted his strapped porphyrin synthesis to prepare a system **199** with a benzene ring above one face (Scheme 59)[160]. Although structurally similar to the "$C_2$-capped" porphyrin **160,** the absence of the two extra linkages resulted in a "floppy" strap which did not prevent six-coordination by ligands such as 1-MeIm or pyridine and which did not prevent μ-oxo bridge formation. An even more bulky strap, incorporating a naphthalene ring as in **200,** was no more successful.

Dolphin and his colleagues[161, 162] have prepared a series of strapped porphyrins with a durene group protecting one face (**201**) (Scheme 60). Incorporation of iron using standard methods[168] gave the corresponding heme complexes **202a–c.** The crystal structure

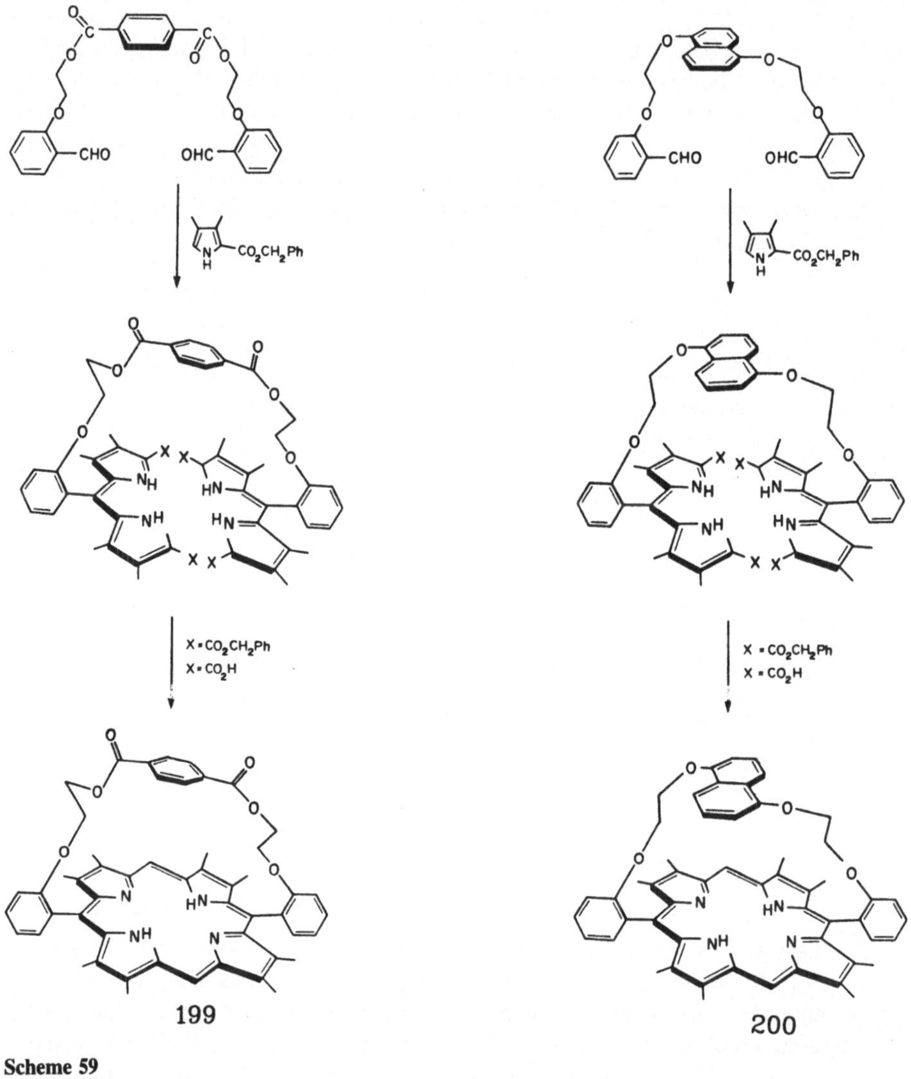

199

200

**Scheme 59**

of the hemin chloride derived from **202a** (Fig. 8) showed considerable distortion of the porphyrin from planarity[169]. The optical spectra for the free base porphyrins indicate that some porphyrin distortion exists also in the durene-5/5 free base **201b**, while the 7/7-derivative **201c** is essentially flat. $^1$H-NMR data suggest that the durene moiety in the 7/7-base **201c** is closer to the porphyrin plane than in the tighter capped 5/5 and -4/4 analogs **201b** and **201a**. The heme derivatives of the durene porphyrins have been studied with respect to their interaction with imidazoles, isocyanides, CO, and $O_2$[114, 170]. The binding constants of 1,5-dicyclohexylimidazole (DcIm) and 1,2-dimethylimidazole (1,2-Me$_2$Im) to the unhindered side of the four-coordinate hemes are similar within the durene series despite differences in the porphyrin plane distortion; for steric reasons, these imidazoles do not coordinate on the capped (or distal) side of the heme. The size of

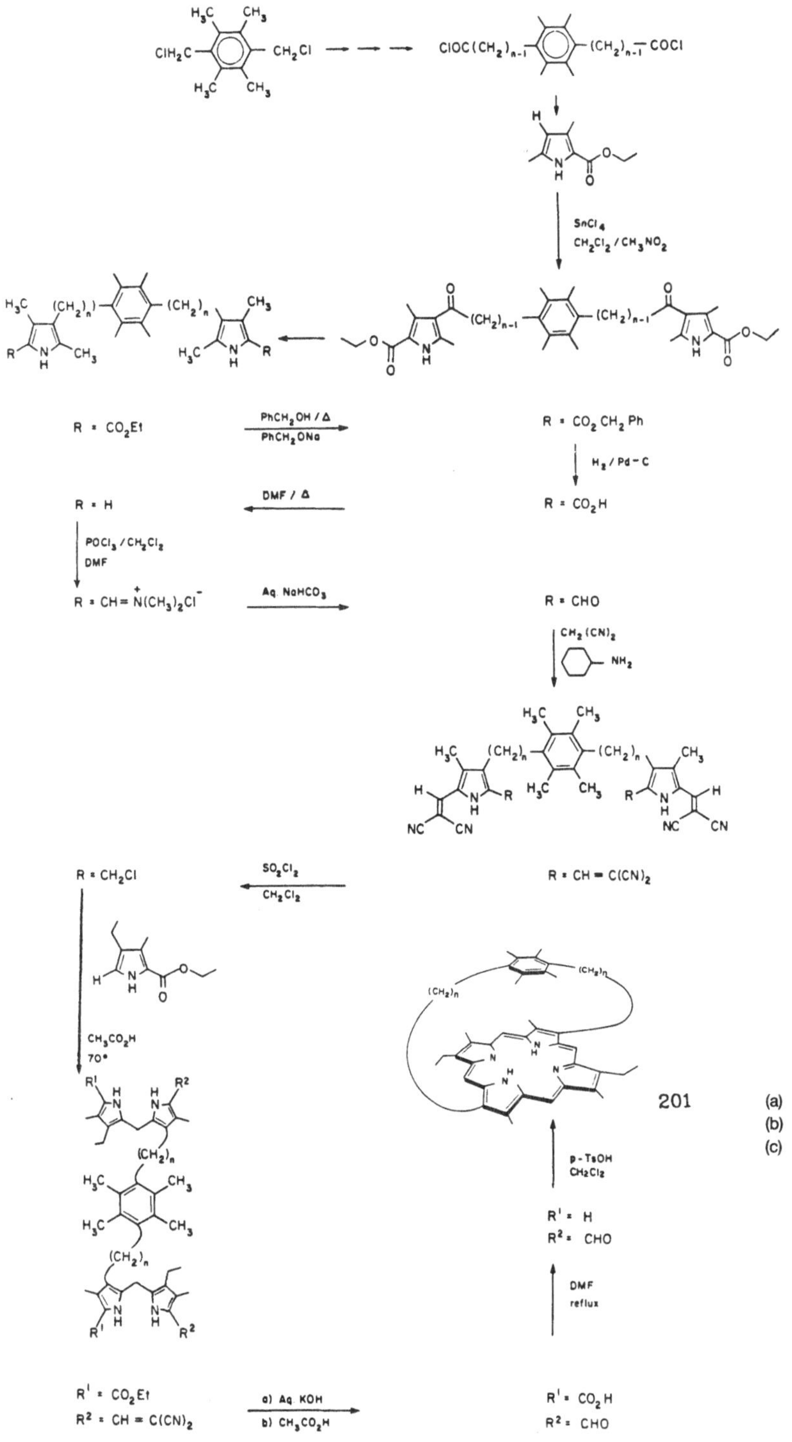

**Scheme 60**

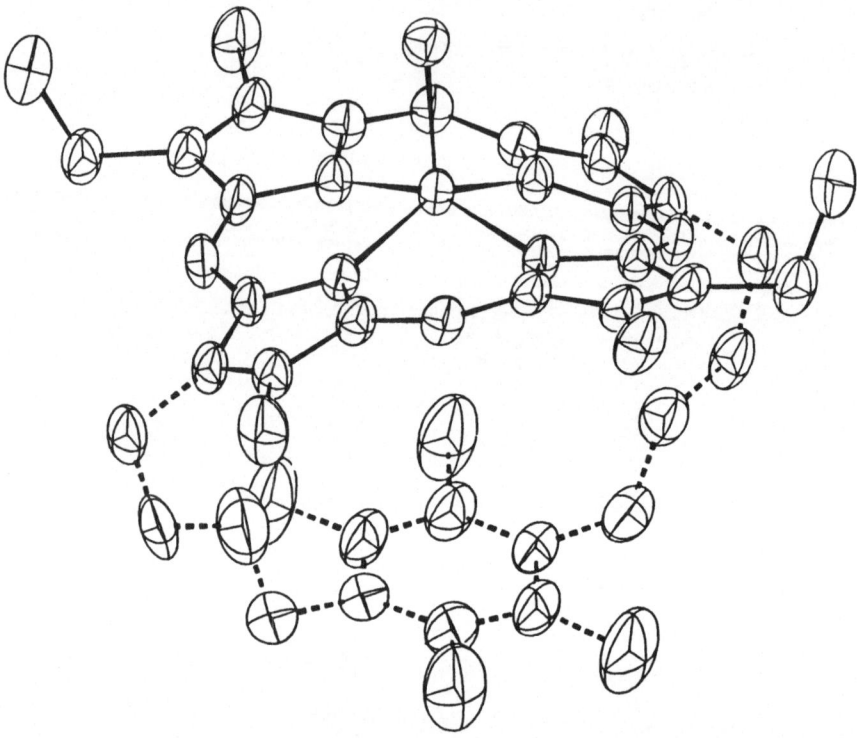

**Fig. 8.** A SNOOPI diagram (E. K. Davies, plotting routine, 1984, Chemical Crystallography Laboratory, 9 Parks Road, Oxford, England) of the hemin chloride of **202a** (50% probability contours for all atoms; hydrogen atoms have been omitted for clarity; the dashed bonds are used to distinguish between the "strap" and the porphyrin skeleton)

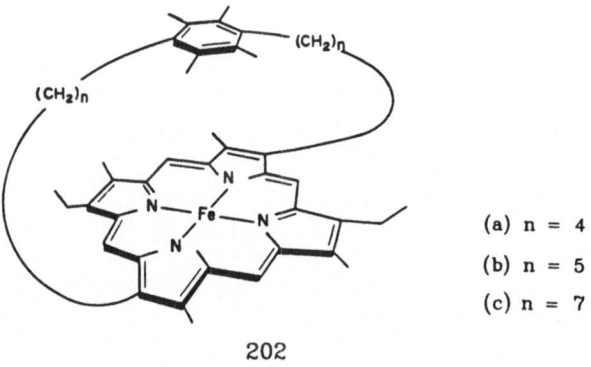

(a) n = 4

(b) n = 5

(c) n = 7

202

the distal cavity was examined using the bulky isocyanides, tosylmethylisocyanide (TMIC) and t-butylisocyanide (t-BuNC), which differ in their spatial requirements. The extremely restrictive distal environment of the durene-4/4 system, however, inhibited the coordination of either isocyanide. The $Fe^{II}$(durene-7/7)(DcIm) complex obtained from **202c** exhibits a reduced overall affinity for CO relative to simple flat, open hemes; this is manifested in a depressed association rate for CO, and was interpreted[170] as a distal steric effect as a result of the durene-cap. The durene-5/5 and -4/4 systems derived from **202a** and **202b** also show reduced CO affinities compared to open hemes, but this results predominantly from increased dissociation rates for CO from the six-coordinate Fe(P)(DcIm)CO complexes because of the proximal steric strain induced by the porphyrin plane distortion. The five-coordinate hemes **202a · B – 202c · B** (B = DcIm or 1,2-Me$_2$Im) bind O$_2$ reversibly to similar extents, implying a negligible effect of porphyrin skeletal distortion. A 10-fold reduced affinity for CO by the **202a · B** complex relative to **202b · B** (or other less distorted hemes) arises from proximal steric discrimination of CO relative to O$_2$ within this severely distorted system. Significantly, the five-coordinate heme complexes **202a · B – 202c · B** all show considerably higher $k^{CO}/k^{O_2}$ ratios (M values) relative to encumbered hemes that incorporate polar amide functions in their distal environments. This is entirely consistent with the concept of electronic interactions within the distal binding pocket stabilizing the Fe–O$_2$ moiety and increasing the affinity of the heme toward O$_2$, relative to CO[114, 170].

The same synthetic strategy (as shown in Scheme 58) was used to strap a bulky adamantane group across one face of a porphyrin to give 1,3-adamantane[6,6] cyclophane. The crystal structure[171] of the free base (Fig. 9) showed no free cavity between the strap and the porphyrin; CO, O$_2$ and isocyanides bind to the Fe$^{II}$ complex **203** with greatly reduced rates. However the adamantane strap displayed no significant steric differentiation between CO and O$_2$ binding. For CO the binding constant was controlled by the association rate, dissociation not being increased by the steric effects of the strap.

In contrast to the ligand behaviour of the pyridine straps of **257** and **259** (see Chap. 6.C.), a singly strapped pyridine-[5,5] cyclophane heme **204** prepared by Traylor et al.[172], showed no evidence of either internal or external iron-pyridine complex formation. Indeed even the binding of CO to **204** failed to induce pyridine ligation even though the binding of CO and bases to iron(II) porphyrins is synergistic. Obviously the shorter strap in this case induces sufficient strain to prevent internal binding and compound **204** behaves more like the anthracene and adamantane cyclophanes **198** and **203**. The

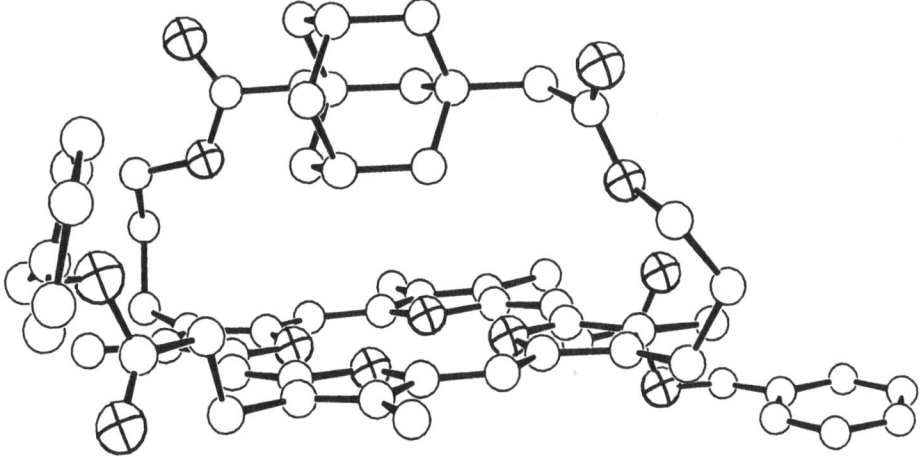

**Fig. 9.** Perspective view of a 1,3-adamantane[6,6] cyclophane porphyrin. Adapted from Ref. 171

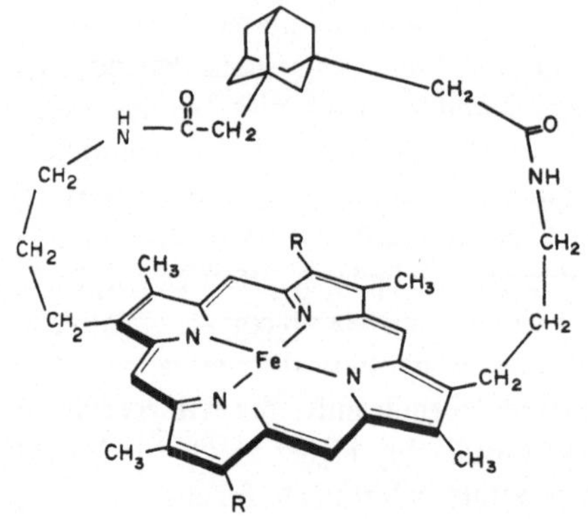

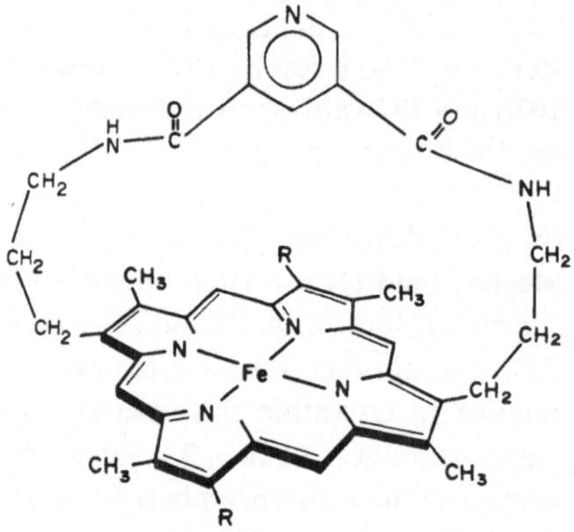

203     R = -CH₂CH₂C-OCH₂-Ph

204     R = -CH₂CH₂C-OCH₂-Ph

pyridine heme **204** displays severe steric hindrance to both $O_2$ and CO binding, which is manifested in lower association rates. Furthermore **204** shows a greater differentiation between CO and $O_2$ binding than other models, the lower binding ratio being possibly due to enhanced binding of $O_2$ in the tight polar pocket[114].

Dolphin and Morgan[173] have prepared the series of strapped porphyrins illustrated in Fig. 10 and Scheme 61. As a consequence of postponing porphyrin ring formation (a

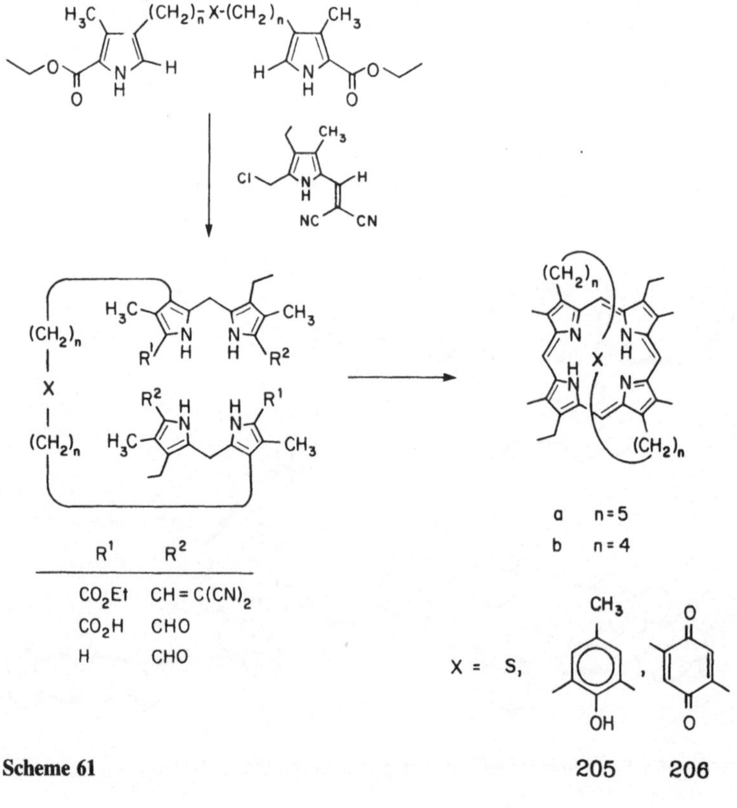

| R¹ | R² |
|---|---|
| CO₂Et | CH=C(CN)₂ |
| CO₂H | CHO |
| H | CHO |

a    n = 5
b    n = 4

X = S,

205    206

**Scheme 61**

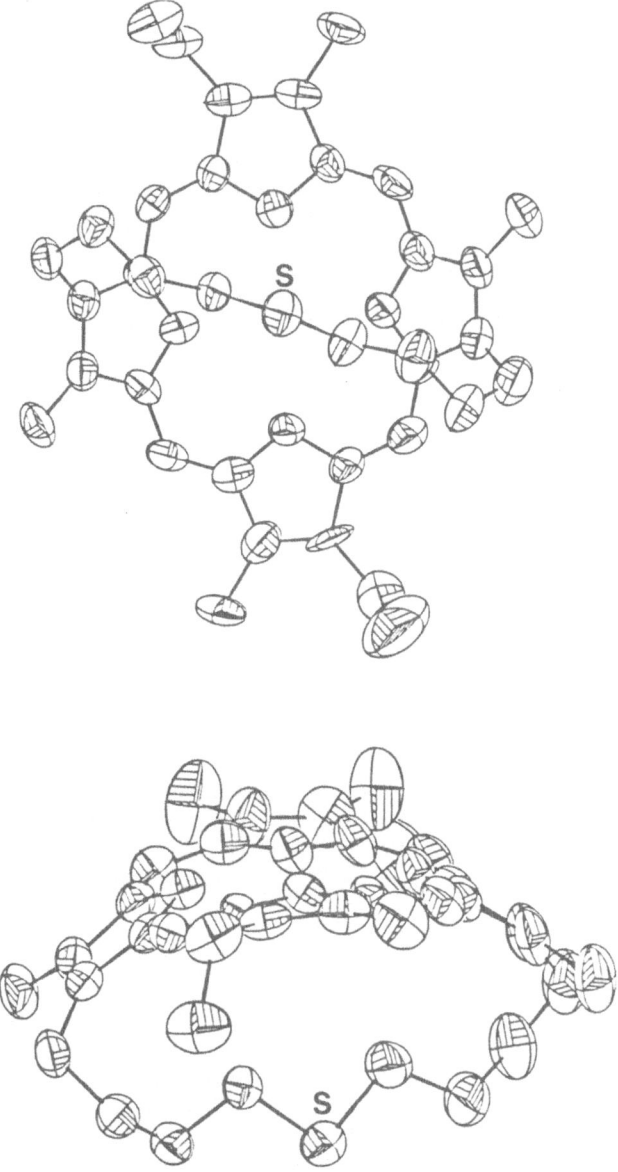

**Fig. 10.** Ortep drawings of a strapped porphyrin containing a thioether linkage. Morgan, B., Dolphin, D., Einstein, F. W. B., Jones, T., manuscript in preparation

specific example is given in Scheme 61) till the later stages on the synthesis, (i) a range of strap lengths are available, some of which may lead to distortion of the porphyrin, and (ii) hydrocarbon straps may be employed, precluding any polar effects due to amide or ester linkages and (iii) increased stability is obtained due to the hydrocarbon strap. Compounds including thioethers (Fig. 10), phenol **205** and quinone **206** groups[174] in the strap have been prepared (Scheme 61) as potential models for cytochrome c, catalase and photosynthetic charge separation, respectively.

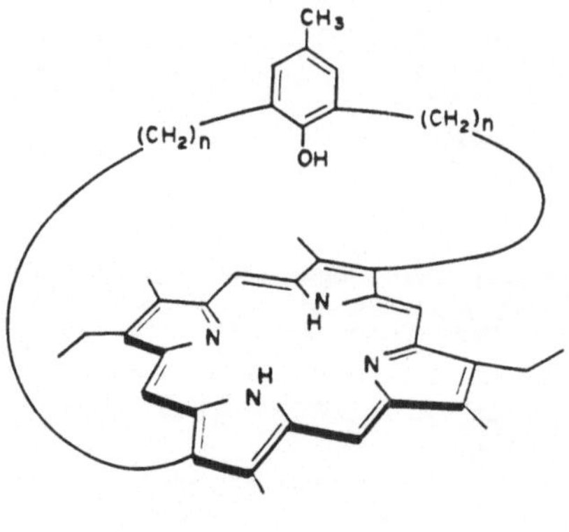

205 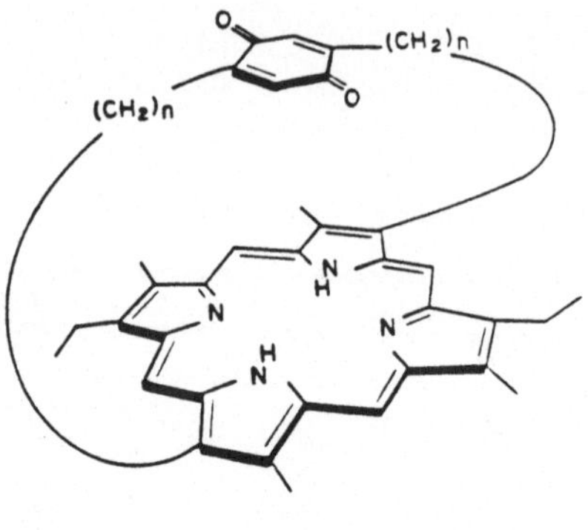                                                  206

## 6.C. Straps Containing Interactive Groups

The incorporation of potential ligands into the porphyrin strap has three advantages.

(i)   A stoichiometric amount of ligand is built into the system, ensuring five-coordina-
      tion without the addition of external ligand. In the case of nitrogen bases, mixtures
      of six- and four-coordinate complexes are avoided.

(ii)  For ligands which bind poorly to iron(II) (e.g. thiolate), coordination would be
      favoured by constraining the ligand into a position suitable for binding to the metal.

(iii) Because the strap is fixed to the porphyrin and the ligand in two positions, compli-
      cations due to ligand replacement or dissociation will be minimized.

The strapped-porphyrin approach is also useful for orientating other interactive
groups (e.g., metal binding sites, electron donor/acceptor groups) into specific geometries with respect to the porphyrin.

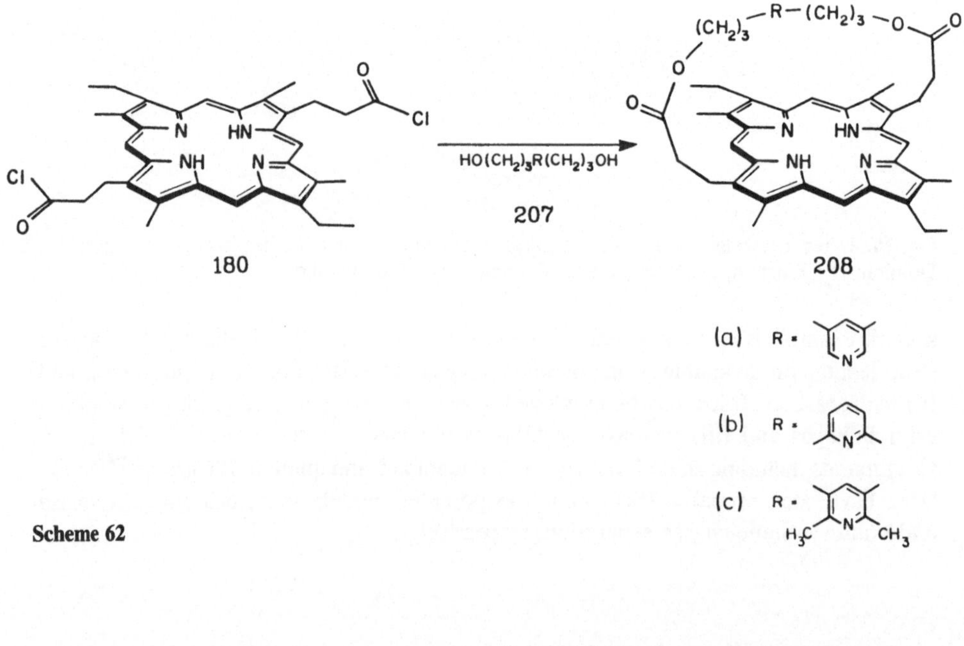

180                                    207                                    208

(a)   R =

(b)   R =

(c)   R =

**Scheme 62**

Battersby et al. have prepared a series of strapped porphyrins bearing variously substituted pyridine ligands[175]. Reaction of the porphyrin bis-acid chloride **180** with the corresponding pyridine diol **207** gave the ester-linked pyridine straps **208 a–c** in up to 38% yield (Scheme 62). The molecular structure of a crystalline side-product of these syntheses is shown in Fig. 11[176]. A cytochrome P450 model was prepared similarly (Scheme 63)[177]. A suitably functionalized diol **211** was reacted under high dilution with the bis-acid chloride of mesoporphyrin II (**180**) to give the strapped porphyrin **212** in 25%

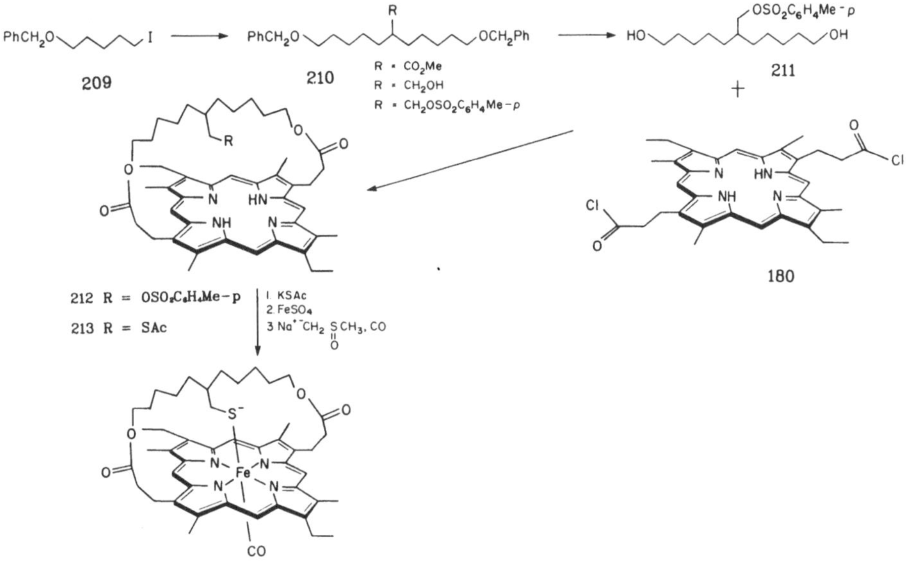

**Scheme 63**          214

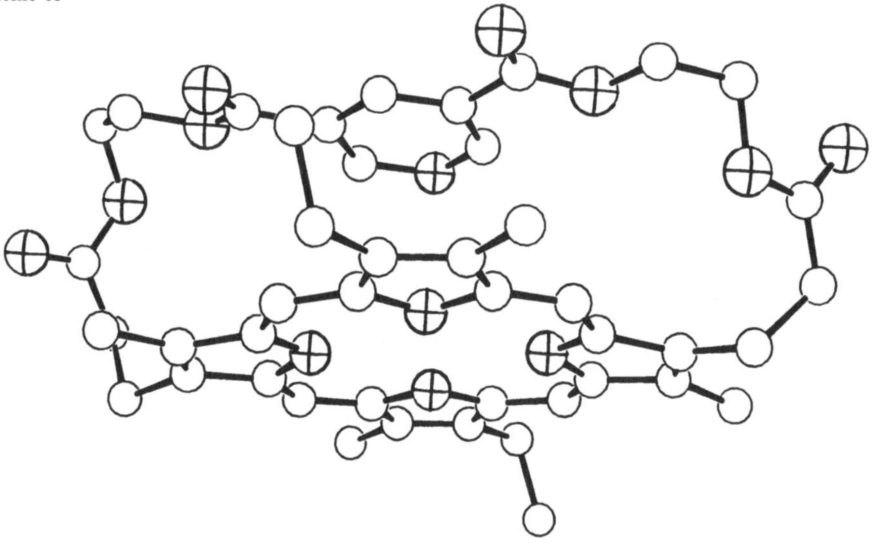

**Fig. 11.** Perspective view of a pyridine strapped porphyrin. Adapted from Ref. 176

yield. The protected-sulfur derivative **213** was obtained by displacement of the tosyloxy group with potassium thioacetate. After iron insertion and reduction to the iron(II) state, the S-acetyl group was cleaved with dimsyl sodium to produce the five-coordinate iron(II) species. On exposure to CO a six-coordinate iron(II) species **214** was formed whose visible absorption spectrum reproduced the major characteristics of the carbon-monoxy cytochrome P450 spectrum – a split hyper Soret band with an intense band at 450 nm. The $^{13}$C-NMR spectrum of the $^{13}$CO complex also supported the $RS^--Fe^{II}-CO$ formulation.

Several binucleating strapped porphyrins have been prepared, which are capable of binding two metal ions in close proximity. The "crowned" porphyrin **217** of Chang[178], formed by the addition of a bis-amino crown ether **216** to the bis-acid chloride **215**, may bind a transition metal ion (in the porphyrin ring) and a group IA or IIA cation (in the diaza-18-crown-6 ring) (Scheme 64). Apart from its metal binding capability the crown ether also exerts some steric control over one face of the porphyrin. For the iron(II) complexes, bulky ligands (e.g., 1-triphenylmethylimidazole) bind only on the unhindered face of the porphyrin to give a five-coordinate species. Oxygen may then bind under the crown to give a reasonably stable oxygen species ($t_{1/2} > 1$ hr at 25 °C in DMA).

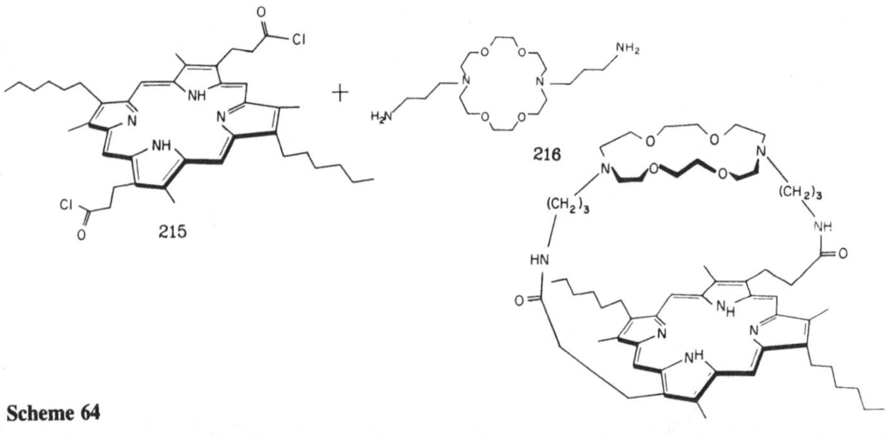

**Scheme 64**

Recently, another crown-strapped porphyrin **219** similar to **217** has been prepared by the condensation of **225b** (see Scheme 67) with the bis-amino crown ether **218** (Scheme 65)[179]. The corresponding metalloporphyrin is a potential host for both anionic and cationic species. Bimetallic complexes were prepared and, for paramagnetic guest cations e.g. $Cu^{II}$, $Fe^{III}$, fluorescence quenching was observed. Studies with the perchlorate salts of diamines indicated attachment of the perchlorate salt at the porphyrin metal site and complexation of the ammonium species at the crown ether.

The combination of crown ether and porphyrin has also been carried out by Bogatskii et al.[180], who have prepared the *meso*-tetra(benzo-18-crown-6)porphyrin **223** by two routes (Scheme 66). Condensation of 4′-formylbenzo-18-crown-6 **220** with pyrrole in refluxing acetic acid or reaction of tetra(3,4-dihydroxyphenyl)porphyrin (**221**) with the dichloride **222** provided the tetra-crown ether porphyrin **223** in 4% and 10% yield respectively. This compound is reported to form complexes with transition and group I, II metals.

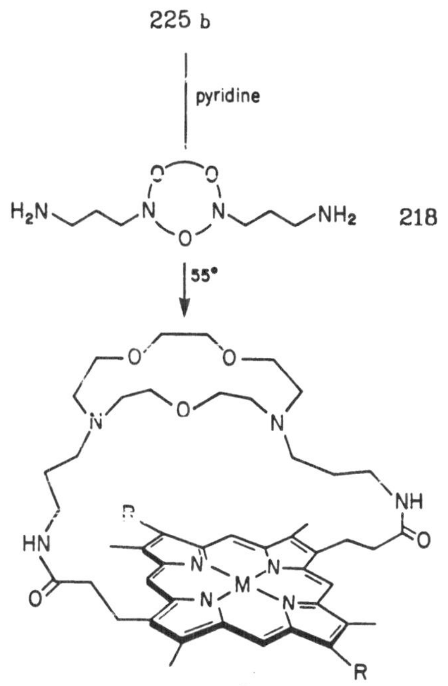

**Scheme 65** 219

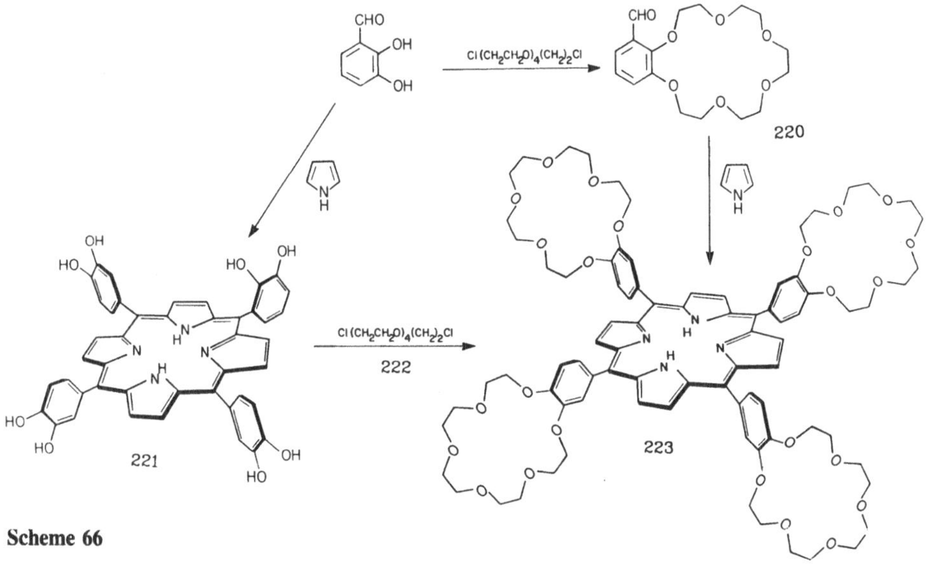

**Scheme 66**

The combination of crown ether and porphyrin has recently been extended by Hamilton et al.[181]. The macrotetracyclic cryptand **226** was prepared by condensation of the biphenyl-linked bis-crown ether **224** with the di-p-nitrophenyl ester porphyrin **225a, b** in pyridine at 55 °C under high dilution conditions (45–54% after chromatography). Reduction to the tetra-amine **227** was effected by treating the zinc complex with diborane

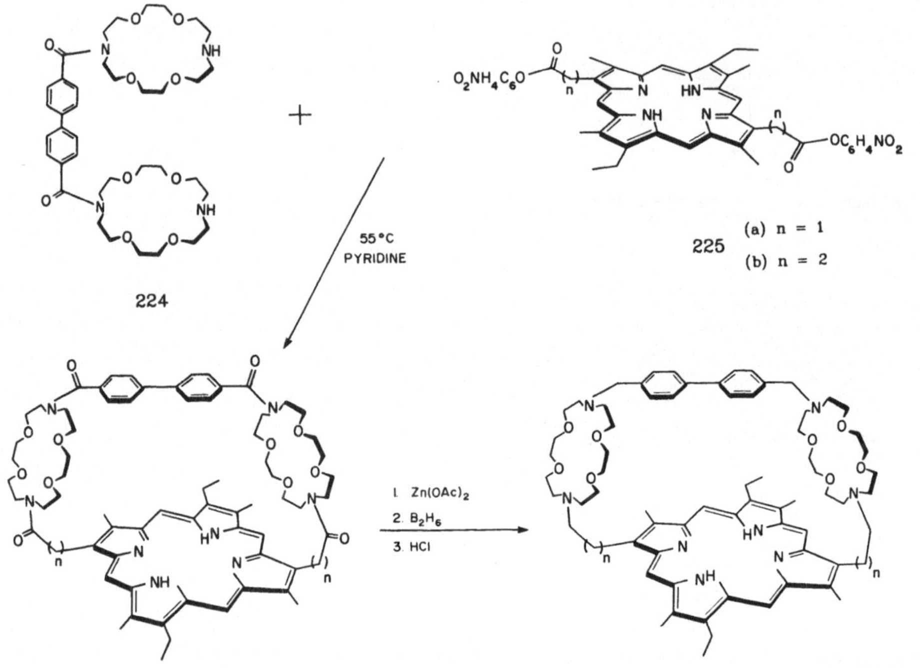

224

225    (a) n = 1
       (b) n = 2

55 °C
PYRIDINE

226          **Scheme 67**          227

followed by demetallation (Scheme 67). This multisite complexing species is capable of selective substrate binding: transition metal cations are bound by the porphyrin and alkyl diammonium salts, $^+H_3N(CH_2)_nNH_3^+$ (n = 8–10), within the central cavity. That binding within the cavity does indeed occur was evidenced by the large upfield shifts of the methylene protons due to the shielding effects of the porphyrin and biphenyl rings.

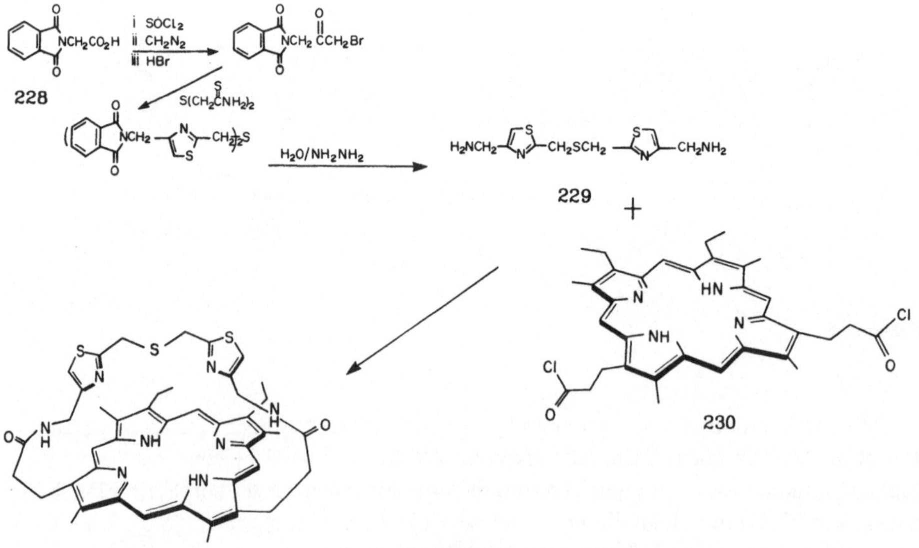

228

229
+

230

231          **Scheme 68**

A copper binding site has been covalently attached to a porphyrin in an attempt to mimic the EPR spectral characteristics of the iron-copper site of cytochrome oxidase[182]. The copper-binding strap was a bis-thiazole derivative **229**, obtainable from phthalylgly-cine **228** in high yield. Because of the need for a non-square-planar copper binding site, the strap was attached to one side of the porphyrin ring rather than to opposite corners. Condensation of the diaminothiazole sulfide **229** with mesoporphyrin XII bis-acid chloride (**230**) under high dilution gave the strapped porphyrin **231** in 72% yield after chromatography (Scheme 68). Metal ions could be differentially introduced into the porphyrin and into the strap to give a dinuclear metal complex containing a high spin iron(III) and a copper(II) ion. The extent of coupling between the two metal centers was investigated by EPR. Gunter et al.[183], used a somewhat different approach to prepare a similar Fe/Cu strapped porphyrin (Scheme 69). Condensation of the tetramethyldipyr-

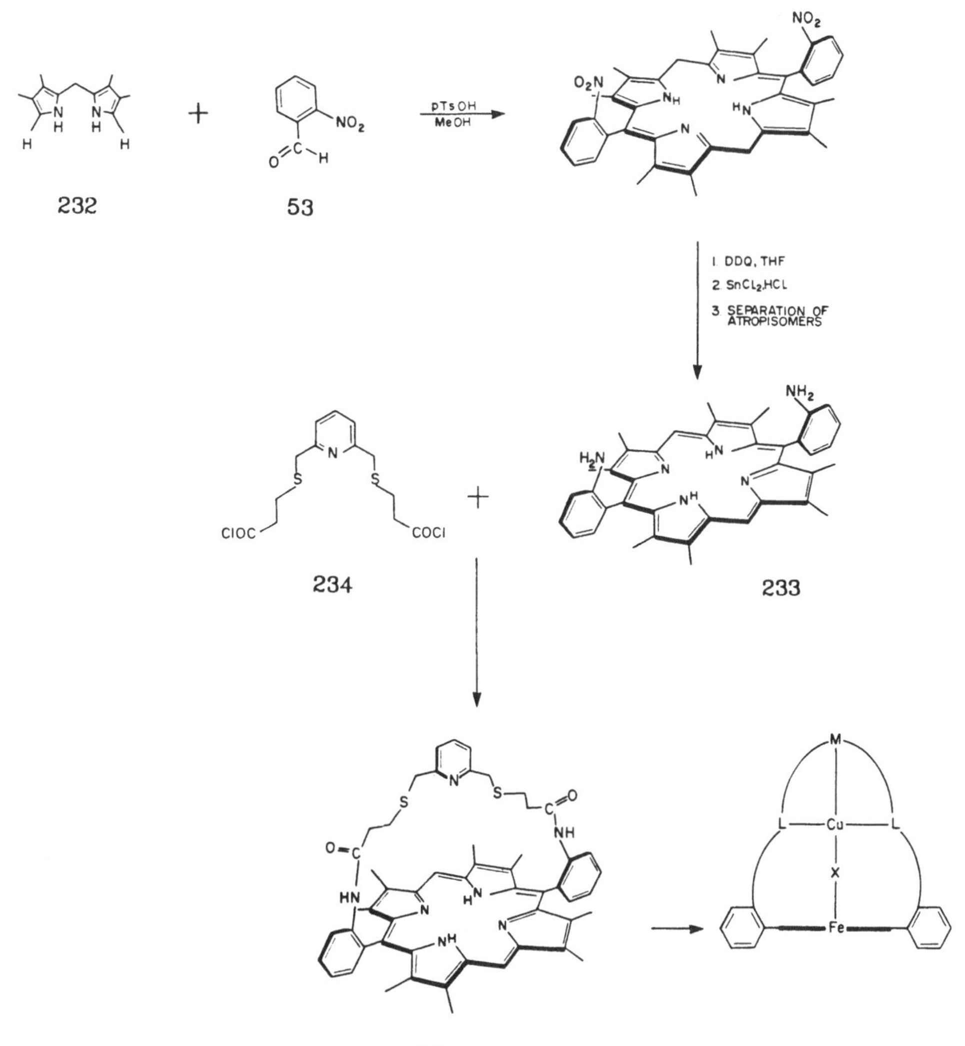

**Scheme 69**

romethane **232** with *o*-nitrobenzaldehyde **(53)**, followed by oxidation afforded the 5,15-*meso*-(*o*-nitrophenyl)porphyrin. After reduction to the amino derivative the atropisomers were separated by chromatography. The α,α-isomer **233** was finally condensed with 2,6-pyridylbis(4'thia-5'-pentanoyl chloride) **(234)**, and yielded the strapped porphyrin **235.** Insertion of iron and copper and the introduction of bridging ligands gave species of the type **236**, whose magnetic properties were investigated[184].

As a model for photosynthetic and electron-transfer systems, Sanders and Ganesh[185, 186] have prepared a quinone-capped porphyrin. Using the approach of the Battersby group, 1,4-dialkoxybenzene derivatives **237** were reacted with mesoporphyrin II bis-acid chloride **(180)** to give the strapped porphyrins **238** in 7 and 15% yield depending on strap length (Scheme 70). Deprotection with boron trichloride afforded the hydroquinones which were oxidized to the quinones **239a, b** with lead dioxide. [1]H-NMR studies on the magnesium complexes **239c, d** suggest that the quinone carbonyl binds to the metal ion and that therefore the quinone and porphyrin chromophores are perpendicular[187].

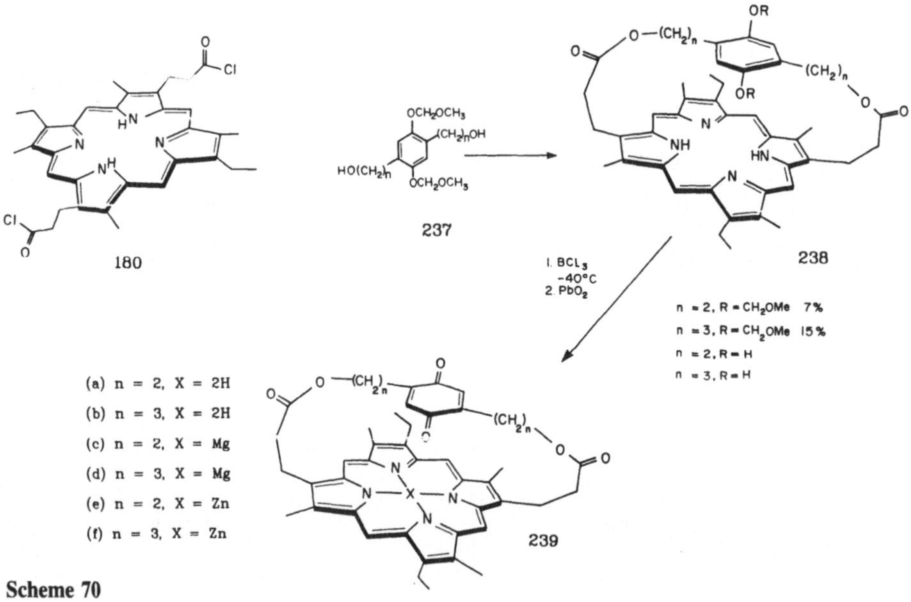

**Scheme 70**

Since zinc is exclusively five-coordinate in porphyrins and has a low affinity for oxygen ligands, intramolecular binding of the chromophores is lessened in the corresponding zinc complexes **239e, f** especially with added ligands present. Fluorescence quenching is observed in the free base **239a, b** and zinc **239e, f** complexes where close, parallel approach of the chromophores is possible, but is much reduced for the magnesium complexes **239c, d** where the chromophores are perpendicular[188]. The quenching is more efficient for the longer chain **239b, f** than for the shorter chain **239a, e**; because of intramolecular binding of the chromophores in **239c, d** the quenching efficiency is similar for both chain lengths.

High dilution coupling of the bipyridyl diols **240** with **180** gave the bipyridyl strapped porphyrins **241** (Scheme 71) in 40–50% yield[189–191]. Treatment with iodomethane fur-

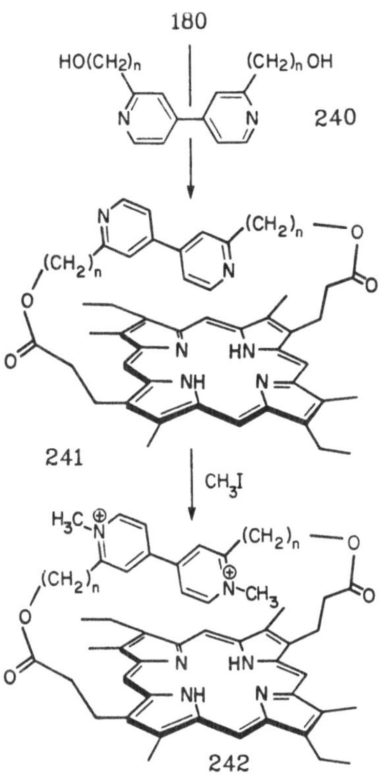

**Scheme 71**

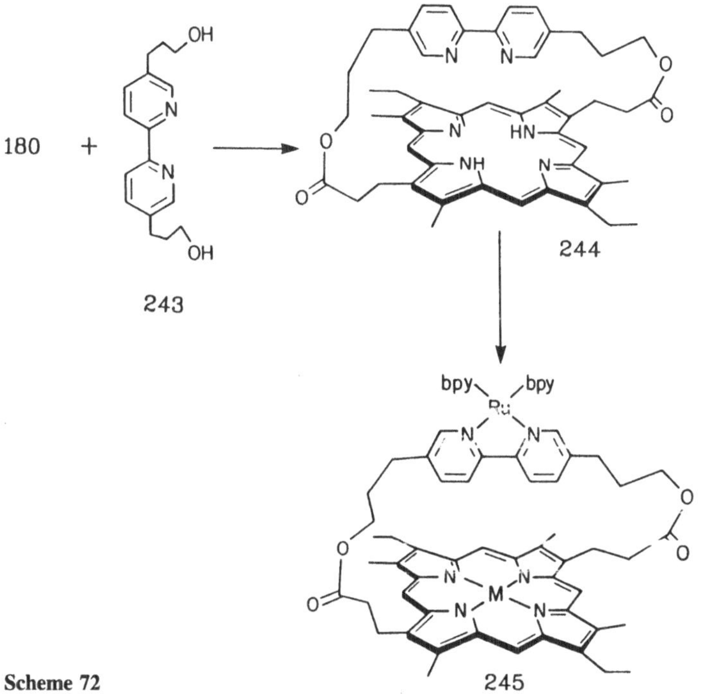

**Scheme 72**

nished the methylviologen strapped porphyrins **242.** The close proximity of an electron donor (porphyrin) and an electron acceptor (methylviologen dication) makes this a potential photosynthetic model. Indeed fluorescence emissions of **241** and **242** were reduced ~200 fold relative to mixtures of methylviologen and unstrapped porphyrin. It was assumed that efficient trapping of an excited electron occurred to give the two radical cations of the connected chromophores.

Hamilton et al.[192], have also prepared a binuclear complex by the condensation of mesoporphyrin II diacid chloride **180** with the bipyridyl diol **243.** Reaction of the strapped porphyrin **244** with $Ru(byp)_2Cl_2$ followed by metal insertion into the porphyrin resulted in **245** (Scheme 72). The close proximity of the two metal centers, estimated at 4 Å, is reflected in both the luminescence and electrochemical properties of the complex.

## 6.D. Doubly-Strapped Porphyrins

We have already seen that sterically encumbered porphyrins may still be susceptible to μ-oxo bridge formation if any four-coordinate species in solution binds oxygen on the open face. Steric encumbrance on both faces of the porphyrin, as in the "bis-pocket" porphyrin systems of Amundsen and Vaska[148] and Suslick[149], may prevent this bimolecular oxidation pathway.

Momenteau and his colleagues have used a combination of approaches to prepare TPP derivatives having two straps on each porphyrin ring[193]. In a strategy reminiscent of Baldwin's "capped" and "strapped" porphyrin syntheses[159, 160], the sodium salt of salicylaldehyde was reacted with a variety of dibromoalkyl and p-(dibromoalkyl)benzene derivatives **246a–e** to give chain-linked dialdehydes **247.** The TPP ring was then formed by condensing the dialdehydes with pyrrole in refluxing propionic acid (Scheme 73).

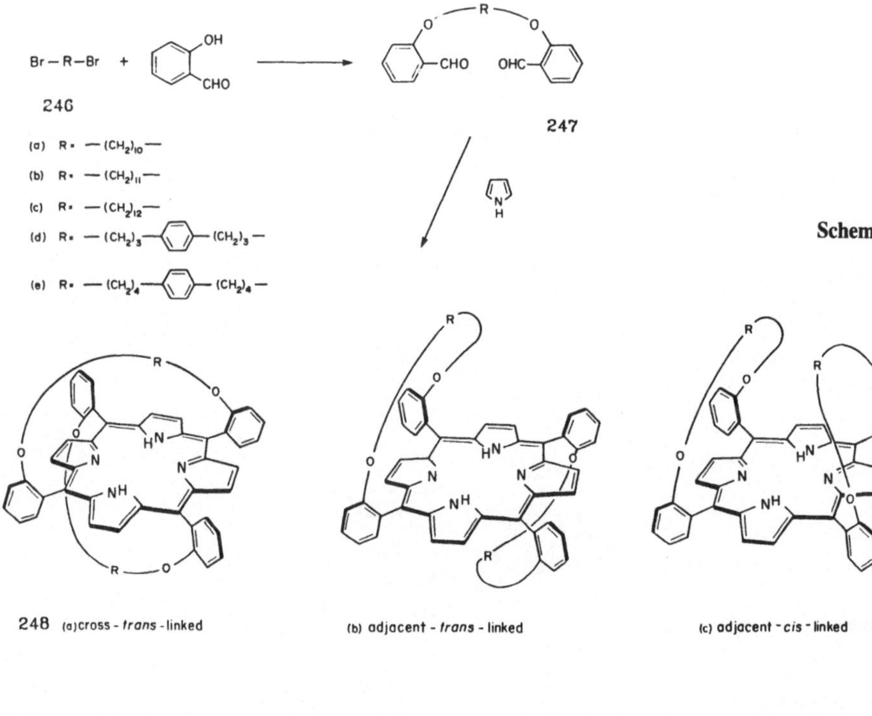
Br—R—Br +
OH
CHO
246
(a) R = —(CH2)10—
(b) R = —(CH2)11—
(c) R = —(CH2)12—
(d) R = —(CH2)3—C6H4—(CH2)3—
(e) R = —(CH2)4—C6H4—(CH2)4—
R
O
CHO
OHC
247
N
H
R
O
HN
N
NH
248 (a) cross-trans-linked
(b) adjacent-trans-linked
(c) adjacent-cis-linked

**Scheme 73**

After removal of polymeric materials, three isomers were obtained by chromatography in low overall yield. The unwanted adjacent cis-linked product **248c** was often the predominant isomer. To increase the yield of the more interesting cross trans-linked isomers, formation of the bridges was delayed until after the porphyrin-forming condensation (Scheme 74). Tetra(*o*-methoxyphenyl)porphyrin (**249**) was obtained from pyrrole and *o*-methoxybenzaldehyde (10% yield) and demethylated to provide tetra(*o*-hydroxyphenyl)porphyrin (**250**). Alkylation with the dibromo derivatives **246** under high dilution in DMF at 100 °C, followed by chromatography led to the isolation of the three porphyrin isomers **248**. In this case the major product of each reaction was the desired cross trans-linked isomer **248a**. The starting porphyrin **250** was used as a mixture of the four possible atropisomers since the conditions of the condensation would lead to equilibration.

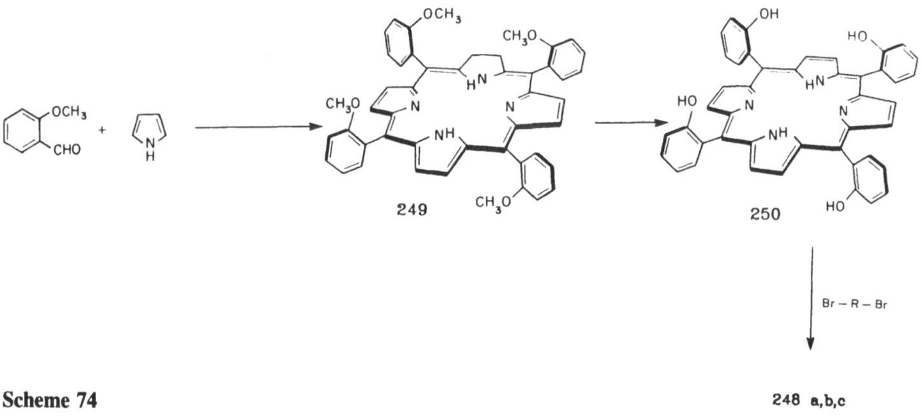

**Scheme 74**                                              248 a,b,c

The degree of steric encumbrance is illustrated by the rates of metallation and oxidation of the various isomers. The adjacent cis-linked isomer **248c** has one face unhindered and is easily metallated. In contrast the other isomers, where both faces are hindered, undergo iron insertion reluctantly. While not preventing ligation of nitrogenous bases or diatomic molecules, the chains do inhibit irreversible oxidation at room temperature. For the four-coordinate iron(II) cross trans-linked isomer **248a** the $t_{1/2}$ for oxidation to the hematin derivative $Fe^{III}(P)OH$ is 1.5–10.5 minutes compared to 7–54 seconds for oxidation of the less hindered isomers to the μ-oxo complex. Similarly, in toluene at 25 °C under $O_2$ (1 atm), $t_{1/2}$ for oxidation of the six-coordinate iron(II) complex is 11–25 min for the cross trans-linked isomer compared to 1.5–12 min for the other two isomers[194].

The basket-handle and picket-fence porphyrins show dramatic effects not only during metallation and binding of small molecules but in their redox and coordination chemistry. Detailed studies on the electrochemistry of the iron complex have been made[195] paralleling the earlier electrochemical studies on the free base, magnesium and zinc complexes of **192** and **201**[196].

A similar doubly-strapped porphyrin has been reported by Zhilina et al.[197]. The α,β,α,β atropisomer of *meso*-tetra(*o*-aminophenyl)porphyrin (**142**) was acylated with triglycolic dichloride at room temperature in the presence of pyridine, conditions that do not cause significant isomerization of the atropisomer. The doubly-strapped porphyrin **251** was obtained in 32% yield after chromatography (Scheme 75).

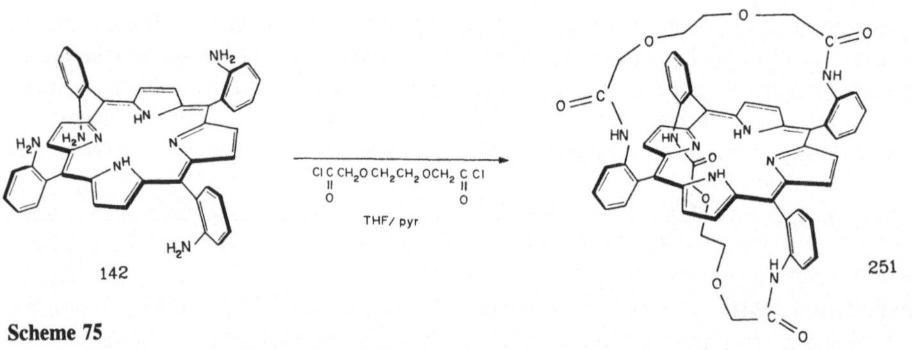

**Scheme 75**

In contrast to the sequential introduction of bridging straps, Weiser and Staab[198] used a one-step synthesis to prepare a porphyrin sandwiched between two parallel quinone units. Thus, condensation of the bis-aldehyde **252** with pyrrole in refluxing propionic acid yielded a mixture of oligomers and polymers, and three doubly-strapped porphyrins. The desired cross trans-linked porphyrin, obtained in ~0.1%, was easily identified by its characteristic symmetric [1]H-NMR spectrum. Demethylation and oxidation then yielded the desired bis-quinone porphyrin **253** (Scheme 76).

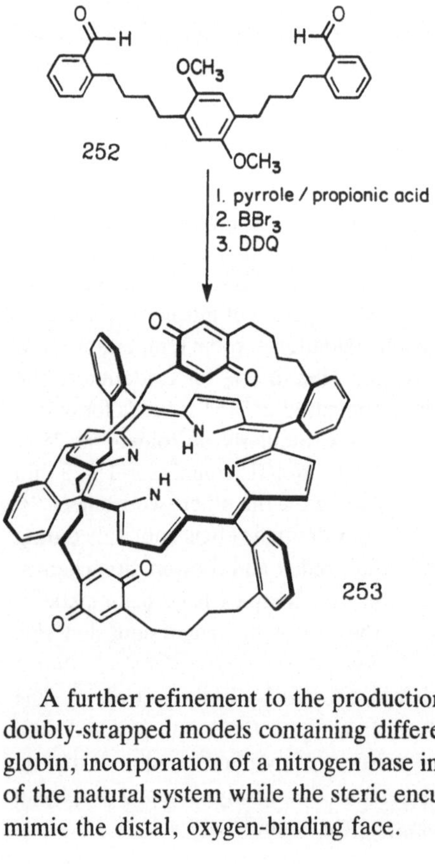

253          **Scheme 76**

A further refinement to the production of heme protein models was the synthesis of doubly-strapped models containing different straps. As models for hemoglobin or myoglobin, incorporation of a nitrogen base into one strap would simulate the proximal face of the natural system while the steric encumbrance provided by the second strap would mimic the distal, oxygen-binding face.

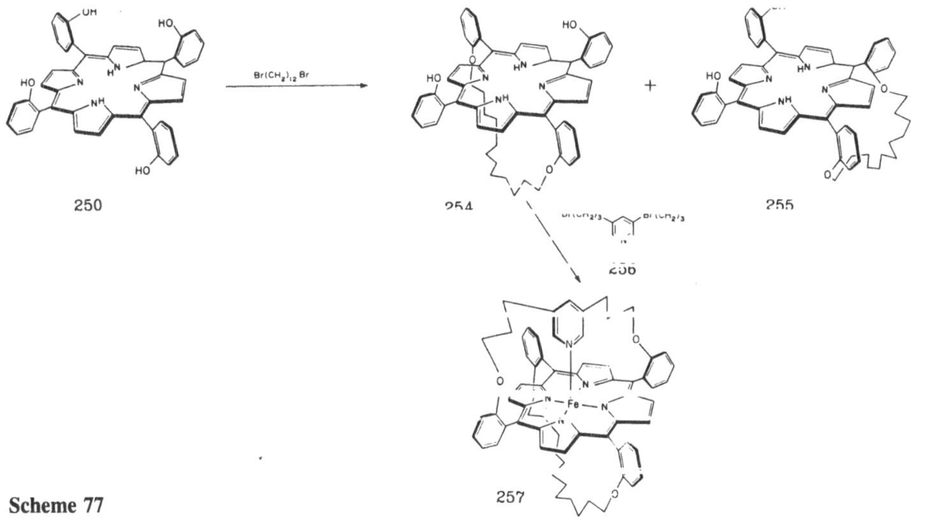

**Scheme 77**

Momenteau's route to doubly-strapped porphyrins was easily adapted to produce compounds in which an axial base was incorporated into one of the straps[199]. Condensation of tetra(o-hydroxyphenyl)porphyrin (250) (mixture of four isomers) with one equivalent of 1,12-dibromododecane gave a mixture of two singly-linked porphyrins, depending on whether adjacent (255), or opposite (254) meso-phenyl groups were linked. This mixture was reacted with 3,5-bis(3-bromopropyl)pyridine 256 and the desired cross trans-linked isomer 257 isolated by preparative tlc (5% overall yield) (Scheme 77). A similar porphyrin 259 was prepared from $\alpha,\beta,\alpha,\beta$-tetra(o-aminophenyl)porphyrin (142); in this case the straps were tied to the porphyrin skeleton by amide linkages (Scheme 78). Following iron insertion and reduction, visible absorption and $^1$H-NMR spectra of both compounds were consistent with a five-coordinate high spin (S = 2) iron(II) complex.

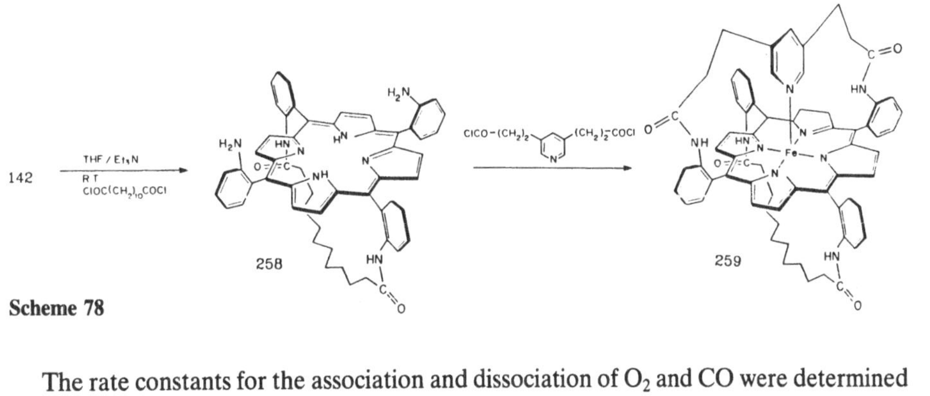

**Scheme 78**

The rate constants for the association and dissociation of $O_2$ and CO were determined by laser flash photolysis. The $O_2$ affinity of the "amide" linked system was higher than that of the "ether" linked compound ($P_{1/2}^{O_2}$ 18.6 vs 2 torr) as a result of a difference of a factor of ca. 10 in the $O_2$ dissociation rates ($10^{-4}$ $k_{off}^{O_2}$ 4 vs 0.5 s$^{-1}$). This increase in stability of the "amide" oxygenated species was attributed to the presence of the N–H group and the possibility of hydrogen-bonding with the terminal oxygen atom. The low

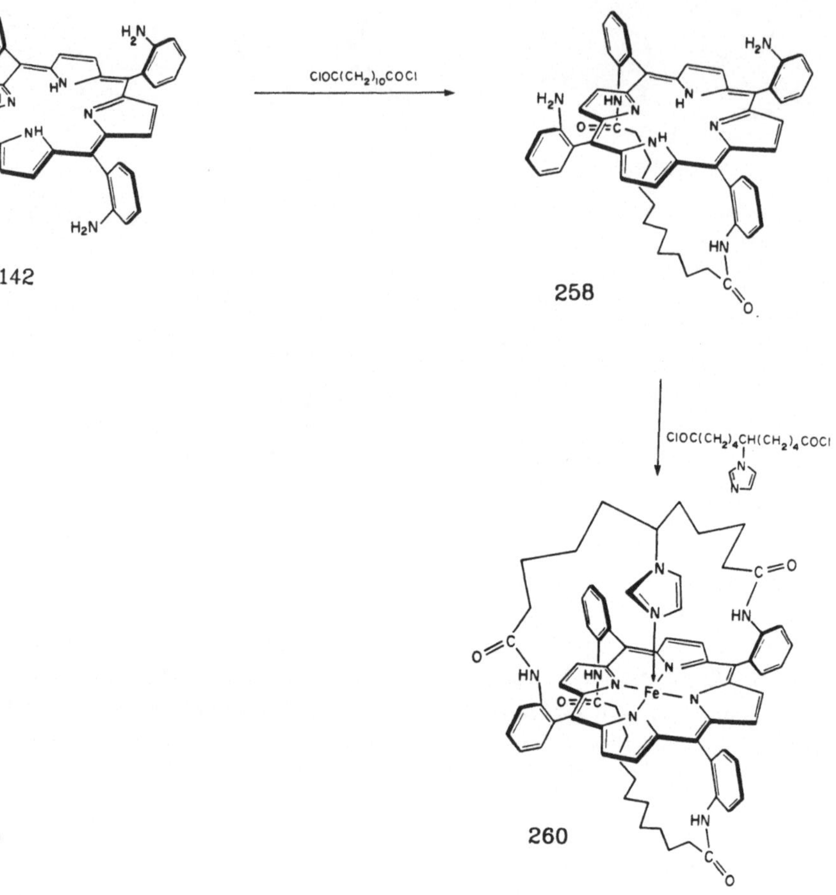

**Scheme 79**

temperature $(-27\,°C)$ $^1$H-NMR spectrum supported this hypothesis, the observed ine-
quivalence of the pyrrole protons as well as the shifts of the amide protons suggesting a
preferred orientation of the oxygen molecule towards the amide N–H groups[200].

To better model the hemoglobin and myoglobin active sites a doubly strapped heme
**260** was prepared incorporating a pendant imidazole (Scheme 79)[201]. **260** was capable of
binding oxygen to give a relatively stable oxygenated species (lifetime was about one day
in dry toluene under 1 atm $O_2$). The kinetics of $O_2$ and CO binding have been determined and initial comparisons with the comparable "pendant pyridine" porphyrins show:
(i)   $O_2$ and CO combination rates are practically constant in the three pendant base
      porphyrins, and
(ii)  a reduction in $k_{off}^{O_2}$ in the imidazole porphyrin due to a combination of hydrogen
      bonding with the amide N–H and the greater basicity of imidazole over pyridine.
Comparison of the pendant imidazole model with myoglobin or isolated hemoglobin
chains shows that the model reacts 10 times faster with $O_2$ and that the dissociation rate is
approximately 100 times faster than in the natural systems.

With the availability of the differentially protected coproporphyrin I **261**, Battersby
and Hamilton adapted their syntheses to the production of doubly-strapped porphyrins

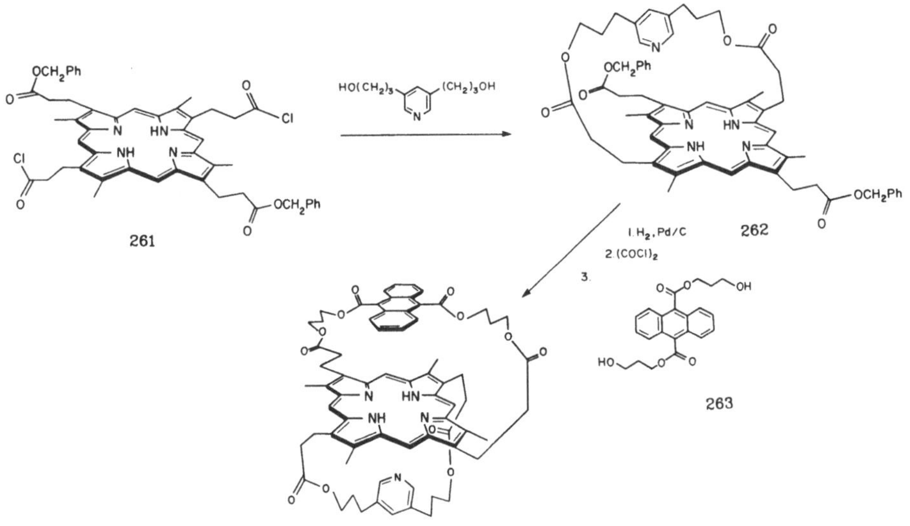

**Scheme 80**                                                264

(Scheme 80)[202]. Reaction of the bis-acid chloride **261** with 3,5-bis(3-hydroxypropyl)py-ridine yielded the pyridine-strapped porphyrin **262** (33%). Hydrogenolysis of the benzyl esters and acid chloride formation was followed by condensation with the anthracene diol **263** to give the doubly-bridged porphyrin **264** (27%). Iron insertion was found to be difficult so the metal was inserted after the introduction of the pyridine strap. Reduction with aqueous dithionite furnished the iron(II) species which on the basis of the visible absorption spectrum was judged to be high spin (S = 2) five-coordinate. Exposure to oxygen gave an oxygenated compound with a $t_{1/2}$ of approximately 15 min at room temperature in $CH_2Cl_2$. In DMF a more stable oxygenated species was formed ($t_{1/2}$ approximately 2 h at 20 °C). The $O_2$ could be displaced by passing CO into the solution, and regeneration of the oxygenated species was accomplished by passage of $O_2$. Such $O_2$–CO cycles could be repeated six times without significant irreversible oxidation. This was in contrast to the resistance of unhindered CO-porphyrin complexes to displacement of CO by $O_2$.

The further refinement of incorporating an imidazole ligand has been recently reported[203]. As before, the differentially protected coproporphyrin **261** was reacted with the anthracene diol **263** to give the porphyrin **265** (57%). After removal of the benzyl esters and treatment with oxalyl chloride, the bis-acid chloride was reacted with the N-substituted imidazole diol **266**. The doubly-bridged porphyrin **267** was obtained in 22% yield (Scheme 81). The iron(II) complex was capable of reversible oxygen binding, four cycles of oxygenation-deoxygenation (by reducing the pressure) being possible before significant irreversible oxidation occurred. The $t_{1/2}$ for the oxygenated species was ca. 24 h at room temperature in DMF solution.

Recognizing that the pendant-imidazole strap was still somewhat floppy, more rigid straps containing a 1,5-disubstituted imidazole **268** were prepared as before (Scheme 81)[203]. Coupling 1,5-bis(4-hydroxybutyl) **268a** or 1,5-bis(3-hydroxypropyl)-imidazole **268b** with the bis-acid chloride of the anthracene-strapped porphyrin gave the doubly-bridged systems **269a, b** in 23% and 6% yield. Distortion of the porphyrin ring

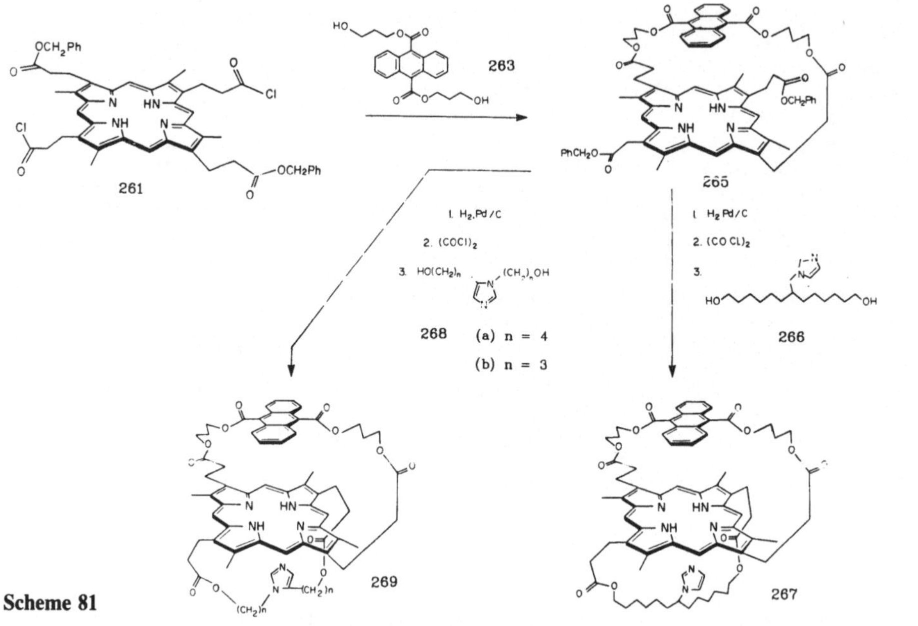

**Scheme 81**

from planarity to accommodate the shorter strap was believed responsible for the low yield of **269b** and also for the lesser stability of the oxygenated iron(II) species. For **269a**, the iron(II) complex could reversibly bind oxygen in DMF solution at ambient temperature. Ten oxygenation-deoxygenation cycles could be performed before irreversible oxidation was significant, and only 20% irreversible oxidation occurred after 2 days in solution.

*Acknowledgements*. This work was supported by the United States National Institute of Health (AM 17989) and the Canadian Natural Sciences and Engineering Research Council.

# 7. List of Abbreviations

| | | | |
|---|---|---|---|
| acac | acetylacetonate | DMF | N,N-Dimethylformamide |
| AQ | Anthraquinone | DMSO | Dimethylsulfoxide |
| bipy | bipyridyl | DNA | Deoxyribonucleic acid |
| 1-n-BuIm | 1-n-Butylimidazole | ee | enantiomeric excess |
| 4-t-BuIm | 4-t-Butylimidazole | Et$_3$N | Triethylamine |
| t-BuNC | t-Butylisocyanide | EXAFS | Extended X-Ray Absorption |
| BQ | Benzoquinone | | Fine Structure |
| CO | Carbon monoxide | H$_2$(OEP) | Octaethylporphyrin |
| Cys | Cysteine | H$_2$(P) | general porphyrin |
| cyt | cytochrome | H$_2$(TAP) | Tetra(*p*-methoxyphenyl)por- |
| DCIM | 1,5-Dicyclohexylimidazole | | phyrin |
| DDQ | 2,3-Dichloro-5,6-dicyano-1,4- | H$_2$(T(OH)PP) | Tetra(*o*-hydroxyphenyl)por- |
| | benzoquinone | | phyrin |
| DMA | N,N-Dimethylacetamide | H$_2$(TPP) | Tetraphenylporphyrin |

| H$_2$(TTP) | Tetra(p-tolyl)porphyrin | phth | phthaloyl |
|---|---|---|---|
| His | Histidine | piv | pivalamido |
| Hm | Heme | THF | Tetrahydrofuran |
| Im | Imidazole | THT | Tetrahydrothiophene |
| MCD | Magnetic Circular Dichroism | TMIC | Tosylmethylisocyanide |
| 1-MeIm | 1-Methylimidazole | Tos | Tosylate |
| 1,2-Me$_2$Im | 1,2-Dimethylimidazole | Tr | Trityl |
| MeOH | Methanol | 1-tritylIm | 1-Tritylimidazole |
| Met | Methionine | Tyr | Tyrosine |
| NQ | Naphthoquinone | | |

# 8. References

1. Hemoglobin and Oxygen Binding (Ho, C., Ed.), Elsevier Biomedical, New York 1982
2. Mathews, F. S.: Prog. Biophys. Mol. Biol. *45*, 1 (1985)
3. Hatefi, Y.: Ann. Rev. Biochem. *54*, 1015 (1985)
4. Frew, J. E., Jones, P.: Adv. Inorg. Bioinorg. Mech. *3*, 175 (1984)
5. Murray, R. T., Fisher, M. T., Debrunner, P. G., Sligar, S. G.: Top. Mol. Struct. Biol. *6* (Metalloproteins, Pt. 1), 157 (1985)
6. The Porphyrins (Dolphin, D., Ed.), Academic Press, New York, Vol. VII, Part B, 1978
8. NATO Adv. Study Inst. Ser., Ser. C., *89* (The Biological Chemistry of Iron) (1981)
8. Kao, O. H. W., Wang, J. H.: Biochemistry *4*, 342 (1965)
9. Hammond, G. S., Wu, C. H. S.: Adv. Chem. Ser. *77*, 186 (1968)
10. Cohen, I. A., Caughey, W. S.: Biochemistry *7*, 636 (1968)
11. Latos-Grazynski, L., Cheng, R.-J., La Mar, G. N., Balch, A. L.: J. Am. Chem. Soc. *104*, 5992 (1982)
12. Holquist, D. E., Saunes, L. J., Juckett, D. A.: Curr. Top. Cell. Regul. *24*, 287 (1984)
13. Brunori, M., Falcioni, G., Fioretti, E., Giardina, B., Rotilio, G.: Eur. J. Biochem. *53*, 99 (1975)
14. Fee, J. A.: Metal Ion Activation of Dioxygen (Ed. Spiro, T. G.), Wiley Interscience, 209–237 (1980)
15. Phillips, S. E. V., Schoenborn, B. P.: Nature (London) *292*, 81 (1981)
16. Shaanan, B.: Nature (London) *296*, 683 (1982)
17. Perutz, M. F.: Scientific American *239*(6), 68 (1978)
18. Brault, D., Rougee, M.: Biochemistry *13*, 4591 (1974)
19. Brault, D., Rougee, M.: ibid. *13*, 4598 (1974)
20. Scheidt, W. R., Reed, C. A.: Chem. Rev. *81*, 543 (1981)
21. Stynes, D. V., Stynes, H. C., James, B. R., Ibers, J. A.: J. Am. Chem. Soc. *95*, 1796 (1973)
22. Anderson, D. L., Weschler, C. J., Basolo, F.: ibid. *96*, 5599 (1974)
23. Almog, J., Baldwin, J. E., Dyer, R. L., Huff, J., Wilkerson, C. J.: ibid. *96*, 5600 (1974)
24. Brinigar, W. S., Chang, C. K.: ibid. *96*, 5595 (1974)
25. Wagner, G. C., Kassner, R. J.: ibid. *96*, 5593 (1974)
26. Traylor, T. G., Chang, C. K., Geibel, J., Berzinis, A., Mincey, T., Cannon, G.: ibid. *101*, 6716 (1979)
27. Gibson, Q. H.: Prog. Biophys., Biophys. Chem. *9*, 1 (1959)
28. Basolo, F., Hoffman, B. M., Ibers, J. A.: Acc. Chem. Res. *8*, 384 (1975)
29. James, B. R., Addison, A. W., Cairns, M., Dolphin, D., Farrell, N. P., Paulson, D. R., Walker, S.: Fundamental Research in Homogeneous Catalysis (Tsutsui, M., Ed.), Plenum Press: New York, Vol. 3, p. 751 (1979)
30. Wang, J. H.: Acc. Chem. Res. *3*, 90 (1970)
31. Leal, O., Anderson, D. L., Bowman, R. G., Basolo, F., Burwell, R. L.: J. Am. Chem. Soc. *97*, 5125 (1975)
32. Collman, J. P.: Acc. Chem. Res. *10*, 265 (1977)
33. Jones, R. D., Summerville, D. A., Basolo, F.: Chem. Rev. *79*, 139 (1979)
34. Smith, P. D., James, B. R., Dolphin, D. H.: Coord. Chem. Rev. *39*, 31 (1981)

35. Traylor, T. G.: Acc. Chem. Res. *14*, 102 (1981)
36. Bogatskii, A. V., Zhilina, Z. I.: Russ. Chem. Rev. *51*, 592 (1982)
37. Collman, J. P., Halpert, T. R., Suslick, K. S.: Metal Ion Activation of Dioxygen (Spiro, T. G., Ed.), Wiley: New York, p. 1 (1980)
38. Lautsch, V. W., Wiemer, B., Zschenderlein, P., Kraege, H. J., Bandel, W., Gunther, D., Schulz, G., Gnichtel, H.: Kolloid. Z. *161*, 36 (1958)
39. Losse, G., Müller, G.: Hoppe-Seyler's Z. Physiol. Chem. *327*, 205 (1962)
40. Van der Heijden, A., Peer, H. G., Van den Oord, A. H. A.: J. Chem. Soc., Chem. Commun., 369 (1971)
41. Warme, P. K., Hager, L. P.: Biochemistry *9*, 1599 (1970)
42. Momenteau, M., Rougee, M., Loock, B.: Eur. J. Biochem. *71*, 63 (1976)
43. Castro, C. E.: Bioinorg. Chem. *4*, 45 (1974)
44. Molokoedov, A. S., Fillippovich, E. I., Mazakova, N. A., Evstigneeva, R. P.: Zhur. Obshch. Khim. Ed. Engl. *47*, 1070 (1977)
45. Kazakova, N. A., Radyukhin, V. A., Luzgiva, V. N., Filippovich, E. I., Kamyshan, N. V., Kudryavtseva, E. V., Evstigneeva, R. P.: ibid. *52*, 1896 (1982)
46. Momenteau, M., Loock, B.: Biochim. Biophys. Acta *343*, 535 (1974)
47. Selve, C., Niedercorn, F., Nacro, M., Castro, B., Gabriel, M.: Tetrahedron *37*, 1893 (1981)
48. Selve, C., Niedercorn, F., Nacro, M., Castro, B., Gabriel, M.: ibid. *37*, 1903 (1981)
49. Gabriel, M., Grange, J., Niedercorn, F., Selve, C., Castro, B.: ibid. *37*, 1913 (1981)
50. Goulon, J., Goulon, C., Niedercorn, F., Selve, C., Castro, B.: ibid. *37*, 2707 (1981)
51. Chang, C. K., Traylor, T. G.: Proc. Natl. Acad. Sci. U.S.A. *70*, 2647 (1973)
52. Chang, C. K., Traylor, T. G.: J. Am. Chem. Soc. *95*, 5810 (1973)
53. Chang, C. K., Traylor, T. G.: ibid. *95*, 8475 (1973)
54. Brinigar, W. S., Chang, C. K., Geibel, J., Traylor, T. G.: ibid. *96*, 5597 (1974)
55. Geibel, J., Chang, C. K., Traylor, T. G.: ibid. *97*, 5924 (1975)
56. Dolphin, D., Hiom, J., Paine III, J. B.: Heterocycles *16*, 417 (1981)
57. Traylor, T. G., Campbell, D., Sharma, V., Geibel, J.: J. Am. Chem. Soc. *101*, 5376 (1979)
58. Traylor, T. G., White, D. K., Campbell, D. H., Berzinis, A. P.: ibid. *103*, 4932 (1981)
59. Traylor, T. G., Mitchell, M. J., Cicone, G. P., Nelson, S.: ibid. *104*, 4986 (1982)
60. Traylor, T. G., Tatsuno, T., Powell, D. W., Cannon, J. B.: J. Chem. Soc. Chem. Commun., 732 (1977)
61. Traylor, T. G., Berzinis, A. P.: J. Am. Chem. Soc. *102*, 2844 (1980)
62. Tabushi, I., Sasaki, T.: Tetrahedron Lett. *23*, 1913 (1982)
63. Tabushi, I., Kugimiyua, S., Kinnaird, M. G., Sasaki, T.: J. Am. Chem. Soc. *107*, 4192 (1985)
64. Tabushi, I., Kugimiya, S., Sasaki, T.: ibid. *107*, 5159 (1985)
65. Denniss, J. S., Sanders, J. K. M.: Tetrahedron Lett., 295 (1978)
66. Boxer, S. G., Wright, K. A.: J. Am. Chem. Soc. *101*, 6791 (1979)
67. Momenteau, M., Loock, B., Bisagni, E., Rougee, M.: Can. J. Chem. *57*, 1804 (1979)
68. Callot, H. J.: Tetrahedron, 899 (1973)
69. Lavalette, D., Tetreau, C., Momenteau, M.: J. Am. Chem. Soc. *101*, 5395 (1979)
70. Lavalette, D., Momenteau, M.: J. Chem. Soc., Perkin Trans. 2, 385 (1982)
71. More, K. M., Eaton, S. S., Eaton, G. R.: Inorg. Chem. *20*, 2641 (1981)
72. Damoder, R., More, K. M., Eaton, G. R., Eaton, S. S.: J. Am. Chem. Soc. *105*, 2147 (1983)
73. Damoder, R., More, K. M., Eaton, G. R., Eaton, S. S.: Inorg. Chem. *22*, 2836 (1983)
74. Collman, J. P., Brauman, J. I., Doxsee, K. M., Halbert, T. R., Bunnenberg, E., Linder, R. E., LaMar, G. N., Del Gaudio, J., Lang, G., Spartalian, K.: J. Am. Chem. Soc. *102*, 4182 (1980)
75. Mashiko, T., Reed, C. A., Haller, K. J., Kastner, M. E., Scheidt, W. R.: ibid. *103*, 5758 (1981)
76. Walker, F. A.: ibid. *102*, 3254 (1980)
77. Walker, F. A., Benson, M.: ibid. *102*, 5530 (1980)
78. Santon, R. J., Wilson, L. J.: J. Chem. Soc., Chem. Commun., 359 (1984)
79. Molinaro, F. S., Little, R. G., Ibers, J. A.: J. Am. Chem. Soc. *99*, 5628 (1977)
80. Goff, H.: ibid. *102*, 3252 (1980)
81. Traylor, T. G., Mincey, T. C., Berzinis, A. P.: ibid. *103*, 7084 (1981)
82. Collman, J. P., Groh, S. E.: ibid. *104*, 1391 (1982)
83. Buckingham, D. A., Rauchfuss, T. B.: J. Chem. Soc., Chem. Commun., 705 (1978)

84. Smith K. M., Bisset, G. M. F.: J. Chem. Soc., Perkin Trans. 1, 2625 (1981)
85. Kong, J. L. Y., Loach, P. A.: J. Heterocycl. Chem. *17*, 737 (1980)
86. McIntosh, A. R., Siemiarczuk, A., Bolton, J. R., Stillman, M. J., Ho, T.-F., Weedon, A. C.: J. Am. Chem. Soc. *105*, 7215 (1983)
87. McIntosh, A. R., Siemiarczuk, A., Bolton, J. R., Stillman, M. J., Ho, T.-F., Weedon, A. C.: ibid. *105*, 7224 (1983)
88. Wang, C.-B., Tien, J. T., Lopez, J. R., Lui, Q.-Y., Joshi, N. B., Hu, Q.-Y.: Photobiochem. Photobiophys. *4*, 177 (1982)
89. Joshi, N. B., Lopez, J. R., Tien, H. T., Wang, C.-B., Lui, Q.-Y.: J. Photochem. *20*, 139 (1982)
90. Tabushi, I., Koga, N., Yanagita, M.: Tetrahedron Lett., 257 (1979)
91. Wasielewski, M. R., Niemczyk, M. P.: J. Am. Chem. Soc. *106*, 5043 (1984)
92. Wasielewski, M. R., Niemczyk, M. P., Svec, W. A., Pewitt, E. B.: ibid. *107*, 1080 (1985)
93. Joran, A. D., Leland, B. A., Geller, G. G., Hopfield, J. J., Dervan, P. B.: ibid. *106*, 6090 (1984)
94. Nishitani, S., Kurata, N., Sakata, Y., Misumi, S., Migita, M., Mataga, N.: Tetrahedron Lett. *22*, 2099 (1981)
95. Mataga, N., Karen, A., Okada, T., Nishitani, S., Kurata, N., Sakata, Y., Misumi, S.: J. Phys. Chem. *22*, 5138 (1984)
96. Nishitani, S., Kurata, N., Sakata, Y., Misumi, S., Karen, A., Okada, T., Mataga, N.: J. Am. Chem. Soc. *105*, 7771 (1983)
97. Moore, T. A., Gust, D., Mathis, P., Mialocq, J.-C., Chachaty, C., Bensasson, R. V., Land, E. J., Doizi, D., Liddel, P. A., Lehman, W. R., Nemeth, G. A., Moore, A. L.: Nature (London) *307*, 630 (1984)
98. Maiya, G. B., Krishnan, V.: Inorg. Chim. Acta *77*, L13 (1983)
99. Kobayashi, N., Akiba, V., Takatori, K., Veno, A., Osa, T.: Heterocycles *19*, 2011 (1982)
100. Eshima, K., Matsushita, Y., Sekine, M., Nishide, H., Tsuchida, E.: Nippon Kayaku Kaishi, 214 (1983)
101. Lown, J. W., Joshua, A. V.: J. Chem. Soc., Chem. Commun., 1298 (1983)
102. Hashimoto, Y., Lee, C.-S., Shudo, K., Okamoto, T.: Tetrahedron Lett. *24*, 1523 (1983)
103. Collman, J. P., Gagne, R. R., Halbert, T. R., Marchon, J.-C., Reed, C. A.: J. Am. Chem. Soc. *95*, 7868 (1973)
104. Collman, J. P., Gagne, R. R., Reed, C. A., Halbert, T. R., Lang, G., Robinson, W. T.: ibid. *96*, 1427 (1975)
105. Anzui, K., Hatano, K.: Chem. Pharm. Bull. *32*, 1273 (1984)
106. Freilag, R. A., Whitten, D. G.: J. Phys. Chem. *87*, 3918 (1983)
107. Freilag, R. A., Whitten, D. G.: J. Am. Chem. Soc. *103*, 1226 (1981)
108. Collman, J. P., Gagne, R. R., Reed, C. A., Robinson, W. T., Rodley, G. A.: Proc. Natl. Acad. Sci. U.S.A. *71*, 1326 (1974)
109. Jameson, G. B., Molinaro, F. S., Ibers, J. A., Collman, J. P., Brauman, J. I., Rose, E., Suslick, K. S.: J. Am. Chem. Soc. *102*, 3224 (1980)
110. Collman, J. P., Gagne, R. R., Gray, H. B., Hare, J. W.: ibid. *96*, 6522 (1974)
111. Collman, J. P., Brauman, J. I., Halbert, T. R., Suslick, K. S.: Proc. Natl. Acad. Sci. U.S.A. *73*, 3333 (1976)
112. Gmai, H., Nakata, K., Nakatsubo, A., Kakagawa, S., Vermori, Y., Kyuno, E.: Synth. React. Inorg. Met.-Org. Chem. *13*, 761 (1983)
113. Baldwin, J. E., Perlmutter, P.: Top. Curr. Chem. *121* (Host Guest Complex Chem. 3), 181 (1984)
114. David, S., Dolphin, D., James, B. R.: in Frontiers of Bioinorganic Chemistry (ed. Xavier, A. V.), VCH Verlagsgesellschaft, Weinheim, pp. 163–182 (1985)
115. Collman, J. P., Gagne, R. R., Reed, C. A.: J. Am. Chem. Soc. *96*, 2629 (1974)
116. Collman, J. P., Brauman, J. I., Suslick, K. S.: ibid. *97*, 7185 (1975)
117. Bogatskii, A. V., Zhilina, Z. I., Danilina, N. I.: Dokl. Akad. Nauk. S.S.S.R. (Engl. Ed.) *252*, 127 (1980)
118. Bogatskii, A. V., Zhilina, Z. I., Krasnoshchekaya, S. P., Zakharova, R. M.: Zh. Org. Khim. Engl. Ed. *18*, 2035 (1982)
119. Valiotti, A., Adeyemo, A., Williams, F. A., Ricks, L., North, J., Hambright, P.: J. Inorg. Nucl. Chem. *43*, 2653 (1981)

120. Buckingham, D. H., Clark, C. R., Webley, W. S.: J. Chem. Soc., Chem. Commun., 192 (1981)
121. Buckingham, D. A., Gunter, M. G., Mander, L. N.: J. Am. Chem. Soc. 100, 2899 (1978)
122. Gunter, M. G., Mander, L. N., McLaughlin, G. M., Murray, K. S., Berry, K. J., Clark, P. E., Buckingham, D. A.: ibid. 102, 1470 (1980)
123. Elliott, C. M., Krebs, R. R.: ibid. 104, 4301 (1982)
124. Groves, J. T., Myers, R. S.: ibid. 105, 5791 (1983)
125. Tabushi, I., Kodera, M., Yokoyama, M.: ibid. 107, 4466 (1985)
126. Lecas-Nawrocka, A., Levisalles, J., Mariacher, C., Renko, Z., Rose, E.: Can. J. Chem. 62, 2054 (1984)
127. Collman, J. P., Brauman, J. I., Doxsee, K. M., Sessler, J. L., Morris, R. M., Gibson, Q. H.: Inorg. Chem. 22, 1427 (1983)
128. Collman, J. P., Brauman, J. I., Doxsee, K. M., Halbert, T. R.,Suslick, K. S.: Proc. Natl. Acad. Sci. U.S.A. 75, 564 (1978)
129. Collman, J. P., Brauman, J. I., Doxsee, K. M.: ibid. 76, 6035 (1979)
130. Almog, J., Baldwin, J. E., Dyer, R. L., Peters, M.: J. Am. Chem. Soc. 97, 226 (1975)
131. Almog, J., Baldwin, J. E., Crossley, MK. J., De Bernardis, J. F., Dyer, R. L., Huff, J. R., Peters, M. K.: Tetrahedron 37, 3589 (1981)
132. Ellis, Jr., P. E., Linard, J. E., Szymanski, T., Jones, R. D., Budge, J. R., Basolo, F.: J. Am. Chem. Soc. 102, 1889 (1980)
133. Budge, J. R., Ellis, Jr., P. E., Jones, R. D., Linard, J. E., Basolo, F., Baldwin, J. E., Dyer, R. L.: ibid. 101, 4760 (1979)
134. Ellis, Jr., P. E., Linard, J. E., Jones, R. D., Budge, J. R., Basolo, F.: ibid. 102, 1896 (1980)
135. Hashimoto, T., Dyer, R. L., Crossley, M. J., Baldwin, J. E., Basolo, F.: ibid. 104, 2101 (1982)
136. Baldwin, J. E., Crossley, M. G., De Bernardis, J.: Tetrahedron 38, 685 (1982)
137. Shimizu, M., Basolo, F., Vallejo, M. N., Baldwin, J. E.: Inorg. Chim. Acta 91, 247 (1984)
138. Shimizu, M., Basolo, F., Vallejo, N. M., Baldwin, J. E.: ibid. 91, 251 (1984)
139. Rose, E. J., Vankatasurbramanian, P. N., Swartz, J. C., Jones, R. D., Basolo, F., Hoffman, B. M.: Proc. Natl. Acad. Sci. U.S.A. 79, 5742 (1982)
140. Jameson, G. B., Ibers, J. A.: J. Am. Chem. Soc. 102, 2823 (1980)
141. Sabat, M., Ibers, J. A.: ibid. 104, 3715 (1982)
142. Clayden, N. J., Moore, G. R., Williams, R. J. P., Baldwin, J. E., Crossley, M. J.: J. Chem. Soc., Perkin Trans. 2, 1693 (1982)
143. Clayden, N. J., Moore, G. R., Williams, R. J. P., Baldwin, J. E., Crossley, M. J.: ibid. 2, 1863 (1983)
144. Baldwin, J. E., Cameron, J. H., Crossley, M. J., Dayley, E. J.: J. Chem. Soc., Dalton Trans., 1739 (1984)
145. Collman, J. P., Brauman, J. I., Collins, T. J., Iverson, B. L., Sessler, J. L.: J. Am. Chem. Soc. 103, 2450 (1981)
146. Collman, J. P., Brauman, J. I., Collins, T. J., Iverson, B. L., Lang, G., Pettman, R. B., Sessler, J. L., Walters, M. A.: ibid. 105, 3038 (1983)
147. Collman, J. P., Brauman, J. I., Iverson, B. L., Sessler, J. L., Morris, R. M., Gibson, Q. H.: ibid. 105, 3052 (1983)
148. Amundsen, A. R., Vaska, L.: Inorg. Chim. Acta. 14, L49 (1975)
149. Suslick, K. S., Fox, M. M.: J. Am. Chem. Soc. 105, 3507 (1983)
150. Lindsey, J. S., Mauzerall, D. C.: ibid. 104, 4498 (1982)
151. Lindsey, J. S., Mauzerall, D. C., Linschitz, H.: ibid. 105, 6528 (1983)
152. Ogoshi, H., Sugimoto, H., Yoshida, Z.: Tetrahedron Lett., 4481 (1976)
153. Ogoshi, H., Sugimoto, H., Yoshida, Z.: ibid. 1515 (1977)
154. Ogoshi, H., Sugimoto, H., Miyake, M., Yoshida, Z. I.: Tetrahedron 40, 579 (1984)
155. Battersby, A. R., Buckley, D. G., Hartley, S. G., Turnbull, M.: J. Chem. Soc., Chem. Commun., 879 (1976)
156. Chang, C. K., Kuo, M.-S.: J. Am. Chem. Soc. 101, 3413 (1979)
157. Ward, B., Wang, C.-B., Chang, C. K.: ibid. 103, 5236 (1981)
158. Yu, N.-T., Kerr, E. A., Ward, B., Chang, C. K.: Biochemistry 22, 4534 (1983)
159. Baldwin, J. E., Klose, T., Peters, M.: J. Chem. Soc., Chem. Commun., 881 (1976)
160. Baldwin, J. E., Crossley, M. J., Klose, T., O'Rear III, E. A., Peters, M. K.: Tetrahedron 38, 27 (1981)

161. Wijesekera, T. P.: Ph.D. Thesis, University of British Columbia 1980
162. Wijesekera, T. P., Paine III, J. B., Dolphin, D., Einstein, F. W. B., Jones, T.: J. Am. Chem. Soc. *105*, 6747 (1983)
163. Diekmann, H., Chang, C. K., Traylor, T. G.: ibid. *93*, 4068 (1971)
164. Traylor, T. G., Campbell, D., Tsuchiya, S.: ibid. *101*, 4748 (1979)
165. Traylor, T. G., Campbell, D., Tsuchiya, S., Mitchell, M., Stynes, D. V.: ibid. *102*, 5939 (1980)
166. Traylor, T. G., Mitchell, M. J., Tsuchiya, S., Campbell, D. H., Stynes, D. V., Koga, N.: ibid. *103*, 5234 (1981)
167. Traylor, T. G., Tsuchiya, S., Campbell, D., Mitchell, M., Stynes, D., Koga, N.: ibid. *107*, 604 (1985)
168. Chang, C. K., DiNello, R. K., Dolphin, D.: in Inorganic Syntheses (ed. Busch, D. H.), John Wiley & Sons, New York, Vol. XX, 147 (1980)
169. David, S., Dolphin, D., James, B. R., Paine, J. B. III, Wijesekera, T. P., Einstein, F. W. B., Jones, T.: Can. J. Chem. *64*, 208 (1986)
170. David, S.: Ph.D. Thesis, University of British Columbia 1985
171. Traylor, T. G., Koga, N., Dearduff, L. A., Swepston, P. N., Ibers, J. A.: J. Am. Chem. Soc. *106*, 5132 (1984)
172. Traylor, T. G., Koga, N., Dearduff, L. A.: ibid. *107*, 6504 (1985)
173. Morgan, B.: Ph.D. Thesis, University of British Columbia 1984
174. Morgan, B., Dolphin, D.: Angew. Chem. Int. Ed. *24*, 1003 (1985)
175. Battersby, A. R., Hartley, S. G., Turnbull, M. D.: Tetrahedron Lett., 3169 (1978)
176. Cruse, W. B., Kennard, O., Sheldrick, G. M., Hamilton, A. D., Hartley, S. G., Battersby, A. R.: J. Chem. Soc., Chem. Commun., 700 (1980)
177. Battersby, A. R., Howson, W., Hamilton, A. D.: ibid. 1266 (1982)
178. Chang, C. K.: J. Am. Chem. Soc. *99*, 2819 (1977)
179. Richardson, N. M., Sutherland, I. O., Camilleri, P., Page, J. A.: Tetrahedron Lett. *26*, 3739 (1985)
180. Bogatskii, A. V., Zhilina, Z. I., Stepanov, D. E.: Zh. Org. Khim. Engl. Ed. *18*, 2039 (1982)
181. Hamilton, A. D., Lehn, J.-M., Sessler, J. L.: J. Chem. Soc., Chem. Commun., 311 (1984)
182. Chang, C. K., Koo, M. S., Ward, B.: ibid. 716 (1982)
183. Gunter, M. J., Mander, L. N.: J. Org. Chem. *46*, 4792 (1981)
184. Gunter, M. J., Mander, L. N., Murray, K. S., Clark, P. E.: J. Am. Chem. Soc. *103*, 6784 (1981)
185. Ganesh, K. N., Sanders, J. K. N.: J. Chem. Soc., Chem. Commun., 1129 (1980)
186. Ganesh, K. N., Sanders, J. K. N.: J. Chem. Soc., Perkin Trans. 1, 1611 (1982)
187. Ganesh, K. N., Sanders, J. K. N., Waterton, J. C.: ibid. 1, 1617 (1982)
188. Sanders, J. K., Leighton, P.: J. Chem. Soc., Chem. Commun., 24 (1985)
189. Leighton, P., Sanders, J. K. M.: ibid. 854 (1984)
190. Abraham, R. J., Leighton, P., Sanders, J. K. M.: J. Am. Chem. Soc. *107*, 3472 (1985)
191. Leighton, P., Sanders, J. K. M.: J. Chem. Soc., Chem. Commun., 856 (1984)
192. Hamilton, A. D., Rubin, H.-D., Bocarsly, A. B.: J. Am. Chem. Soc. *106*, 7255 (1984)
193. Momenteau, M., Mispelter, J., Loock, B., Bisagni, E.: J. Chem. Soc., Perkin Trans. 1, 189 (1983)
194. Momenteau, M., Loock, B., Mispelter, J., Bisagni, E.: Nouv. J. Chem. *3*, 77 (1979)
195. Lexa, D., Momenteau, M., Rentien, P., Rytz, G., Saveant, J. M., Xu, F.: J. Am. Chem. Soc. *106*, 4755 (1984)
196. Becker, J. Y., Dolphin, D., Paine, J. B. III, Wijeskera, T. P.: J. Electroanal. Chem. *164*, 335 (1984)
197. Zhilina, Z. I., Bogatskii, A. V., Vodzinskii, S. V., Abramovich, A. E.: Zh. Org. Khim. Eng. Ed. *18*, 2271 (1982)
198. Weiser, J., Staab, H. A.: Angew. Chem. Int. Ed. *23*, 623 (1984)
199. Momenteau, M., Lavalette, D.: J. Chem. Soc., Chem. Commun., 341 (1982)
200. Mispelter, J., Momenteau, M., Lavalette, D. Lhoste, J.-M.: J. Am. Chem. Soc. *105*, 5165 (1983)
201. Momenteau, M., Loock, B., Lavalette, D., Tetreau, C., Mispelter, J.: J. Chem. Soc., Chem. Commun., 962 (1983)
202. Battersby, A. R., Hamilton, A. D.: ibid. 117 (1980)
203. Battersby, A. R., Bartholomew, S. A. J., Nitta, J.: ibid. 1291 (1983)

# Synthesis, Electrochemistry, and Structural Properties of Porphyrins with Metal-Carbon Single Bonds and Metal-Metal Bonds

**R. Guilard[1], C. Lecomte[2] and K. M. Kadish[3]**

1 Laboratoire de Synthèse et d'Electrosynthèse Organométallique associé au C.N.R.S. (UA 33), Faculté des Sciences "Gabriel", 6, Boulevard Gabriel, 21100 Dijon (France)
2 Laboratoire de Minéralogie et Cristallographie (UA CNRS 809), Faculté des Sciences, Centre de 2ème Cycle, Boite Postale n° 239, 54506 Vandœuvre les Nancy Cedex (France)
3 University of Houston, Department of Chemistry, Houston, Texas 77004 (U.S.A.)

Porphyrins are known to coordinate with most of the metallic or pseudo-metallic elements. However, in spite of the large number of isolated metalloporphyrin derivatives, metal-carbon σ-bonded complexes are limited only to the following series: Fe, Ru, Co, Rh, Ir, Ti, Al, Ga, In, Tl, Si, Ge, Sn, and Zn.

These complexes are of great interest as model compounds for understanding the functions and relationships of several biological macromolecules, as well as for their chemical reactivity. In the latter case the insertion of small molecules between the metal ion and the carbon atom may result in activation of the inserted molecule or may generate new monomeric or polymeric materials. In addition, metal-carbon σ-bonded porphyrins can act as precursors in the synthesis of metal-metal bonded derivatives.

This review will cover the general strategies for synthesizing organometallic porphyrin complexes. Furthermore, we will also discuss the electrochemical and structural properties of metalloporphyrins from the viewpoint of organometallic chemistry. In doing this we will focus on the following points:

(i) We will give a summary of the synthetic procedures and characteristics of metal-alkyl (aryl) σ-bonded porphyrins. Special emphasis will be placed on the magnetic properties of the iron-carbon porphyrin derivatives.

(ii) The insertion of carbon dioxide and sulfur dioxide into the metal-carbon bond of metalloporphyrins containing main group metals such as Ga, In, and Al will be discussed.

(iii) The synthesis of metal-metal bonded metalloporphyrins containing a σ-bond or a donor-acceptor bond will be presented; we will also focus on the structural and electrochemical properties of some typical compounds.

Structure and Bonding 64
© Springer-Verlag Berlin Heidelberg 1987

# A) Introduction

A considerable number of publications have concentrated on the characterization and reactivity of synthetic and naturally occuring σ-bonded alkyl or aryl iron(III) porphyrins. In some cases, the goal of these studies was devoted to investigations of various substrates occuring in the redox processes of cytochrome P450. Preparation of numerous synthetic models and in vivo experiments have provided proof that the formation of cytochrome P450 complexes involves an iron-carbon bonded species upon metabolic reduction of derivatives like polyhalogenated compounds[1-3]. Especially noteworthy are the carbene complexes, e.g. Fe(TPP)(CCl₂)(H₂O) first prepared by Mansuy et al.[4]. More recently, several laboratories have postulated that cytochrome P450, myoglobin, and hemoglobin give rise to intermediate iron(III) σ-bonded alkyl or aryl derivatives upon metabolic oxidation of various monosubstituted hydrazines[5-7]. The biochemical role of hemes has lead in large part to the synthesis of σ-bonded alkyl or aryl iron(III) porphyrins, but similar σ-bonded derivatives with other transition or non-transition metals have been prepared in order to demonstrate the specific role of the porphyrin ligand and the metal ion in reactions and properties of these organometallic complexes. Electrochemical techniques have been used to monitor the mechanism of specific oxidation or reduction reactions. One such mechanism is the migration of a σ-bonded alkyl or aryl group from the cobalt or iron of a metalloporphyrin to one of the four nitrogens of the porphyrin ring after electrooxidation of the complex. While the carbene complexes contain metal-carbon multiple bonds with important contributions of metal-to-carbon backbonding, the σ-bonded alkyl and aryl derivatives have metal-to-carbon single bonds. Only these will be covered in detail by this review.

From an organometallic point of view, the σ-bonded alkyl and aryl complexes are good precursors in the synthesis of metal-metal bonded derivatives. These latter compounds have been studied in order to determine the potentials for oxidation or reduction, electron transfer rates, and electron transfer mechanisms of metalloporphyrins as a function of solvent, axial ligand coordination, and of the macrocycle. Recently, bimetallic compounds have attracted growing interest due to their potential applications as starting materials for synthesizing polymeric conductors[8]. Some aspects of the reactivity for this family of compounds have been studied and will be discussed in this review as well.

# B) Metal Alkyl (Aryl) σ-Bonded Porphyrins

## B I) General Synthetic Procedures

Most σ-bonded metalloporphyrins are obtained from the reaction of chloro and perchlorato metalloporphyrins M(Por)Xₙ (n = 1 or 2) with Grignard reagents or with alkyl (aryl) lithium (Scheme 1). The corresponding mono or dialkyl (aryl) derivatives are generally isolated in good yield but this will depend upon the nature of the metal.

The above method has been used for the synthesis of metal alkyl (or aryl) σ-bonded porphyrins of iron[9-17], cobalt[9, 10], rhodium[18], titanium[19], iridium[20], gallium[21], indium[22, 23], thallium[24], silicon[25], germanium[26-30], and tin[28, 29].

$$M(Por)X_n + n\,RM'Y \rightarrow M(Por)R_n + n\,XM'Y$$

n = 1: M = Fe, Co, Rh, Ti, Ir, Ga, In, Tl          X   = halide, $ClO_4^-$
n = 2: M = Si, Ge, Sn                              M'  = Mg;    Y = halide
                                                   M'  = Li;    Y = none
                                                   M'  = Ag;    Y = $B(C_6H_5)_4$

**Scheme 1**

An alternative route used in organometallic chemistry is the reaction of low valent organometallic derivatives with alkyl (aryl) halides. The two electron oxidative addition of alkyl (aryl) halides or cyclopropane derivatives to metalloporphyrins such as $[M^I(Por)]^-$ leads to metal alkyl (aryl) σ-bonded porphyrins of cobalt[31–35], rhodium[36–38], and iridium[20] (Scheme 2). Substitution of aryl and vinyl halides by electrochemically generated iron(I) porphyrins also leads to σ-bonded $Fe^{III}$ complexes[39, 40].

$$[M^I(Por)]^- + R(Ar)X \rightarrow M^{III}(Por)[R(Ar)] + X^-$$

M = Fe, Co, Rh, Ir

$$[M^I(Por)]^- + \overset{\ddots}{X}\!\!\diagdown \rightarrow M^{III}(Por)(CH_2CH_2X)$$

M = Rh

**Scheme 2**

Oxidative addition of various organic substrates RX with $Rh^I$ porphyrins also yields σ-bonded organometallic porphyrin complexes[41–44] (Eq. 1).

$$(Por)[Rh^I(CO)_2]_2 + RX \rightarrow Rh^{III}(Por)(R) + Rh^I(CO)_2^+ + X^- + 2\,CO \qquad (1)$$

The iron(II) or rhodium(II) complexes undergo oxidative addition with alkylating agents via a monoelectronic process[45] (Scheme 3).

$$M^{II}(Por) + RX \xrightarrow{\text{slow}} M^{III}(Por)X + R^{\cdot}$$

$$R^{\cdot} + M^{II}(Por) \xrightarrow{\text{fast}} M^{III}(Por)(R)$$

M = Fe, Rh

**Scheme 3**

Steady-state pulse radiolysis of methylene chloride solutions containing ferric deuteroporphyrin or the chemical reduction of ferric deuteroporphyrin solutions containing methyl iodide led to iron methyl σ-bonded species which were spectrophotometrically characterized[46, 47].

Other synthetic procedures have also been reported for specific metalloporphyrins. The reactivity of rhodium(III) and cobalt(III) porphyrins towards diazoalkanes have been described by Callot and Schaeffer[48–51] where R or R' are H, $CH_3$, or alkyl carboxylate groups. The first step in this reaction is given by Eq. 2.

$$M(Por)X + N_2C\overset{R}{\underset{R'-H}{\diagdown}} \rightarrow M(Por)(C=R'R) + N_2 + HX \tag{2}$$

M = Rh, Co

The reactions of various diazo derivatives with rhodium(III) porphyrins leads to high yields of alkyl rhodium(III) porphyrins. For the corresponding $Co^{III}$ series, catalytic decomposition of the reagents and fragmentation of primary olefinic products give a methylcarboxylate cobalt(III) compound if R is a methylcarboxylate group. Nucleophilic attack of the CO ligand by the ethoxy anion leads to an ethylcarboxylate rhodium(III) complex as shown below[52].

$$Rh(TPP)(CO)Cl + EtO^- \rightarrow Rh(TPP)(COOEt) + Cl^- \tag{3}$$

Chloro(aquo) rhodium(III) octaethylporphyrin reacts with ethyl vinyl ether in the presence of ethanol to give a rhodium(III) porphyrin which is readily hydrolysed to produce a (formylmethyl) rhodium(III) complex (Scheme 4)[38].

$$Rh(OEP)(H_2O)Cl + CH_2 = CHOEt \xrightarrow{EtOH} Rh(OEP)[CH_2CH(OEt)_2] \xrightarrow{H_3O^+} Rh(OEP)(CH_2CHO)$$

**Scheme 4**

If the same complex reacts with acetylenic compounds various derivatives can be obtained, such as vinyl or acyl rhodium(III) porphyrins (Scheme 5)[38].

$$Rh(OEP)(H_2O)Cl + R-C\equiv CH \nearrow \begin{array}{l} Rh(OEP)(CH=CClR) \\ R = H, C_6H_5 \end{array}$$
$$\searrow \begin{array}{l} Rh(OEP)(COCH_2R) \\ R = C_6H_5, n\text{-}C_4H_9, n\text{-}(CH_2)_2CH_2OH \end{array}$$

**Scheme 5**

A neutral metalloformyl complex is formed by the apparent insertion of carbon monoxide between the rhodium hydrogen bond of rhodium(III) hydride, Rh(OEP)H (Eq. 4)[53].

$$Rh(OEP)H + CO \rightarrow Rh(OEP)(CHO) \tag{4}$$

The mechanism for this reaction has been determined by Halpern[54] on the basis of $^1$H NMR and UV-visible studies. A radical chain process is initiated by an equilibrium between the hydride Rh(OEP)H and the dimer $[Rh(OEP)]_2$. The different steps of the process are summarized in Scheme 6.

Initiation/termination
$[Rh(OEP)]_2 \rightleftharpoons 2 \cdot Rh(OEP)$

Propagation
$\cdot Rh(OEP) + CO \rightleftharpoons \cdot(CO)Rh(OEP)$

$\cdot(CO)Rh(OEP) + HRh(OEP) \rightleftharpoons Rh(OEP)(CHO) + \cdot Rh(OEP)$

**Scheme 6**

Aluminum[55, 56)], thallium[57, 58)], and zinc[59)] porphyrins containing a metal carbon bond are synthesized by treating the porphyrins with organometallic compounds such as $Al(Et)_3$ or $Al(CH_3)Cl_2$.

Recently, Collman has shown that the ruthenium porphyrin dimer $[Ru(OEP)]_2^{2+}$ produces an alkyl complex according to Eq. 5[60, 61)].

$$[Ru(OEP)]_2^{2+} + 2\,EtMgX \rightarrow 2\,Ru(OEP)Et \tag{5}$$

The great synthetic ability of the oxidized Ru metalloporphyrin dimers permits the development of a general synthesis of σ-bonded ruthenium(III) porphyrins.

## B II) NMR and UV-Visible Characteristics

Most σ-bonded alkyl (or aryl) metalloporphyrins exhibit typical $^1H$ NMR features of diamagnetic metalloporphyrins. The σ-alkyl (or aryl) iron(III) complexes have characteristic spectra of paramagnetic species. In addition, the NMR data for some σ-bonded complexes correlate with the UV-visible data. The presence of an alkyl or aryl group coordinated to the metal is demonstrated by the appearance of a signal at very high fields.

### B II 1) NMR Data

#### B II 1 a) Diamagnetic Complexes

The chemical shift of a methyl group bound to a metal(III) porphyrin depends on the nature of the metal. Comparison of values in Table 1 shows that complexes with more electronegative metals normally give signals at higher fields. For example, the $Rh^{III}$-carbon bond has a higher covalent bonding character than the $Co^{III}$-carbon bond. Consequently, chemical shifts for the σ-bonded methyl are $-5.00$ and $-6.47$ ppm for the $Co^{III}$ and $Rh^{III}$ octaethylporphyrins respectively. Characteristic shifts of the $^1H$ NMR signal of the alkyl or aryl groups which are directly bonded to the central metal (see Tables 1 and 2) are primarily due to the diamagnetic anisotropy of the ring current. In order to evaluate the diamagnetic ring current, calculations of the shifts have been made using a modified Johnson-Bovey equation[18)].

**Table 1.** Chemical shifts of the alkyl protons of various $M(Por)R_n$ complexes where n = 1 or 2

| M | Por | R | $H_\alpha$ | $H_\beta$ | $H_\gamma$ | $H_\delta$ | Ref. |
|---|-----|---|-----------|-----------|------------|------------|------|
| Co | OEP | $CH_3$ | $-5.00$ | | | | 62 |
| Co | TPP | $CH_3$ | $-4.75$ | | | | 33 |
| Co | TPP | $(CH_2)_2CH_3$ | $-3.56$ | $-4.64$ | $-1.35$ | $-0.73$ | 33 |
| Rh | OEP | $CH_3$ | $-6.47$ | | | | 18 |
| Rh | TPP | $CH_3$ | $-5.80$ | | | | 62 |
| Rh | OEP | $(CH_2)_3CH_3$ | $-5.58$ | $-5.04$ | $-1.93$ | $-1.08$ | 18 |
| Ge | TPP | $CH_2CH_3$ | $-6.66$ | $-4.28$ | | | 29 |
| Ge | TPP | $(CH_2)_3CH_3$ | $-6.55$ | $-4.59$ | $-1.42$ | $-0.77$ | 29 |

**Table 2.** Comparative chemical shifts for alkyl protons of Ga(Por)(R) complexes where Por = OEP or TPP[?]

| Por | R | $H_\alpha$ | $H_\beta$ | $H_\gamma$ | $H_\delta$ |
|-----|---|------|------|------|------|
| OEP | $CH_3$ | − 6.71 | | | |
| OEP | $CH_2CH_3$ | − 6.15 | − 3.74 | | |
| OEP | $(CH_2)_3CH_3$ | − 6.15 | − 3.79 | − 1.79 | − 0.80 |
| TPP | $CH_3$ | − 6.22 | | | |
| TPP | $CH_2CH_3$ | − 5.66 | − 3.34 | | |
| TPP | $(CH_2)_3CH_3$ | − 5.64 | − 3.34 | − 1.50 | − 0.60 |

**Table 3.** Chemical shifts of the methinic protons and characteristic absorption maxima (nm) and molar absorptivities ($\varepsilon$) of Soret bands for the In(OEP)(R) complexes[23, 63]

| R | $\delta(-CH=)$ | $\lambda_{max}$, nm ($\varepsilon \times 10^{-3}$) | | $\varepsilon(II)/\varepsilon(III)$ |
|---|---|---|---|---|
| | | Band III | Band II | |
| $C(CH_3)_3$ | 10.17 | 377(95.3) | 443(61.4) | 0.64 |
| $CH(CH_3)_2$ | 10.18 | 372(88.8) | 439(96.8) | 1.09 |
| $C_4H_9$ | 10.17 | 365(59.3) | 435(142) | 2.39 |
| $C_2H_5$ | 10.18 | 365(58.5) | 435(145) | 2.48 |
| $CH_3$ | 10.19 | 354(49.0) | 428(247) | 5.04 |
| $C_6H_5$ | 10.21 | 354(45.1) | 424(264) | 5.85 |
| $C_2H_2(C_6H_5)$ | 10.22 | 354(39.8) | 426(246) | 6.18 |
| $C_2(C_6H_5)$ | 10.37 | 345(30.7) | 415(426) | 13.88 |

Since shifts of the $^1H$ NMR signal from the protons in close proximity to the central metal vary from metal to metal (see Table 1) the constants for the Johnson-Bovey equation must be readjusted for each metalloporphyrin series. However, this work demonstrates that proton shifts for the axial ligand far from the central metal appear at approximately the same magnetic field as observed for other systems. The $\alpha$ protons generally appear in the range of − 4.5 to − 7.0 ppm whereas the $\beta$ protons are observed in the region of − 3 to − 5 ppm. However, for the Co series the more shielded protons are the $\beta$ protons. In all of the porphyrin series the shielding of the $\gamma$ and $\delta$ protons decreases when the proton-metal distance increases. Furthermore, the resonance signal of an R group is more shielded for octaethylporphyrin derivatives than for tetraphenylporphyrin derivatives. This observation can be easily explained since the ring current of the tetraphenylporphyrins is smaller than that of the octaethylporphyrins.

It is well known that chemical shifts for meso protons of the octaethylporphyrin complexes depend upon the oxidation state of the central metal. Divalent metals have a resonance in the region of 9.75–10.08 ppm while for tri- and tetravalent metals this resonance is in the range of 10.13–10.39 and 10.30–10.58 ppm respectively. For the In(OEP)(R) series[23, 63], the indium oxidation state is III and the observed methine proton shifts of 10.17–10.37 ppm correspond to those of a typical trivalent metal (Table 3). The exact position of the In(OEP)(R) methinic protons resonance varies systematically with the electron-donating ability of the axial ligand: i.e. the more basic the axial ligand, the higher the field. Similar results are observed for σ-bonded complexes of the rhodium series[18].

## B II 1 b) Paramagnetic Complexes

On the basis of NMR data most of the iron aryl (or iron-alkyl) σ-bonded porphyrins are described as low spin complexes[14, 16, 39]. The signal from the pyrrole protons of tetraphenylporphyrin derivatives is at a high field position in the region of $-17$ to $-20$ ppm (see Fig. 1). Signals of the $C_6H_5$ axial ligand for the $Fe(Por)(C_6H_5)$ complexes appear as three singlets close to $-80$ ppm ($o$-H), $-25$ ppm ($p$-H) and $+15$ ppm ($m$-H). The line width is logically increased with respect to the distance of the protons from the paramagnetic center (110, 10.5 and 8–10 Hz respectively). The phenyl protons of the porphyrin macrocycle are shielded, compared to a diamagnetic species and the ortho and meta protons of the phenyl groups appear as two singlets for the $o$- and $o'$-H and a singlet for the $m$-H.

## B II 2) UV-Visible Data

Electronic absorption spectra for σ-bonded complexes of cobalt[33], rhodium[18], thallium[24], silicon[25], germanium[26, 30], and tin[28, 29] are characteristic of regular porphyrins.

Other σ-bonded metal porphyrins have electronic absorption spectra belonging to the hyperclass. The σ-bonded complexes of iron[16, 17], indium[23, 63], gallium[21, 64] exhibit such typical spectra. Thus, replacement of the anionic axial ligand on the halogenated complexes $M(Por)X_n$ by an alkyl (or aryl) σ-bonded ligand results in a splitting of the Soret band into two bands. One of the bands is red shifted and the other is blue shifted with respect to the same porphyrin complex with anionic axial ligands. This is illustrated in Fig. 2 for $In(OEP)Cl$ and $In(OEP)(C(CH_3)_3)$.

The ratio of the molar absorptivities ($\varepsilon(II)/\varepsilon(III)$) for the split Soret peaks varies as a function of the R groups (see Table 3, which gives molar absorptivities for $In(OEP)(R)$ complexes). For example, the indium(III) ions of $In(Por)(R)$ have filled d orbitals, and consequently the blue shifted band III may be attributed to a 5 $p_z \rightarrow eg(\pi^*)$ transition. On the other hand, the red shifted band II may be attributed to a $\pi \rightarrow \pi^*$ electronic transition of the porphyrin ring. The $\varepsilon(II)/\varepsilon(III)$ molar absorptivity ratio of these two bands should be systematically related to the electron-donating character of the bound R group. Complexes with more electron-donating R groups should have a higher extinction of transition III and a smaller $\varepsilon(II)/\varepsilon(III)$ ratio. This is generally true for all the $In(Por)(R)$ (see Table 3) and $Ga(Por)(R)$ complexes[64]. A dependence of the molar absorptivities on the R donor ligand is observed independent of the nature of the porphyrin macrocycle (i.e. OEP or TPP ligand). However, for a given R group, the ratio of the molar absorptivities for an octaethylporphyrin complex is about half that of the corresponding tetraphenylporphyrin derivative. This is due to a difference in electron donor capacity of the two porphyrin rings. One would expect the hyper character of the spectra to be more pronounced for the OEP complexes than for the TPP complexes, and this appears to be the case. The data for the indium series as well as that for the gallium series suggest that the greater the electron donor properties of the bound R group, the larger the electron density on the metal and on the conjugated porphyrin π system. As described in the preceeding section, NMR studies on the indium complexes suggest the same interpretation.

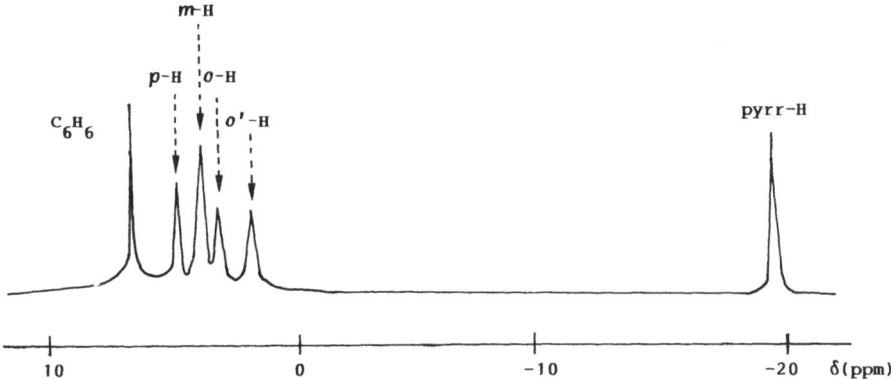

**Fig. 1.** $^1H$ NMR spectrum of Fe(TPP)(CH$_3$) recorded at 294 K in C$_6$D$_6$

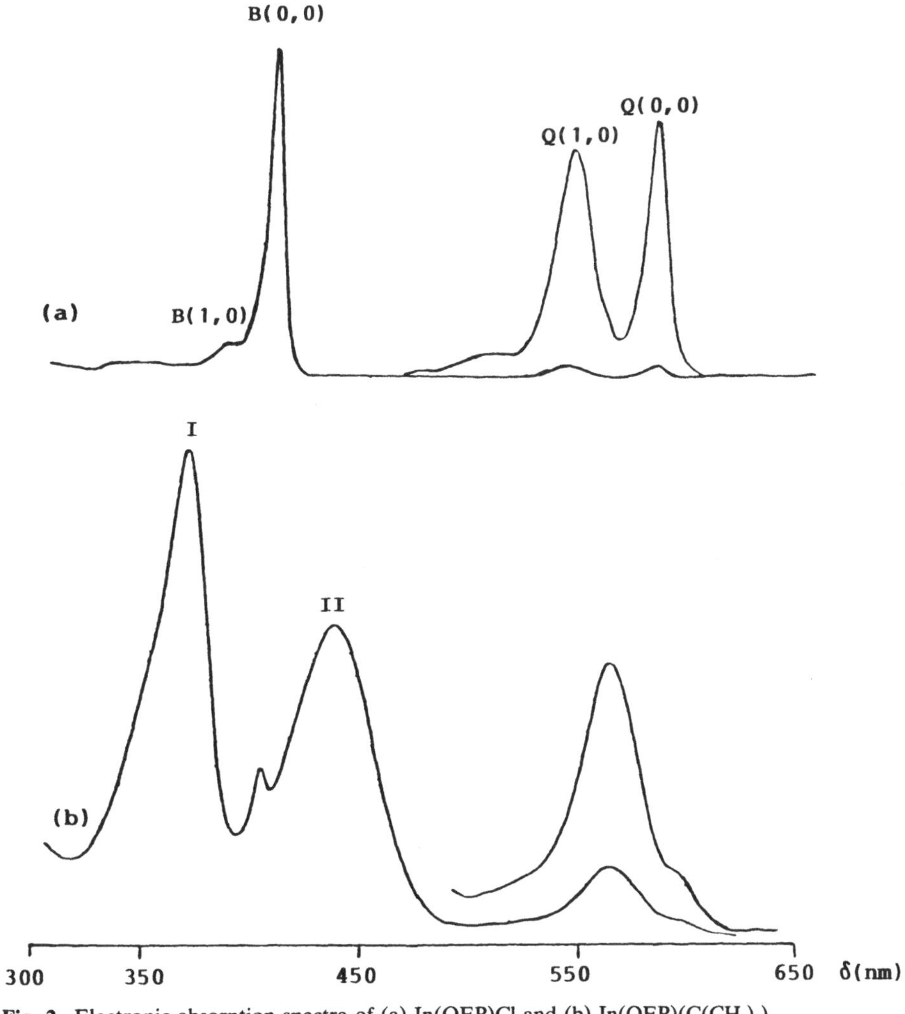

**Fig. 2.** Electronic absorption spectra of (a) In(OEP)Cl and (b) In(OEP)(C(CH$_3$)$_3$)

## B III) ESR and Magnetic Properties of Iron Complexes

### B III 1) Spin States of Alkyl(Aryl) Iron(III) Porphyrins

As mentioned above, the iron-alkyl or iron-aryl σ-bonded porphyrins are generally described as low spin complexes[14, 16, 39]. However, the iron atom of these alkyl or aryl σ-bonded systems may also exhibit a mixture of spin states or a high spin state depending on the nature of the alkyl or aryl group and the equatorial ligand. As shown in Fig. 3, the ESR spectrum of Fe(TpTP)(C₆H₅) in toluene at 115 K has three lines at g values close to 2 which are attributable to a low spin state and two lines centered at g = 1.96 and 6.08 which are typical of a high spin iron(III) species. This spin mixture depends on numerous parameters such as solvent, method of sample preparation, type of porphyrin ligand and type of axial ligand.

Since the formation of complexes with σ-bonded iron species seems to play a fundamental role in the biological activity of cytochrome P 450[1-5], a definition of the molecular parameters that determine a given spin state model is of interest. The nature and the number of axial ligands bound to the iron atom are two major factors to consider[66, 67]. It has also been shown that variation of the axial and macrocyclic ligands can affect the magnetic behaviour of synthetic iron porphyrins. Changes in either the organic σ-bonded axial ligand or the equatorial porphyrin ring may lead to complexes with either pure high spin or pure low spin character[68] (Scheme 7).

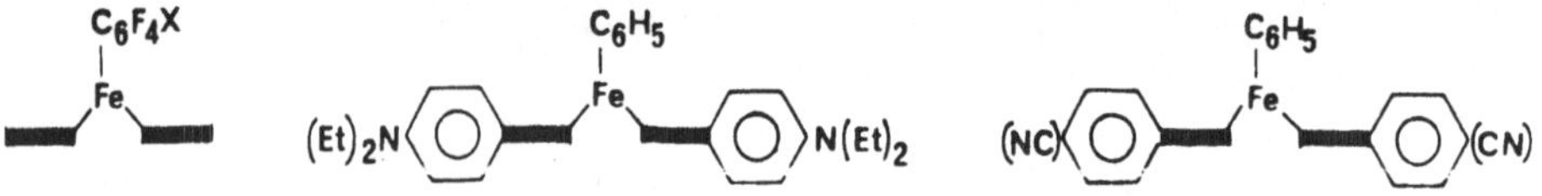

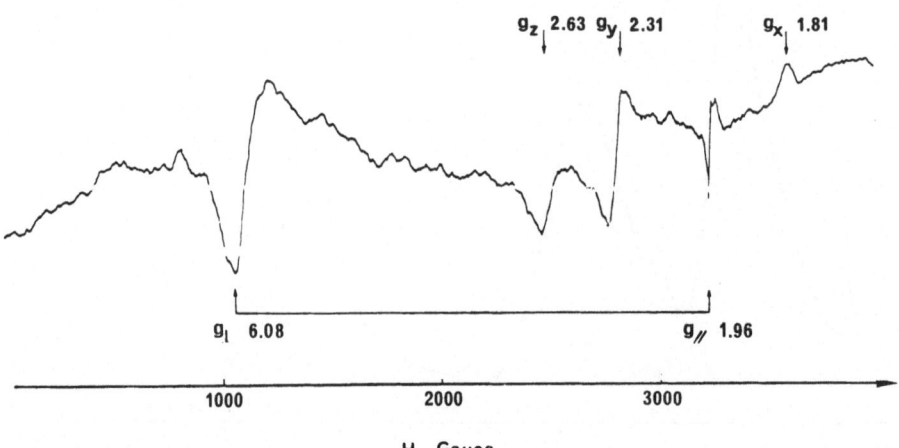

$C_6F_4X$

Fe

Fe(Por)(C₆F₄X)
with X = H or F

S = 5/2

$C_6H_5$

Fe

(Et)₂N〈O〉━━〈O〉N(Et)₂

Fe(TpEt₂NPP)(C₆H₅)

S = 5/2, 1/2

$C_6H_5$

Fe

(NC)〈O〉━━〈O〉(CN)

Fe((CN)₄TPP)(C₆H₅)

S = 1/2

**Scheme 7**

$g_z$ 2.63   $g_y$ 2.31                    $g_x$ 1.81

$g_⊥$ 6.08                              $g_∥$ 1.96

1000              2000              3000

H , Gauss

**Fig. 3.** ESR spectrum of Fe(TpTP)(C₆H₅) in toluene at 115 K[65]

High spin iron(III) is obtained for porphyrin complexes bound with electrophilic axial ligands such as perfluoro groups. Complexes with a more basic porphyrin ligand such as TpEt$_2$NPP, and a coordinated σ-bonded iron-phenyl moiety may exist in either the high or low spin state depending upon the experimental conditions. In contrast, the electron attractor (CN)$_4$TPP ligand gives an iron-phenyl species with pure low spin character.

## B III 2) Factors Governing the Spin State in Solution

The factors governing the spin state of the alkyl or aryl ferriporphyrins are summarized below.

### B III 2 a) Effect of the Nature of the Axial Ligand

Complexes with bound perfluoro groups are well defined high spin derivatives independent of the porphyrin macrocycle, the temperature, the solvent, and the sample preparation. This behavior is very close to that described for halogenated ferriporphyrins[69] and demonstrates the electron withdrawing character of this type of ligand.

Alkyl or aryl iron(III) porphyrins behave differently. The room temperature NMR data are characteristic of low spin state ferriporphyrin systems, even though the pentacoordination scheme is usually associated with high spin derivatives[66]. Alkyl and aryl groups behave as strong field ligands. The nature of the metal-porphyrin and metal-ligand bonding was determined by calculation of contact and dipolar isotropic chemical shifts. Normal charge transfers are noted for low spin iron(III) derivatives, but a diminished polarization of the σ electron system initiated by the porphyrin π system is also observed. This latter phenomenon which increases with the inductive character of the ligand, demonstrates the relationship between the magnetic properties and the nature of the axial ligand. At low temperature, the role of the hydrocarbon ligands is more important.

The replacement of an alkyl ligand by an aryl ligand on Fe(Por)(R) for a given porphyrin macrocycle increases the fraction of the complex present in a high spin form. This is shown in Table 4 and explains why the only high spin states are seen for Fe(Por)(R) with perfluoro ligands.

### B III 2 b) Effect of the Porphyrin Macrocycle

The ESR data in Table 4 demonstrate that a basic porphyrin ligand such as TpEt$_2$NPP preferentially gives a high spin Fe(III) complex. In contrast, the electron withdrawing ligand (CN)$_4$TPP gives a pure low spin state for the aryl iron(III) species. For the other macrocycles, two spin states are observed (see Table 4). The spin state is critically dependent on the basicity of the macrocycle. The metal-ligand bond strength decreases when the metal electron density or donor properties of the macrocycle increase. Therefore, it is then possible to order the porphyrin ligands with increasing basicity as follows: (CN)$_4$TPP < TPP ≃ TpTP ≃ TmTP < OEP < TpEt$_2$NPP.

**Table 4.** ESR data of Fe(Por)(R) in toluene at 115 K

| Compound | $g_x$ | $g_y$ | $g_z$ | $\dfrac{\Lambda}{\lambda}$ | $\dfrac{V}{\lambda}$ | $\dfrac{V}{\Delta}$ | $g_\perp$ | $g_\parallel$ |
|---|---|---|---|---|---|---|---|---|
| | | | S = 1/2 | | | | S = 5/2 | |
| Fe(OEP)(n-C$_4$H$_9$) | | | | | | | 5.99 | 1.99 |
| Fe(OEP)(CH$_3$) | | | | | | | 6.17 | 2.01 |
| Fe(OEP)(C$_6$H$_5$) | | | | | | | 6.31 | 1.99 |
| Fe(OEP)(p-MeC$_6$H$_4$) | | | | | | | 5.99 | 2.02 |
| Fe(TPP)(n-C$_4$H$_9$) | 1.90 | 2.28 | 2.56 | 5.20 | 3.85 | 0.74 | 6.10 | 1.98 |
| Fe(TPP)(CH$_3$) | 1.89 | 2.27 | 2.60 | 5.44 | 3.56 | 0.65 | 6.08 | |
| Fe(TPP)(C$_6$H$_5$) | 1.92 | 2.39 | 2.60 | 4.14 | 3.51 | 0.85 | 5.96 | |
| Fe(TPP)(p-MeC$_6$H$_4$) | 1.86 | 2.23 | 2.48 | 5.98 | 3.99 | 0.67 | 5.97 | |
| Fe(TmTP)(n-C$_4$H$_9$) | 1.89 | 2.27 | 2.52 | 5.03 | 3.99 | 0.79 | 6.45 | 1.99 |
| Fe(TmTP)(CH$_3$) | 1.87 | 2.28 | 2.61 | 5.02 | 3.46 | 0.69 | 6.02 | 1.99 |
| Fe(TmTP)(C$_6$H$_5$) | 1.86 | 2.25 | 2.55 | 5.11 | 3.65 | 0.71 | 6.02 | 1.98 |
| Fe(TpTP)(n-C$_4$H$_9$) | 1.88 | 2.27 | 2.53 | 4.94 | 3.88 | 0.78 | 6.26 | 1.98 |
| Fe(TpTP)(CH$_3$) | 1.88 | 2.30 | 2.57 | 4.65 | 3.72 | 0.80 | 6.08 | |
| Fe(TpTP)(C$_6$H$_5$) | 1.81 | 2.31 | 2.63 | 5.04 | 3.18 | 0.63 | 6.08 | 1.96 |
| Fe(TpEt$_2$NPP)(C$_6$H$_5$) | | | | | | | 5.96 | 2.02 |
| Fe((CN)$_4$TPP)(C$_6$H$_5$) | 1.97 | 2.06 | 2.25 | 21.55 | 7.81 | 0.36 | | |
| Fe(OEP)(C$_6$F$_4$H) | | | | | | | 5.88 | 1.98 |
| Fe(OEP)(C$_6$F$_5$) | | | | | | | 6.08[a] | 1.97[a] |
| Fe(TPP)(C$_6$F$_4$H) | | | | | | | 5.86 | 1.98 |
| Fe(TPP)(C$_6$F$_5$) | | | | | | | 5.86 | 1.99 |
| Fe(TmTP)(C$_6$F$_4$H) | | | | | | | 5.86 | 1.98 |
| Fe(TmTP)(C$_6$F$_5$) | | | | | | | 5.86 | 1.98 |
| Fe(TpTP)(C$_6$F$_4$H) | | | | | | | 5.81 | 1.98 |
| Fe(TpTP)(C$_6$F$_5$) | | | | | | | 5.85 | 1.98 |

[a] In 2 : 1 toluene-methylene chloride mixtures

The room temperature NMR data are in agreement with low spin entities for all Fe(Por)(R) complexes with the exception of the perfluoro derivatives. These contradictory results suggest a low spin-high spin conversion at low temperature:

$$Fe(Por)R \rightleftharpoons Fe(Por)R \tag{6}$$
$$S = 1/2, T > 273 \text{ K} \quad S = 5/2, T < 273 \text{ K}$$

Thermodynamics alone cannot explain this conversion since low temperature favors a low spin species and other parameters must be invoked.

## B III 2 c) Solvent Effects

The effect of solvent is illustrated by the ESR spectrum of Fe(OEP)(CH$_3$) recorded in toluene and carbon disulfide (Fig. 4). At a low temperature, carbon disulfide induces a shift of the spin mixture towards the low spin state. However, the axial and equatorial ligand effects are predominant[65]. Furthermore, results from Raman spectroscopy indicate that the solvent is not coordinated to the metal[70].

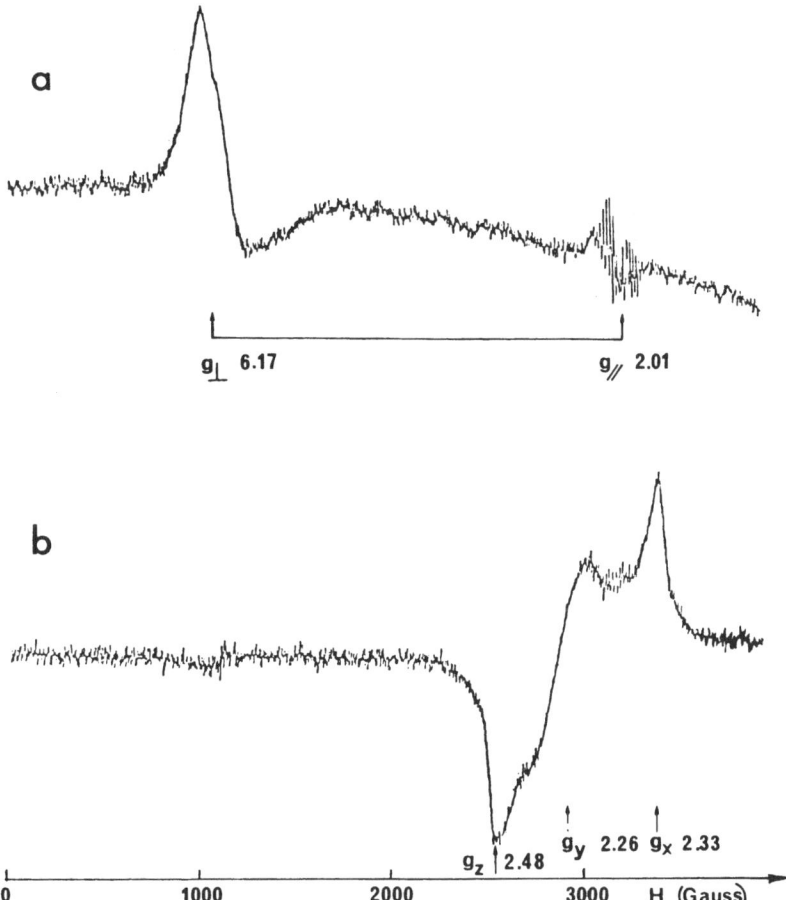

**Fig. 4a, b.** ESR spectrum of Fe(OEP)(CH$_3$) recorded at 115 K (**a**) in toluene and (**b**) in carbon disulfide[65]

## B III 3) Solid State Measurements[65]

The ESR spectrum of Fe(Por)(R) greatly depends on the method of sample preparation. Samples which are ground give different ESR spectra. The intensity of the high spin signals increases with more grinding and that of the low spin signals decrease. This change can be related to the ligand-metal center bond strength since the compressibility is anisotropic. The ESR data demonstrate that a higher pressure results in high spin multiplicity. The stabilization of the high spin state must be due to a modification of the molecular structure resulting in a decrease of the radius of the central hole and a larger displacement of the metal ion out of the macrocyclic plane[71]. However, the shift of the spin mixture is small, since it is not clearly demonstrated by either magnetic susceptibility measurements (Fig. 5) or Mössbauer spectroscopy. The grinding would modify the molecular structure only at the surface.

The Mössbauer data unambigously demonstrate the high spin state of the perfluoro-aryl porphyrins and are in good agreement with a low spin state of the alkyl(aryl) ferriporphyrins independent of the grinding (Table 5).

## B IV) Electrochemistry

The electrochemistry of metalloporphyrins with σ-bonded alkyl or aryl groups has been reported for complexes with nine different central metals. These include transition metal complexes of iron[39, 40, 68, 77–79)], cobalt[33, 35, 81, 82)], rhodium[83–86)], and iridium[87, 88)], and metalloporphyrins with gallium[64)], indium[89)], thallium[24)], silicon[25)], and germanium[90)] main group metals. The first seven types of metalloporphyrins are σ-bonded to a single

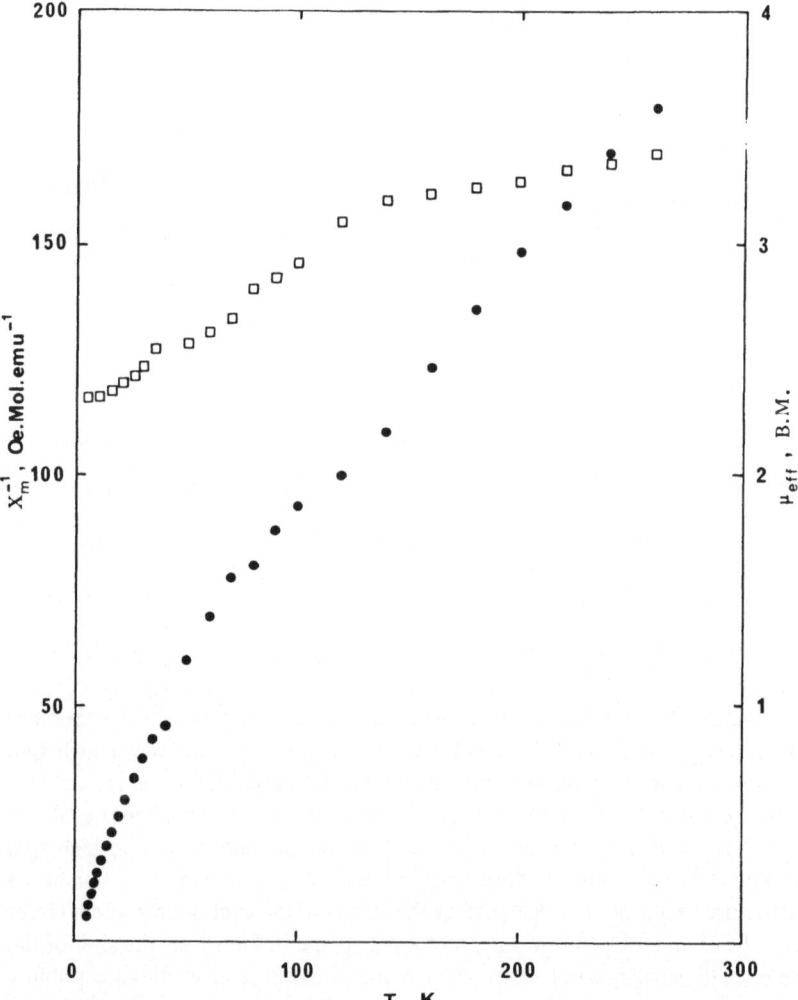

**Fig. 5.** A plot of $1/\chi_m$ vs. T (●●●) and $\mu_{eff}$ vs. T (□□□) for Fe(OEP)(C$_6$H$_5$)

**Table 5.** Mössbauer data for porphyrinatoiron(III) complexes

| Compound | T (K) | $\delta^a$ (mms$^{-1}$) | $\Delta E_Q$ (mms$^{-1}$) | Spin state | Ref. |
|---|---|---|---|---|---|
| Fe(TPP)Cl | 4.2 | 0.30 | 0.48 | 5/2 | 72 |
| Fe(TPP)(C$_6$F$_4$H) | 85 | 0.53 | 0.48 | 5/2 | 65 |
| Fe(Me$_2$CPS$_2$)$_3$ | RT | 0.51 | 0.18 | 5/2 | 73 |
| Fe(i-PrCPS$_2$)$_3$ | RT | 0.48 | 0.29 | 5/2 | 73 |
| Fe(Et$_2$NCS$_2$)$_3$ | RT | 0.23 | 0.91 | 1/2 | 73 |
| Fe(PhCH$_2$CS$_2$)$_3$ | RT | 0.35 | 1.72 | 1/2 | 73 |
| Fe(PhCS$_2$)$_3$ | RT | 0.25 | 1.87 | 1/2 | 73 |
| Hemoglobin azide | 195 | 0.26 | 2.30 | 1/2 | 74 |
| Hemoglobin cyanide | 195 | 0.29 | 1.40 | 1/2 | 74 |
| Fe(TPP)(Im)$_2$Cl | 298 | 0.13 | 2.11 | 1/2 | 75 |
| | 77 | 0.23 | 2.23 | | |
| Fe(OEP)(CH$_3$) | 295 | 0.25 | 2.13 | 1/2 | 65 |
| | 80 | 0.34 | 2.16 | | |
| Fe(OEP)(C$_6$H$_5$) | 295 | 0.20 | 2.10 | 1/2 | 65 |
| | 230 | 0.22 | 2.18 | | |
| | 144 | 0.27 | 2.27 | | |
| | 95 | 0.27 | 2.30 | | |
| Fe(OEP)(C$_6$H$_5$)$^b$ | 295 | 0.20 | 2.16 | 1/2 | 65 |
| | 143 | 0.26 | 2.28 | | |
| | 85 | 0.28 | 2.30 | | |
| Fe(OEP)(C$_6$H$_5$)$^c$ | 85 | 0.28 | 2.32 | 1/2 | 65 |
| Fe(OEP)ClO$_4$ | 295 | 0.29 | 3.16 | spin mixed with | 76 |
| | 115 | 0.37 | 3.52 | predominantly | |
| | 4.2 | 0.37 | 3.57 | S = 3/2 spin state | |

$^a$  Relative to metallic iron
$^b$  In magnesium sulfate (1 : 1) without grinding
$^c$  In magnesium sulfate (1 : 1) with grinding

alkyl or aryl group and have a metal ion in the + 3 oxidation state. The latter two compounds contain Si$^{IV}$ and Ge$^{IV}$ and are σ-bonded to two alkyl or aryl groups.

Electroreduction of the σ-bonded metalloporphyrins proceeds in either one or two single electron addition steps depending upon the solvent, the type of porphyrin ring, and the nature of the σ-bonded R group. Almost all of the reductions are reversible and occur at potentials negatively shifted from those for reduction of the same metalloporphyrin with bound ionic ligands. In contrast, oxidations of the σ-bonded complexes may be irreversible and shifted either positively or negatively from potentials for oxidation of the same metalloporphyrin containing bound ionic ligands. The sign and the magnitude of the potential shift is related to the stability of the singly oxidized species and the presence or absence of coupled chemical reactions. Fe(Por)(R) is oxidized at much more negative potentials than Fe(Por)Cl[77] but Co(Por)(R) is oxidized at much more positive potentials than Co(Por)Cl[35]. The R group in both iron[77] and cobalt[35] complexes undergoes a metal to nitrogen migration after electro-oxidation. Complexes of In(Por)(R) and Ga(Por)(R)[64] as well as Ge(Por)(R)$_2$[90] may undergo rapid loss of an R group after electro-oxidation and when this occurs an irreversible peak is observed at potentials much more negative than potentials for the reversible oxidation of In(Por)X and Ga(Por)X. Complexes of In(Por)(R) where R = $C_6F_5$ or $C_6F_4H$ are reversibly oxidized without loss of the σ-bonded ligand[89] and under these conditions, the potentials are only slightly more negative than for oxidation of the same indium porphyrin with ionic ligands[91]. Electro-oxidized Tl(Por)(R)[24] complexes are also relatively stable and potentials for the reversible abstraction of one electron from Tl(Por)(R) occur at $E_{1/2}$ values which are negatively shifted by 170 to 420 mV from those of Tl(Por)Cl[24]. This is discussed in the following sections.


## B IV 1) Iron Porphyrins

The electrochemistry of σ-bonded iron porphyrins has been carried out almost exclusively by the groups of Lexa and Saveant[39, 40] and Kadish and Guilard[68, 77–80]. The latter group studied complexes which were initially generated and characterized as stable $Fe^{III}$ complexes[16]. In contrast, Lexa and Savéant[39] in-situ generated several σ-alkyl-iron porphyrins which were then electrochemically investigated in DMF. One reduction and one oxidation of Fe(Por)(R) were reported in DMF. The reductions occurred between − 0.76 V and − 1.06 V vs. SCE (depending on the porphyrin) and were attributed to the reaction $Fe^{III} \rightleftarrows Fe^{II}$. Oxidation potentials were only reported for Fe(OEP)(n-$C_4H_9$) ($E_{1/2}$ = 0.25 V) and Fe($C_{12}$TPP)(n-$C_4H_9$) ($E_{1/2}$ = 0.39 V) where $C_{12}$TPP represents the cross trans-linked basket-handle porphyrin[92, 93].

Dramatic cathodic shifts of half-wave potentials are observed for both oxidation and reduction of the low spin Fe(Por)(R) complexes with respect to the corresponding high spin Fe(Por)X derivatives (where X is an anionic ligand) in the same media[77, 78, 80, 94]. For example, the first reduction and first oxidation of Fe(TPP)($C_6H_5$) and Fe(OEP)($C_6H_5$) are shifted by up to 800 mV with respect to the first reduction and first oxidation of Fe(TPP)$ClO_4$[94]. All electrode reactions of Fe(TPP)($C_6H_5$) and Fe(OEP)($C_6H_5$) are reversible to quasi-reversible on the cyclic voltammetric time scale but, at longer time scales, a migration of the phenyl group on Fe(Por)($C_6H_5$) occurs[77].

Fe(TPP)($C_6H_5$) has a $Fe^{III}$ ion which is displaced by 0.17 Å from the mean porphyrin plane[95] and this suggests that $C_6H_5^-$ is a very strong field ligand. In solution Fe(Por)(R) complexes have generally been described as five-coordinate, low-spin iron(III) derivatives (see Sect. B III 1). However, Fe(Por)($C_6F_4H$) and Fe(Por)($C_6F_5$) contain high spin $Fe^{III}$ for complexes, where Por = OEP, TPP, TmTP, and TpTP[68]. The Fe(TpEt$_2$ NPP)($C_6H_5$) complex contains high spin $Fe^{III}$[68] at low temperature but low spin $Fe^{III}$ at room temperature. In addition, lowering the temperature in noncoordinating solvents (such as toluene or benzene) shifts the spin state from low to high spin for Fe(Por)($C_6H_5$)[65]. Thus complexes of Fe(Por)(R) may exist with high spin $Fe^{III}$, with low spin $Fe^{III}$, or with $Fe^{III}$ in two discrete spin states depending upon temperature.

A direct correlation exists between the spin state of $Fe^{III}$ and the stability of the oxidized or reduced species[68]. All investigated high spin σ-bonded iron porphyrins are unstable upon undergoing a one-electron reduction but are moderately stable upon being oxidized by one electron[68]. This is in contrast to the low spin σ-bonded $C_6H_5$ complexes where the singly reduced species is extremely stable but the singly oxidized species undergoes a rapid migration of the aryl group[77] from metal to nitrogen.

Half-wave potentials for the reversible oxidation or reduction of $Fe(Por)(R)$ are directly influenced by the nature of the electron donating or electron withdrawing group on the porphyrin ring. This was initially pointed out by Laviron and Guilard[13] for the reactions of $Fe(OEP)(CH_3)$ and $Fe(TPP)(CH_3)$ and by Lexa and Saveant[39, 40] for the reactions of $Fe(Por)(R)$ where $Por = C_{12}TPP$, OEP, DP, and TPP and R was one of several different alkyl groups. For a given R group (such as $CH_3$) the $E_{1/2}$ values become progressively more negative and follow the order $TPP > C_{12}TPP > DP > OEP$[39].

It was originally suggested that there was very little dependence of $E_{1/2}$ on the nature of the R group of $Fe(Por)(R)$[39]. This is true for $Fe(Por)(R)$ where R is a simple alkyl group but not for complexes where R is an aryl group such as $C_6H_5$, $C_6F_5$ or $C_6F_4H$. Replacing either four or five of the protons on $C_6H_5$ by fluoro groups contributes to a 70–80 mV positive shift of the half-wave potential for each F atom. A similar positive potential shift is observed upon adding electron withdrawing CN groups to the TPP of $Fe(TPP)(C_6H_5)$[68]. For example, the difference between $Fe(Por)(C_6H_5)$ reduction where $Por = (CN)_4TPP$ and $Por = TPP$ is 670 mV, with the former complex being the most easy to reduce ($E_{1/2} = -0.03$ V in PhCN, 0.1 M $TBA(PF_6)$).

The neutral $Fe(Por)(R)$ complex contains a formal $Fe^{III}$ oxidation state and addition of one electron will generate a formal $Fe^{II}$ species. Reduced $[Fe(TPP)(C_6H_5)]^-$ and $[Fe(OEP)(C_6H_5)]^-$ complexes have optical absorption spectra with a Soret region characteristic of a $Fe^{II}$ porphyrin[77] and these data seem to support the proposed $Fe^{III} \rightleftharpoons Fe^{II}$ transition. On the other hand, the electronic absorption spectra of $[Fe(OEP)(C_6H_5)]^-$ and $[Fe(TPP)(C_6H_5)]^-$ have maxima close to 760 nm. Similar absorption bands are found for a number of anion radicals[96] but these bands are also observed in the electronic absorption spectra of oxyheme complexes[97]. These bands may be distinguished by their extinction coefficients.

Because of the dual spectral properties of $Fe^{II}$ and a porphyrin $Fe^{III}$ anion radical, it was suggested that the reduction product of Eq. (7) could be described by the resonance hybrid shown below[77]:

$$Fe^{III}(Por)(C_6H_5) \quad \underset{\rightleftharpoons}{\overset{e}{}} \quad \begin{array}{c} [Fe^{II}(Por)(C_6H_5)]^- \\ \updownarrow \\ [Fe^{III}(Por)(C_6H_5)]^{-\cdot} \end{array} \qquad (7)$$

Pyridine can coordinate to the oxidized, neutral, and reduced iron(III)-phenyl σ-bonded porphyrins producing the corresponding six coordinate derivatives[78]. In this case the exact sequence of electron transfer steps will depend upon the concentration of pyridine in solution. The overall possible electron transfer sequences in $CH_2Cl_2$/pyridine mixtures is given by the series of reactions in Scheme 8. At low pyridine concentration the oxidation and reduction of $Fe(TPP)(C_6H_5)$ proceeds via reaction pathways 2 and 4. As the pyridine concentration is increased, $Fe(TPP)(C_6H_5)(py)$ is formed in solution and under these conditions, oxidation and reduction of $Fe(TPP)(C_6H_5)(py)$ is via reaction pathways 3 and 5. Finally, at high pyridine concentration $[Fe(TPP)(C_6H_5)(py)]^-$ can be formed in solution and under these conditions the reactions proceed via pathways 3 and 6[78].

It was noted that oxidized six coordinate σ-bonded complexes exist only as transient intermediates[78]. The formation of $Fe(Por)(C_6H_5)(py)$ from $Fe(Por)(C_6H_5)$ was monitored both spectrally and electrochemically and formation constants of $10^{1.6}$ to $10^{2.5}$ were

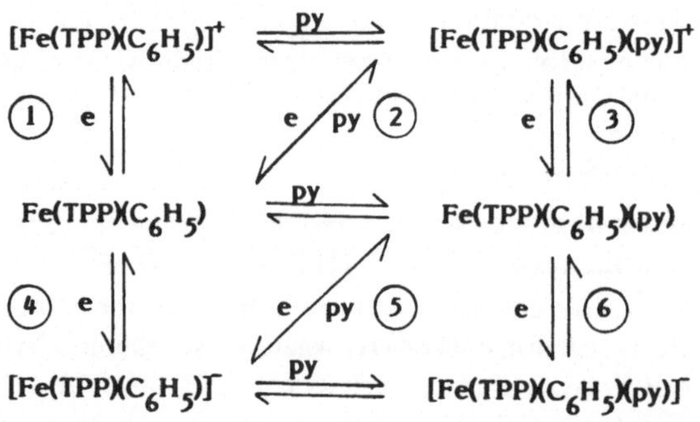

$$[\text{Fe(TPP)}(C_6H_5)]^+ \xrightleftharpoons{\text{py}} [\text{Fe(TPP)}(C_6H_5)(\text{py})]^+$$

(1) e          e  py (2)          e (3)

$$\text{Fe(TPP)}(C_6H_5) \xrightarrow{\text{py}} \text{Fe(TPP)}(C_6H_5)(\text{py})$$

(4) e          e  py (5)          e (6)

$$[\text{Fe(TPP)}(C_6H_5)]^- \xrightleftharpoons{\text{py}} [\text{Fe(TPP)}(C_6H_5)(\text{py})]^-$$          **Scheme 8**

calculated for pyridine binding by the five coordinate $Fe^{III}$ complexes[78]. The reduction of $Fe(Por)(C_6H_5)$ is electrochemically reversible and stable solution products could be identified spectroscopically. Potentials for the reduction are shifted in negative direction from those of $Fe(Por)(C_6H_5)$, consistent with stabilization of the six coordinate species. The electron addition to $Fe(Por)(C_6H_5)(py)$ was also proposed to yield a resonance hybrid involving an $Fe^{II}$ and an $Fe^{III}$ anion radical[78] similar to that of $[Fe(Por)(C_6H_5)]^-$ (Eq. 7) but the spectral data indicated a smaller contribution of the anion radical form to the resonance hybrid.

## B IV 1 a)  Oxidation of Iron Complexes and Electrochemically Initiated Migration of Alkyl and Aryl Groups

Electrooxidation mechanisms of $Fe(TPP)(C_6H_5)$ and $Fe(OEP)(C_6H_5)$ have been studied in nonbonding solvents[77] and in mixed solvent systems containing pyridine and $CH_2Cl_2$[78]. The initial electro-oxidation proceeds via an one electron abstraction to yield $[Fe(Por)(C_6H_5)]^+$ or $[Fe(Por)(C_6H_5)(py)]^+$. These singly oxidized complexes are not stable and undergo a series of chemical and electrochemical reactions leading ultimately to the iron(III) N-phenyl porphyrins $[Fe^{III}(N-C_6H_5TPP)]^{+2}$ and $[Fe^{III}(N-C_6H_5OEP)]^{+2}$. These derivatives can be reversibly reduced by a single electron to quantitatively give $[Fe^{II}(N-C_6H_5TPP)]^+$ and $[Fe^{II}(N-C_6H_5OEP)]^+$. Further reductions are also possible, leading to a transient species tentatively identified as an N-phenyl $Fe^I$ porphyrin[77].

The overall two electron oxidation of $Fe(Por)(C_6H_5)$ involves a chemical reaction coupled between two reversible electron transfer steps[78]. This electrochemical ECE type mechanism is shown below:

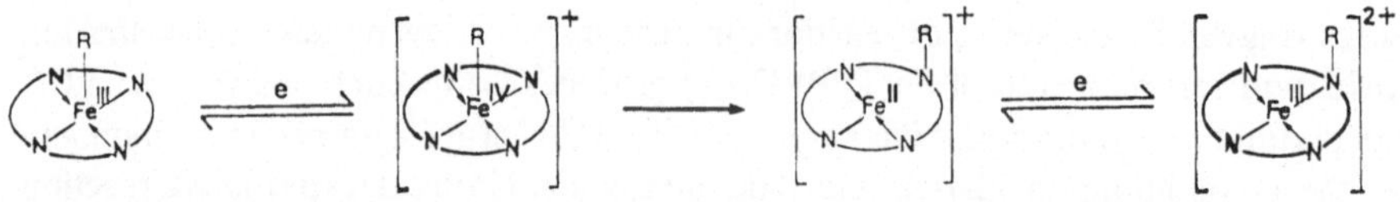

$$\tag{8}$$

A time resolved spectrum of singly oxidized $[Fe(TPP)(C_6H_5)]^+$ has not been obtained but spectral identification of $[Fe(OEP)(C_6H_5)]^+$ was possible on the thin-layer time scale[77]. $[Fe(OEP)(C_6H_5)]^+$ does not exhibit any absorption bands above 600 nm and values of molar extinction coefficients for this complex are very close to the corresponding values of the neutral species. In addition, there is a sharp blue shifted $\beta$ band in the spectrum. This absorbance is not found for any reported porphyrin cation radical spectra and on this basis the oxidized species was characterized as containing $Fe^{IV}$ [77].

The reduction of electrogenerated $[Fe^{II}(N-RPor)]^+$ was also investigated and found to proceed via an ECE mechanism in which the chemical step is a migration of the R group from the nitrogen of the porphyrin ring to the iron center, thus regenerating the initial σ-bonded complex. This overall oxidation/reduction sequence involving "reversible" migrations of the alkyl or aryl groups is shown in Scheme 9 [77].

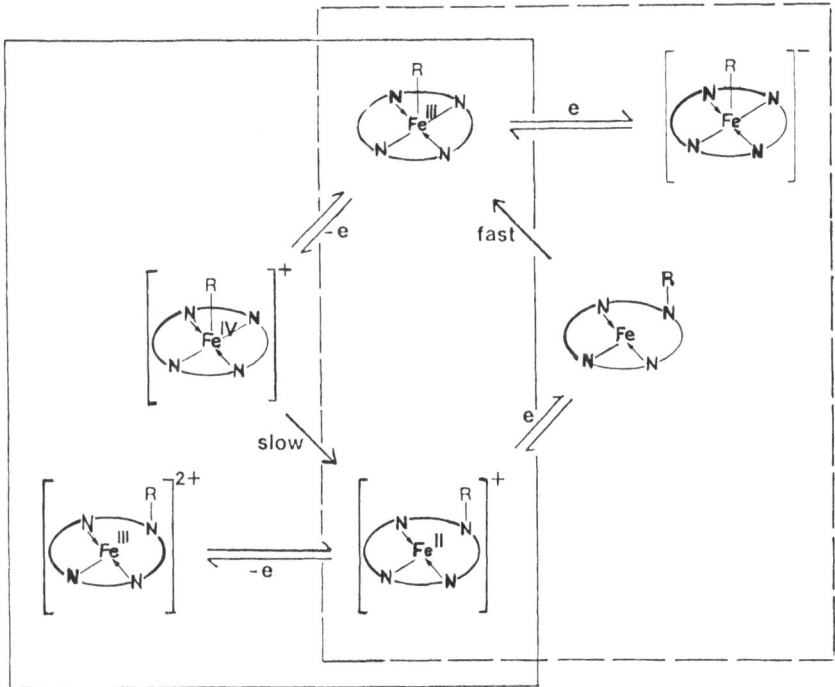

**Scheme 9**

## B IV 1 b) Iron-Alkyl and Iron-Aryl Porphyrins with a Bound NO Molecule

The formation of six coordinate Fe(Por)(R)(NO) is limited to studies under an NO atmosphere. Yields of 50–100% Fe(Por)(R)(NO) are obtained from Fe(Por)(R) depending upon the NO pressure and the specific alkyl or aryl group [98]. A series of Fe(Por)(R)(NO) have been prepared in fairly pure form and were electrochemically [79, 80] and chemically characterized [80]. Iron-NO complexes containing OEP and TPP porphyrin rings and σ-bonded $CH_3$, n-$C_4H_9$, $C_6H_5$, pMeC$_6H_4$, pOMeC$_6H_4$, and $C_6F_4H$ groups have been characterized by $^1H$ NMR, IR, and UV-visible spectroscopy [80]. Based on these data the central metal was assigned as being in the $Fe^{II}$ oxidation state [80, 98].

Reversible oxidations of Fe(Por)(R)(NO) are obtained at potentials positively shifted with respect to those of Fe(Por)(R), Fe(Por)(R)(py), and Fe(Por)(NO) [94]. For example, Fe(TPP)($C_6H_5$) is oxidized at 0.61 V in PhCN while Fe(TPP)($C_6H_5$)(py) has an oxidation potential of 0.53 V in PhCN containing 1 M pyridine. In the same solvent system Fe

(TPP)(NO) is reversibly oxidized at 0.75 V and Fe(TPP)(C$_6$H$_5$)(NO) under a partial pressure of 72 mm HgNO, is reversibly oxidized at 0.86 V[79, 80]. This latter potential is shifted over 1.5 V from the formal Fe$^{II}$/Fe$^{III}$ potential of Fe(Por)(C$_6$H$_5$) and is consistent with the strong stabilization of the Fe$^{II}$ oxidation state by NO, even in the presence of a σ-bonded phenyl group.

## B IV 2) Cobalt, Rhodium, and Iridium Porphyrins

The electrochemistry of cobalt porphyrins with σ-bonded alkyl or aryl groups is limited to three studies by Dolphin[35], Gross[81], and Gaudemer et al.[33]. The electro-oxidation of Co(TPP)(R) where R = C$_2$H$_5$ or C$_2$H$_5$O$_2$C was reported by Dolphin[35]. Both σ-bonded Co$^{III}$ complexes were oxidized by an electrochemical EC mechanism which led to the formation of an N-ethyl or N-ethoxycarbonyl cobalt(II) porphyrin. The first oxidation of neutral Co(TPP)(R) occurs at 0.98 V (R = C$_2$H$_5$) or 1.04 V (R = C$_2$H$_5$O$_2$C) vs. Ag/AgCl in CH$_2$Cl$_2$, 0.1 M TBA(PF$_6$). A second oxidation of Co(TPP)(C$_2$H$_5$O$_2$C) is also observed on the cyclic voltammetric time scale. This reaction occurs at 1.27 V vs. Ag/AgCl. The second oxidation of Co(TPP)(C$_2$H$_5$) is only reversible at − 40 °C, and this reaction occurs at − 1.19 V vs. Ag/AgCl[35].

Migration of the σ-bonded R group from Co$^{III}$ to one of the four nitrogens of the porphyrin ring is given by Eq. 9.

$$Co^{III}(TPP)(R) \ \overset{-e}{\rightleftarrows} \ [Co^{III}(TPP)(R)]^{+\cdot} \rightarrow [Co^{II}(N\text{–}RTPP)]^+ \tag{9}$$

**Table 6.** Half-wave potentials (V vs. SCE) for oxidation and reduction of Co(TPP)(R)

| R Group | Oxidation | | Reduction | Solvent | Ref. |
|---|---|---|---|---|---|
| | 1st | 2nd | 1st | | |
| C$_6$H$_4$OCH$_3$ | 1.04 | 1.30 | − 1.35 | CH$_2$Cl$_2$[b] | 81 |
| C$_6$H$_4$CH$_3$ | 1.07 | 1.30 | − 1.37 | CH$_2$Cl$_2$[b] | 81 |
| C$_6$H$_5$ | 1.09 | 1.41 | − 1.30 | CH$_2$Cl$_2$[b] | 81 |
| C$_6$H$_4$Br | 1.09 | 1.37 | − 1.36 | CH$_2$Cl$_2$[b] | 81 |
| C$_6$H$_4$Cl | 1.07 | 1.32 | − 1.34 | CH$_2$Cl$_2$[b] | 81 |
| C$_6$H$_4$NO$_2$ | 1.15 | 1.33 | (− 1.10)[a] | CH$_2$Cl$_2$[b] | 81 |
| C$_2$H$_5$ | 0.98[d, e] | 1.19[d, e] | | CH$_2$Cl$_2$[c] | 35 |
| C$_2$H$_5$O$_2$C | 1.04[e] | 1.27[e] | | CH$_2$Cl$_2$[c] | 35 |
| CH$_3$ | | | − 1.330 | Me$_2$SO | 33 |
| C$_2$H$_5$ | | | − 1.300 | Me$_2$SO | 33 |
| C$_3$H$_7$ | 0.795 | | − 1.310 | Me$_2$SO | 33 |
| C$_4$H$_9$ | 0.780 | | − 1.320 | Me$_2$SO | 33 |
| iso-C$_4$H$_9$ | 0.785 | | − 1.305 | Me$_2$SO | 33 |
| C$_6$H$_{11}$ | 0.725 | | − 1.320 | Me$_2$SO | 33 |
| C$_2$H$_4$(C$_6$H$_4$) | 0.775 | | − 1.318 | Me$_2$SO | 33 |

[a]  Reduction involves nitro group
[b]  0.1 M TBAP supporting electrolyte
[c]  0.1 M TBA(PF$_6$) supporting electrolyte
[d]  At − 40 °C
[e]  Measured vs. Ag/AgCl reference electrode

The generated $[Co^{II}(N-RTPP)]^+$ is also electroactive and oxidation waves have been reported at 1.10 and 1.70 V vs. Ag/AgCl for electrogenerated $[Co^{II}(N-C_2H_5TPP)]^{+\,35)}$. These latter two reactions correspond to the successive formation of a $Co^{III}$ N-alkyl and a $Co^{III}$ N-alkyl porphyrin cation radical.

Callot and Gross investigated the conversion of σ-bonded $Co^{III}(TPP)(R)$ to N-substituted $[Co(N-RTPP)]^+$ using chemical[81, 82)] and electrochemical[81)] oxidation. Five Co(TPP)(R) compounds were synthesized where $R = C_6H_4X$ and $X = OCH_3$, H, Br, Cl, or $NO_2$[82)]. Two reversible oxidations of each complex are obtained by cyclic voltammetry in $CH_2Cl_2$[81)]. These potentials are listed in Table 6. Values of $E_{1/2}$ for the first oxidation in $CH_2Cl_2$ vary between 1.04 and 1.15 V vs. SCE while the second oxidation of Co(Por)(R) has potentials which vary between 1.30 and 1.41 vs. SCE in $CH_2Cl_2$. A migration was not observed by Gross using cyclic voltammetry[81)] but chemical oxidation of the σ-bonded $Co^{II}$ complexes yielded chlorocobalt(II) N-aryl porphyrins[81, 82)].

The oxidation of Co(TPP)(R) has also been investigated by rotating disk voltammetry in $Me_2SO$[33)]. In this coordinating solvent, potentials are shifted in a negative direction by 200–300 mV from values in $CH_2Cl_2$. A similar shift of potential is observed upon adding concentrations of $Me_2SO$ to $CH_2Cl_2$ containing Co(Por)(R)[33b)] and this suggests the coordination of $Me_2SO$ to oxidized $[Co(TPP)(R)]^+$. The reduction potential of Co(TPP)(R) is not shifted on going from $CH_2Cl_2$ to $Me_2SO$ and this suggests only a small interaction of $Me_2SO$ with the neutral form of the complex.

The reduction of Co(TPP)(R) is reversible on the cyclic voltammetric time scale[33, 81)] and yields $[Co(TPP)(R)]^-$ before cleavage of the cobalt-carbon bond (Eq. 10).

$$Co(TPP)(R) + e \rightleftarrows [Co(TPP)(R)]^- \rightarrow Co(TPP) + R^- \qquad (10)$$

Potentials for reduction of different Co(TPP)(R) complexes vary little with the nature of R and values of $E_{1/2}$ are all close to $-1.30$ V (see Table 6). The potential for reduction of Co(TPP) to $[Co(TPP)]^{-1}$ is $-0.81$ V in $Me_2SO$[99, 100)]. Thus, the Co(TPP) formed after cleavage of the cobalt-carbon bond is immediately reduced to $[Co(TPP)]^{-1}$. Under these conditions, the overall electrode reaction is given by Eq. 11[33)].

$$Co(TPP)(R) + 2e \rightarrow [Co(TPP)]^- + R^- \qquad (11)$$

**Table 7.** Half-wave potentials (V vs. SCE) for oxidation and reduction of Rh(Por)(R) complexes in nonaqueous solvents containing 0.1 M TBAP

| Compound | Oxidation | | Reduction | | Solvent | Ref. |
|---|---|---|---|---|---|---|
| | 2nd | 1st | 1st | 2nd | | |
| Rh(TPP)(CH$_3$) | 1.24 | 0.94 | −1.44 | – | CH$_2$Cl$_2$ | 86 |
| | 1.34 | 0.96 | −1.39 | −1.87 | PhCN | 86 |
| | – | – | −1.43 | −1.85 | THF | 86 |
| Rh(TPP)(COCH$_3$) | 1.33$^a$ | 1.03 | −1.36 | −1.84 | PhCN | 84 |
| | | | −1.37 | −1.78 | THF | 84 |
| Rh(TPP)(CH$_2$Cl) | 1.38 | 0.92 | −1.50 | – | CH$_2$Cl$_2$ | 85 |
| Rh(OEP)(CH$_3$) | – | 0.85 | −1.69 | – | THF | 86 |

$^a$ Irreversible reaction. Value presented is $E_{pa}$ at scan rate = 0.1 V/s

Electrochemistry of σ-bonded Rh porphyrins has been limited to several studies by Kadish and Guilard[84-86]. Rh(Por)(R) is reduced by one or two one-electron transfer steps in nonaqueous media[86]. These reactions were monitored by UV-visible and ESR spectroscopy and the products characterized as anion radicals and dianions which are formed by the stepwise addition of one or two electrons to the complexes. The first reduction of Rh(TPP)(CH₃) varies between − 1.39 and − 1.44 V vs. SCE depending on solvent while the second reduction occurs at − 1.85 V (in THF) or − 1.87 V vs. SCE (in PhCN) (see Table 7). Two reductions also occur for Rh(TPP)(COCH₃)[84] but Rh(TPP)(CH₂Cl) and Rh(OEP)(CH₃) undergo only a single electroreduction within the potential range of the solvent[85, 86].

The reduction of [Rh(TPP)(L)₂]⁺Cl⁻ where L = dimethylamine generates a highly reactive Rh$^{II}$ species[83]. In PhCN, THF, and py, this complex dimerizes to generate [Rh(TPP)]₂. However, in CH₂Cl₂, the Rh$^{II}$ radical does not dimerize but instead reacts with the solvent to form Rh(TPP)(CH₂Cl)[85]. This is the first example of a σ-bonded carbon rhodium(III) species whose synthesis is electrochemically initiated. Rh(TPP) (CH₂Cl) has been isolated after bulk electrolysis of [Rh(TPP)(L)₂]⁺Cl⁻ in CH₂Cl₂ and characterized by infrared, ¹H NMR, UV-visible spectroscopy and cyclic voltammetry[85]. The properties of Rh(TPP)(CH₂Cl) are virtually identical to those of Rh(TPP)(CH₃).

The first and second oxidations of Rh(TPP)(CH₃)[86] and Rh(TPP)(CH₂Cl)[85] occur by reversible single electron transfer steps. Both oxidations involve abstraction of an electron from the porphyrin π ring system to yield cation radicals and dications. In contrast, the second oxidation of Rh(TPP)(COCH₃) is irreversible and leads to cleavage of the Rh-carbon bond with formation of [Rh(TPP)]⁺ or [Rh(TPP)]⁺² depending upon the applied potential[84]. The first oxidation seems to involve an electron abstraction involving the acetyl group rather than the porphyrin π ring system.

The electrochemistry of Ir(OEP)(CH₃) and Ir(OEP)(CH₃)(L), where L = CO, CN⁻, py, NH₃ and N-MeIm has also been reported[87]. A quasireversible wave was reported at positive potentials using cyclic voltammetry. The potential for this reaction in CH₂Cl₂ was 0.68 V for Ir(Por)(CH₃). Other potentials varied as a function of the L ligand in Ir(OEP)(CH₃)(L) (see Table 8). The cyclic voltammetric waves were assigned as due to a reduction of the Ir(OEP)(CH₃) or Ir(OEP)(CH₃)(L) complex[87], despite the positive

**Table 8.** Half-wave potentials (V vs. SCE) for oxidation and reduction of Ir(OEP)(R) complexes in CH₂Cl₂ and THF containing 0.1 M TBAP

| Compound | Oxidation | Reduction | Solvent | Ref. |
|---|---|---|---|---|
| Ir(OEP)(CH₃) | 0.68ᵃ | – | CH₂Cl₂ | 87 |
| Ir(OEP)(CH₃)(CO) | 0.90ᵃ | – | CH₂Cl₂ | 87 |
| Ir(OEP)(CH₃)(N–MeIm) | 0.56ᵃ | – | CH₂Cl₂ | 87 |
| [Ir(OEP)(CH₃)(CN)]⁻ | < 0.43ᵃ | – | CH₂Cl₂ | 87 |
| Ir(OEP)(CH₃)(py) | 0.65ᵃ | – | CH₂Cl₂ | 87 |
| Ir(OEP)(CH₃)(NH₃) | 0.58ᵃ | – | CH₂Cl₂ | 87 |
| Ir(OEP)(C₈H₁₃) | 0.68 | − 1.79 | THF | 88 |
| Ir(OEP)(C₈H₁₃)(CO) | 0.80ᵇ | − 1.65ᵇ | THF | 88 |

ᵃ Waves erroneously reported as reductions in original Ref. 87.
ᵇ $P_{CO}$ = 1 atmosphere

potential observed. However, later studies of other σ-bonded Ir$^{III}$ complexes[88] suggest that these waves are actually due to an oxidation of the Ir$^{III}$ complexes.

Two reversible one electron processes are observed for oxidation and reduction of Ir(OEP)(C$_8$H$_{13}$)[88]. These occur at 0.68 V and − 1.79 V vs. SCE in THF, 0.1 M TBAP. The wave at 0.68 V was shown by rotating disk voltammetry to be an oxidation while the wave at − 1.79 V was shown to be a reduction. The oxidation is reversible on both cyclic voltammetric and bulk electrolysis time scales (1–5 min) and was shown to involve the abstraction of an electron from the bound axial ligand.

The reduction of Ir(OEP)(C$_8$H$_{13}$) also proceeds via a reversible one electron process on the cyclic voltammetric time scale. However, the generated [Ir(OEP)(C$_8$H$_{13}$)]$^-$ is not sufficiently stable to be characterized and on longer time scales [Ir(OEP)]$^-$ is formed after a global two electron reduction and loss of the σ-bonded ligand[88].

Two reversible waves are observed for oxidation and reduction of Ir (OEP)(C$_8$H$_{13}$)(CO). These occur at 0.80 V and − 1.65 V vs. SCE in THF, 0.1 M TBAP. The wave at 0.80 V was ascertained to be an oxidation and the wave at − 1.65 V to be a reduction by rotating disk voltammetry. The oxidation of Ir(OEP)(C$_8$H$_{13}$)(CO) was reversible only on cyclic voltammetric time scales and bulk oxidation resulted in the formation of Ir(OEP)(CO)(ClO$_4$). The reduction of Ir(OEP)(C$_8$H$_{13}$)(CO) is reversible on bulk electrolysis time scales. For this complex the reduction occured at the porphyrin ligand and led to the proposed formation of [Ir(OEP)(C$_8$H$_{13}$)(CO)]$^-$ [88]. The reported potentials for all of the compounds are presented in Table 8.

## B IV 3) Gallium, Indium, and Thallium Porphyrins

The electrochemistry of gallium, indium, and thallium porphyrins with σ-bonded aryl and alkyl groups has been reported by Kadish and Guilard[24, 63, 64, 89]. Complexes of Ga(Por)(R), In(Por)(R) and Tl(Por)(R) are reversibly reduced by up to two single electron transfer additions. These electroreductions are shown in Eq. 12 and 13 where M = Ga, In, or Tl.

$$M(Por)(R) + e \rightleftarrows [M(Por)(R)]^- \tag{12}$$
$$[M(Por)(R)]^- + e \rightleftarrows [M(Por)(R)]^{-2} \tag{13}$$

Half-wave potentials for these reversible reductions are shown in Tables 9 and 10.

Values of E$_{1/2}$ for reduction of a given M(Por)(R) complex containing the same porphyrin ring and the same σ-bonded alkyl or aryl group shift only slightly upon changing from M = Ga to M = Tl. For example the reduction of M(OEP)(C$_6$H$_5$) in CH$_2$Cl$_2$ occurs at − 1.49 V vs. SCE for M = Ga[64], at − 1.47 V vs. SCE for M = In[89] and at − 1.45 V vs. SCE for M = Tl[24]. A similar positive shift of E$_{1/2}$ occurs on going from Ga(TPP)(CH$_3$) to In(TPP)(CH$_3$) to Tl(TPP)(CH$_3$). For example, the reduction of M(TPP)(CH$_3$) occurs at − 1.29 V vs. SCE for M = Ga[64], at − 1.24 V vs. SCE for M = In[89], and at − 1.23 V vs. SCE for M = Tl[24].

Reduction of the gallium and indium complexes involves the porphyrin π ring system[64, 89]. The site of electron transfer in Tl(Por)(R) is not clear. A reduction of Tl$^{III}$ to Tl$^I$ is possible in Tl(Por)(R) but the similarity in potentials between Ga(Por)(R) and In(Por)(R) suggests that reduction of this complex occurs at the porphyrin π ring system[24].

**Table 9.** Half-wave potentials (V vs. SCE) for the reversible reduction of M(OEP)(R) (M = Ga, In, or Tl) in $CH_2Cl_2$, 0.1 M TBA($PF_6$)

| R | Central Metal | | |
|---|---|---|---|
| | Ga[a] | In[b] | Tl[c] |
| $C(CH_3)_3$ | − 1.54 | − 1.54 | − |
| $CH(CH_3)_2$ | − | − 1.53 | − |
| $C_4H_9$ | − 1.53 | − 1.50 | − |
| $C_2H_5$ | − 1.51 | − 1.48 | − |
| $CH_3$ | − 1.53 | − 1.50 | − 1.48 |
| $C_6H_5$ | − 1.49 | − 1.47 | − 1.45 |
| $C_2H_2C_6H_5$ | − 1.47 | − 1.48 | − |
| $C_2C_6H_5$ | − 1.43 | − | − |
| $C_6F_4H$ | − | − 1.45 | − 1.39 |
| $C_6F_5$ | − | − 1.46 | − |

[a] Ref. 64;  [b] Ref. 63;  [c] Ref. 24

**Table 10.** Half-wave potentials (V vs. SCE) for the reversible reduction of M(TPP)(R) (M = Ga, In, or Tl) in $CH_2Cl_2$, 0.1 M TBA($PF_6$)

| R | Central metal | | | | | |
|---|---|---|---|---|---|---|
| | Ga[a] | | In[b] | | Tl[c] | |
| | 1st | 2nd | 1st | 2nd | 1st | 2nd |
| $C(CH_3)_3$ | − 1.31 | − 1.74 | − 1.27 | − 1.71 | − | − |
| $CH(CH_3)_2$ | − | − | − 1.26 | − 1.67 | − | − |
| $C_4H_9$ | − 1.29 | − 1.73 | − 1.26 | − 1.69 | − | − |
| $C_2H_5$ | − 1.27 | − 1.70 | − 1.25 | − 1.66 | − | − |
| $CH_3$ | − 1.29 | − 1.71 | − 1.24 | − 1.64 | − 1.23 | − 1.64 |
| $C_2H_2C_6H_5$ | − 1.24 | − 1.67 | − 1.24 | − 1.64 | − | − |
| $C_6H_5$ | − 1.22 | − 1.66 | − 1.22 | − 1.62 | − 1.21 | − 1.63 |
| $C_2C_6H_5$ | − 1.19 | − 1.59 | − | − | − | − |
| $C_6F_4H$ | − | − | − 1.20 | − 1.60 | − 1.17 | − |
| $C_6F_5$ | − | − | − 1.20 | − 1.60 | − | − |

[a] Ref. 64;  [b] Ref. 63;  [c] Ref. 24

The oxidation of Ga(Por)(R), In(Por)(R), and Tl(Por)(R) proceeds in one to three steps depending on the metal ion, the nature of the R group, and the solvent system. Ga(Por)(R) and In(Por)(R) complexes with alkyl groups undergo an electrochemically reversible one electron oxidation which is followed by rapid cleavage of the metal-carbon bond[64, 89].

$$M(Por)(R) \underset{}{\overset{-e}{\rightleftarrows}} [M(Por)(R)]^+ \rightarrow [M(Por)]^+ + R^{\cdot} \tag{14}$$

This results in an overall irreversible oxidation peak which occurs at an $E_p$ between 0.64 and 1.05 V depending upon the nature of the alkyl group and the type of porphyrin ring (OEP or TPP). The generated $[Ga(Por)]^+$ and $[In(Por)]^+$ are also electroactive and two

**Table 11.** Peak potentials (V vs. SCE) for the irreversible oxidation of Ga(Por)(R) and In(Por)(R) by cyclic voltammetry in $CH_2Cl_2$, 0.1 M TBA $(PF_6)$[a]

| R | OEP | | TPP | |
|---|---|---|---|---|
| | Ga[b] | In[c] | Ga[b] | In[c] |
| $C(CH_3)_3$ | 0.68 | 0.53 | 0.76 | 0.64 |
| $CH(CH_3)_2$ | – | 0.62 | – | 0.74 |
| $C_4H_9$ | 0.71 | 0.67 | 0.79 | 0.78 |
| $C_2H_5$ | 0.72 | 0.67 | 0.80 | 0.83 |
| $CH_3$ | 0.76 | 0.70 | 0.86 | 0.88 |
| $C_2H_2C_6H_5$ | 0.98 | 1.81 | 0.99 | 0.92 |
| $C_6H_5$ | 0.89[d, e] | 0.79[d, e] | 1.05 | 0.94[d, e] |
| $C_2C_6H_5$ | 0.93[e] | 1.01[e, f] | 1.27[e] | 1.15[e, f] |

[a] $E_{pa}$ obtained at scan rate = 0.1 V/s
[b] Ref. 64
[c] Ref. 63
[d] Measured at $-10\,°C$
[e] Reversible $E_{1/2}$ value
[f] Measured in PhCN, 0.1 M TBA($PF_6$)

**Table 12.** Half-wave potentials (V vs. SCE) of In(Por)($C_6F_4H$) and In(Por)($C_6F_5$) complexes in $CH_2Cl_2$, PhCN and pyridine containing 0.1 M (TBA)$PF_6$[a]

| Compound | Solvent | Oxidation | | Reduction | |
|---|---|---|---|---|---|
| | | 2nd | 1st | 1st | 2nd |
| In(OEP)($C_6F_4H$) | PhCN | 1.37 | 0.96 | $-1.38$ | – |
| In(TPP)($C_6F_4H$) | $CH_2Cl_2$ | 1.52 | 1.03 | $-1.20$ | $-1.60$ |
| | PhCN | 1.54 | 1.09 | $-1.11$ | $-1.57$ |
| | Pyridine | – | – | $-1.05$ | $-1.50$ |
| In(TmTP)($C_6F_4H$) | PhCN | 1.52 | 1.06 | $-1.12$ | $-1.59$ |
| | Pyridine | – | – | $-1.08$ | $-1.54$ |
| In(TpTP)($C_6F_4H$) | $CH_2Cl_2$ | 1.48 | 0.96 | $-1.24$ | $-1.64$ |
| | PhCN | 1.50 | 1.03 | $-1.14$ | $-1.60$ |
| | Pyridine | – | – | $-1.08$ | $-1.54$ |
| In(OEP)($C_6F_5$) | $CH_2Cl_2$ | 1.43 | 0.90 | $-1.46$ | – |
| | PhCN | 1.39 | 0.96 | $-1.38$ | – |
| | Pyridine | – | – | $-1.30$ | $-1.80$ |
| In(TPP)($C_6F_5$) | $CH_2Cl_2$ | 1.53 | 1.05 | $-1.20$ | $-1.60$ |
| | PhCN | 1.53 | 1.10 | $-1.10$ | $-1.55$ |
| | Pyridine | – | – | $-1.04$ | $-1.49$ |
| In(TmTP)($C_6F_5$) | $CH_2Cl_2$ | 1.50 | 1.01 | $-1.22$ | $-1.63$ |
| | PhCN | 1.51 | 1.09 | $-1.10$ | $-1.56$ |
| | Pyridine | – | – | $-1.06$ | $-1.52$ |
| In(TpTP)($C_6F_5$) | $CH_2Cl_2$ | 1.49 | 0.99 | $-1.22$ | $-1.62$ |
| | PhCN | 1.48 | 1.04 | $-1.13$ | $-1.59$ |
| | Pyridine | – | – | $-1.06$ | $-1.52$ |

[a] Taken from Ref. 89

well-defined oxidations of these complexes are observed[64, 89]. Potentials for the first oxidation of Ga(Por)(R) and In(Por)(R) are summarized in Table 11.

Both Ga(Por)(C$_2$C$_6$H$_5$)[64] and In(Por)(C$_2$H$_2$C$_6$H$_5$)[89] give reversible waves for oxidation. The latter In$^{III}$ complexes are not stable in CH$_2$Cl$_2$ but are well defined in PhCN. Complexes of In(Por)(R) where R = C$_6$F$_4$H or C$_6$F$_5$ are also stable after electro-oxidation and two well defined oxidations of the porphyrin ring are observed. These reactions are given by Eq. 15 and 16 and potentials for the oxidation and reduction are given in Table 12.

$$In(Por)(R) \rightleftarrows [In(Por)(R)]^+ + e \tag{15}$$

$$[In(Por)(R)]^+ \rightleftarrows [In(Por)(R)]^{+2} + e \tag{16}$$

Likewise, all of the investigated Tl(Por)(R) complexes are reversibly oxidized by two electrons without cleavage of the Tl-carbon bond on the time scale of cyclic voltammetry[24]. Potentials for these reactions are given in Table 13.

### B IV 4) Germanium and Silicon Porphyrins

The electrochemistry of Ge$^{IV}$ and Si$^{IV}$ porphyrins with σ-bonded alkyl and aryl groups has been investigated by Kadish and Guilard[25, 90]. The reactions are complicated and must be carried out both in the dark and in the strict absence of O$_2$. An insertion of O$_2$ into the Ge-carbon bond of Ge(Por)(R)$_2$ has been postulated[101] but later studies[25, 29] show that these reactions are actually more complicated than initially suggested[101].

The reduction of Ge(Por)(R)$_2$ and Si(Por)(R)$_2$ occurs in either one or two single electron transfer steps. $E_{1/2}$ values of $-1.45$ V and $-1.48$ V are obtained for the first reduction of Ge(OEP)(CH$_3$)$_2$ and Si(OEP)(CH$_3$)$_2$ in PhCN and no second reduction is observed in this solvent[25]. In contrast, Ge(TPP)(C$_6$H$_5$)$_2$ is reduced in two steps which occur at $-1.10$ and $-1.65$ V in PhCN. Similar values of $-1.17$ V and $-1.72$ vs. SCE are measured for the reduction of Ge(TPP)(CH$_2$C$_6$H$_5$)$_2$. First and second reduction poten-

**Table 13.** Half-wave potentials (V vs. SCE) for the reversible oxidation and reduction of Tl(Por)Cl and Tl(Por)(R) in CH$_2$Cl$_2$, 0.1 M TBA(PF$_6$)[a]

| Compound | Oxidation | | Reduction | |
|---|---|---|---|---|
| | 2nd | 1st | 1st | 2nd |
| Tl(OEP)Cl | 1.55 | 1.07 | $-1.12^b$ | – |
| Tl(OEP)(C$_6$F$_4$H) | 1.39 | 0.90 | $-1.39$ | – |
| Tl(OEP)(C$_6$H$_5$) | 1.27 | 0.80 | $-1.45$ | – |
| Tl(OEP)(CH$_3$) | 1.18 | 0.75 | $-1.48$ | – |
| Tl(TPP)Cl | 1.57 | 1.21 | $-1.04$ | $-1.43$ |
| Tl(TPP)(C$_6$F$_4$H) | 1.49 | 1.01 | $-1.17$ | – |
| Tl(TPP)(C$_6$H$_5$) | 1.41 | 0.90 | $-1.21$ | $-1.63$ |
| Tl(TPP)(CH$_3$) | 1.34 | 0.86 | $-1.23$ | $-1.64$ |

[a]  Taken from Ref. 24
[b]  $E_{pa}$ measured at 0.10 V/s. A 0.03 V cathodic shift in $E_p$ was observed for each 10-fold increase in scan rate

**Table 14.** Half-wave potentials (V vs. SCE) for the oxidation and reduction of $Ge(Por)(R)_2$ and $Si(Por)(R)_2$ in PhCN, 0.1 M TBAP[a]

| Compound | Oxidation | | Reduction | |
|---|---|---|---|---|
| | 2nd[b] | 1st[c] | 1st | 2nd |
| $Ge(OEP)(CH_3)_2$ | 1.37 | 0.75 | − 1.48 | |
| $Ge(OEP)(CH_2C_6H_5)_2$ | | 0.64 | − 1.54 | |
| $Ge(OEP)(C_6H_5)_2$ | 1.39 | 0.88 | − 1.40 | |
| $Ge(TPP)(CH_2C_6H_5)_2$ | | 0.73 | − 1.17 | − 1.72 |
| $Ge(TPP)(C_6H_5)_2$ | 1.45 | 0.95 | − 1.10 | − 1.65 |
| $Si(OEP)(CH_3)_2$ | 1.22 | 0.72 | − 1.45 | |

[a]  Taken from Ref. 25
[b]  Oxidation is of species generated after cleavage of the metal-carbon bond
[c]  $E_{pa}$ measured at 0.10 V/s

tials of $Ge(TPP)(R)_2$ are given in Table 14 and are shifted negatively with respect to $Ge(TPP)X_2$ where X is an anionic ligand[90]. ·

The oxidation of $Ge(Por)(R)_2$ is complicated and involves several chemical and electrochemical reactions[90]. In the dark, $Ge(Por)(R)_2$ loses a single electron followed by cleavage of one or more of the σ-bonded groups. The generated complex is electroactive and depending upon scan rate will be reversibly or irreversibly reduced by a single electron at potentials between − 0.6 V and − 0.8 V vs. SCE. Studies of this reaction are still in progress.

## B V) Crystallography of σ-Bonded Metalloporphyrins

Relatively a few σ-bonded metalloporphyrins have been studied by X-ray diffraction methods. Complexes with $Rh^{III}$, $Rh^{IV}$, $Tl^{III}$, $In^{III}$, $Ge^{IV}$, $Fe^{III}$, and $Mo^{IV}$ [27c, 57, 95, 102–105] form five or six coordinated species whose structural parameters are dependent on the electronic state of the metal. Two classes of σ-bonded metalloporphyrins may be defined on the basis of structural data. These are the transition metal and the non transition metal porphyrins.

### B V 1) Five Coordinated Compounds

#### B V 1 a) Transition Metal Complexes

The first reported crystal structure of a metalloporphyrin with a σ-bonded methyl group is $Rh(OEP)(CH_3)$ [102] shown in Fig. 6.

The rhodium atom lies at the center of a square pyramid whose basal plane is defined by the four nitrogen atoms ($\langle Rh-N \rangle$ = 2.031 ± 0.01 Å) and the apex by the methyl group ($\langle CH_3-Rh-N \rangle$ = 91.4 ± 1.5°). The methyl rhodium(III) covalent bond (2.031(6) Å) is equal to the average rhodium-nitrogen distances and agrees well with measured distances in other square pyramidal rhodium(III) derivatives[106–108]. The distances of the metal atom from the plane of the four nitrogens and from the average

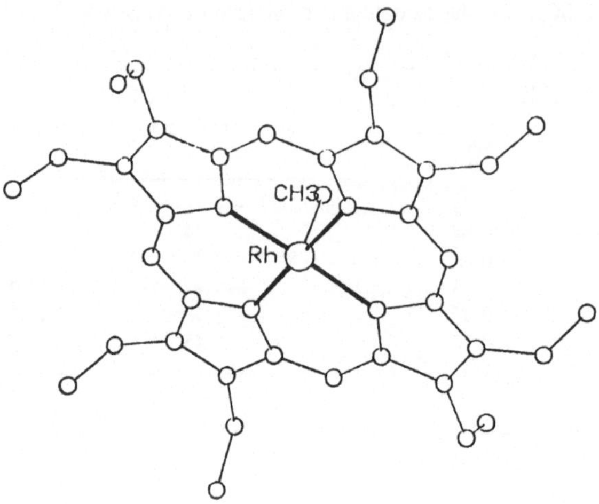

CH3

Rh

**Fig. 6.** ORTEP view of
Rh(OEP)(CH₃)
(coordinates from Ref. 102)

macrocycle are 0.051 and 0.013 Å, respectively. This leads to a 0.038 Å doming of the
porphyrin moiety towards the metal atom. Furthermore, a small out-of-plane displacement results from the low spin character of the diamagnetic (Rh(OEP)(CH₃) complex.
The empty $d_{x^2-y^2}$ orbital allows the rhodium atom to lie close to the plane of the four
nitrogens without a large expansion of the macrocycle (N–$C_t$ = 2.030 Å). As proposed in
Ref. 102, the bonding scheme in valence theory consists of a $[d_{x^2-y^2} sp^2]$ hybridization and
an almost pure $d_{z^2}$ axial bond which allows the metal to be in the plane of the porphyrin.
Furthermore a coplanar location of the rhodium atom favors metalloporphyrin backbonding which is important in all noble-metal porphyrins[109].

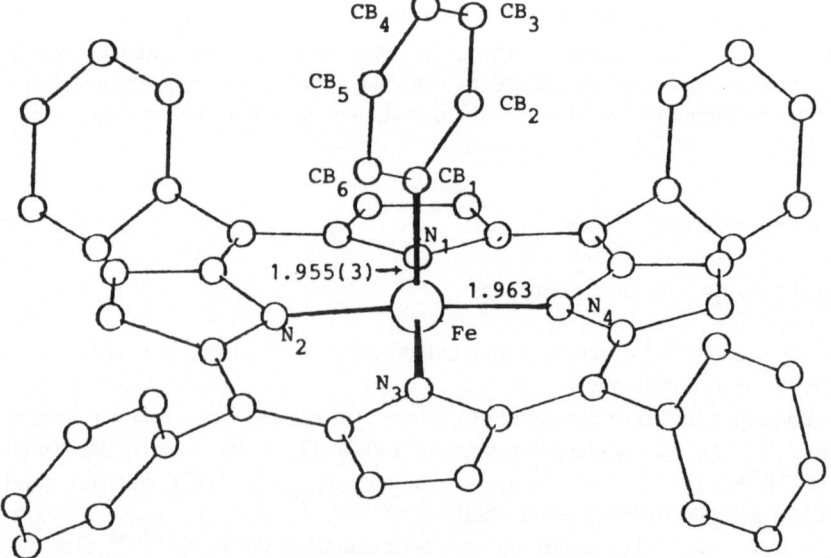

**Fig. 7.** Crystal structure of Fe(TPP)(C₆H₅) (coordinates from Ref. 95)

The covalent metal-carbon bond in Fe(TPP)(C$_6$H$_5$)[95] is a Fe–C$_{sp^2}$ bond; Fig. 7 gives an ORTEP view of this complex.

The five coordinated iron atom lies close to the porphyrin plane as expected for a low spin complex (see above). The Fe$^{III}$ ion is 0.175 Å from the plane of the four nitrogens and 0.165 Å from the macrocycle ($\Delta 4N = 0.175$ Å and $\Delta Por = 0.165$ Å). These values are shorter than those usually observed in other iron(II) or iron(III) low spin five coordinated complexes ($\Delta 4N = 0.29$ Å in Fe$^{III}$(OEP)(NO)ClO$_4$[110] and 0.22 Å in Fe$^{II}$(TPP)(NO)[111]. The small out-of-plane distance results in slight contacts between the hydrogen atoms of the σ-bonded phenyl ring and the four nitrogen atoms of the macrocycle (N$_2$–HB$_6$ = 2.71 Å, N$_3$–HB$_6$ = 2.86 Å, N$_4$–HB$_2$ = 2.68 Å, N$_1$–HB$_2$ = 2.93 Å). The phenyl ring is 10° from a perfect staggered conformation. The dihedral angles of the average phenyl ring with the N$_i$–Fe–CB$_1$ planes (i = 1–4) are respectively 55.0°, 35.5°, 54.5°, and 36.0°. The conformation minimizes the steric interactions with the nitrogen atoms.

The average iron-nitrogen bond length (1.963 ± 0.008 Å) is slightly shorter than those observed in other five or six coordinated low spin iron complexes (see for example Ref. 66). The radius of the macrocycle (1.953(1) Å) indicates a large contraction of the core when the coordination number changes from 6 to 5 (in the six coordinated iron(III) low-spin species the average radius of the core is 1.99(10) Å[66]. This contraction imposes a severe ruffling of the macrocycle as shown on the stereoview of the complex (Fig. 8) and the doming parameter[112] is 0.01 Å.

The metal-sp$^2$ carbon bond length of 1.955(3) Å in Fe(TPP)(C$_6$H$_5$) is longer than those found in the carbenic complexes[4, 113] (1.83(3) Å in Fe(TPP)(CCl$_2$)(H$_2$O) and 1.914(7) Å in Fe(TPP)[C(C$_6$H$_4$Cl)$_2$]Cl).

Comparisons may be made between the above two transition metal σ-bonded metalloporphyrins. In the rhodium complex the metal atom is almost in the plane of the macrocycle and this is not exactly the case for the ferric complex. It is not clear if this conformation difference is due only to a difference between the metal atoms. One explanation is that the σ-bonded ligands are not the same (C$_6$H$_5$ or CH$_3$ with two different hybridization schemes). As shown above, this leads to different magnetic behaviour and in this case the out-of-plane distance could be different. One would expect a slightly larger out-of-plane distance for the methyl species because of its spin state. However, the steric interactions of a methyl group are far less than those of the phenyl ring and Fe(TPP)(CH$_3$) could have the iron atom more in the porphyrin plane, thus favoring hexacoordination[78].

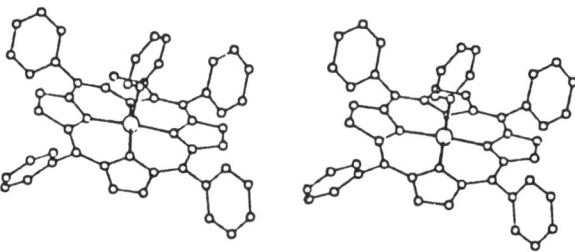

**Fig. 8.** Stereoview of
Fe(TPP)(C$_6$H$_5$)
(coordinates from Ref. 95)

## B V 1 b) Group 13 σ-Bonded Complexes

Group 13 methyl σ-bonded metalloporphyrins differ significantly in stereochemistry from the transition metal porphyrins described above. This is illustrated by Figure 9 which gives the ORTEP view of $In(TPP)(CH_3)$[104]. The main difference between σ-bonded complexes in the transition and non-transition metal series is the metal out-of-plane distance $\Delta 4N = 0.98$ Å and $0.78$ Å for group 13 metals and $0.17$ Å and $0.05$ Å for the transition metals (Table 15). These values cannot be explained by ionic radii considerations, because $In(TPP)SO_3CH_3$ is polymeric and the indium atom is exactly in the plane of the four nitrogens. With the exception of seven and six coordinated niobium(V) porphyrins[117-119] and eight coordinated hafnium(IV) and zirconium(IV) porphyrins[112] for which the displacement is between $0.9$ and $1.1$ Å, these out-of-plane distances are the largest yet observed (excluding rare earth metalloporphyrins). Furthermore, compared to the halogenated metalloporphyrins (Table 15), $\Delta 4N$ and $\Delta Por$ are $0.15$ to $0.20$ Å larger. This difference cannot be explained by crystal packing effects because, as also shown in Table 15, $In(TPP)(CH_3)$[104], $In(TPP)Cl$[114], $Tl(TPP)(CH_3)$[57], and $Tl(TPP)Cl$[57] are isotypic. Furthermore, this observation is exactly opposite to that of the transition metal porphyrins. The R groups are very strong field ligands and lead to an iron or

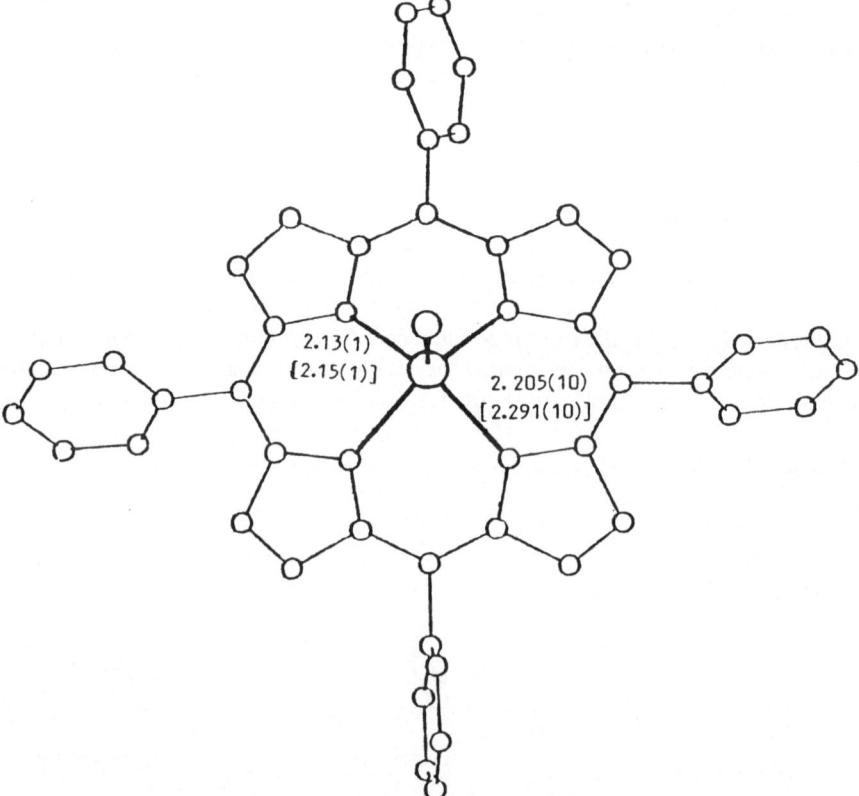

**Fig. 9.** Projection of $In(TPP)(CH_3)$[104]; the values given in *square brackets* are those of $Tl(TPP)(CH_3)$[57]

Table 15. Main characteristics of five-coordinated σ-bonded metalloporphyrins and five-coordinated chloro derivatives with the same porphyrin ring

| Compound | a (Å) | b (Å) | c (Å) | β (°) | Δ4N (Å) | ΔPor (Å) | (M–N) (Å) | M–L (Å) | Ref. |
|---|---|---|---|---|---|---|---|---|---|
| In(TPP)(CH$_3$) | 10.064(1) | 16.221(3) | 23.355(2) | 115.47(2) | 0.78(2) | 0.92(2) | 2.205(10) | 2.132(15) | 104 |
| Tl(TPP)(CH$_3$) | 10.046(2) | 16.244(3) | 23.373(3) | 115.5(1) | 0.98 | 1.11 | 2.291(10) | 2.147(12) | 57 |
| In(TPP)Cl | 10.099(1) | 16.117(2) | 23.382(2) | 115.6(1) | 0.61 | 0.71 | 2.15(1) | 2.369(2) | 114 |
| Tl(TPP)Cl | 10.064(2) | 16.177(2) | 23.354(5) | 115.3(1) | 0.74 | 0.86 | 2.21(1) | 2.420(4) | 57 |
| Tl(OEP)Cl | – | – | – | – | 0.69 | 0.75 | 2.212(6) | 2.449(2) | 115 |
| Rh(OEP)(CH$_3$) | – | – | – | – | 0.05 | 0.01 | 2.031(10) | 2.031(6) | 102 |
| Fe(TPP)(C$_6$H$_5$) | – | – | – | – | 0.175 | 0.165 | 1.963(11) | 1.955(3) | 95 |
| Fe(TPP)Cl | – | – | – | – | 0.38 | 0.38 | 2.049(9) | 2.192(12) | 116 |

rhodium atom in a low spin state. This allows depopulation of the $d_{x^2-y^2}$ orbital which may then come more closely to the plane of the four nitrogens leading to an almost pure $dsp^2$ hybridization and a $d_{z^2}$ σ ligand bonding. In the Group 13 complexes, the 3d orbitals are filled and the covalent interaction seems to be an $sp^3d$ (s, $p_x$, $p_y$, $p_z$, $d_{x^2-y^2}$) hybridization leading (because of the $p_z$ orbital) to a large out-of-plane position of Tl or In atoms.

## B V 2) Six Coordinated Species

Stereochemical parameters have been very briefly reported for Rh(TPP)(C₆H₅)Cl[103] and recently the crystal structures of Mo(TPP)(C₆H₅)Cl[105] and Ge(MGP)(CH₃)₂ [27c] have been described where MGP = dimethyl-5,10,15,20-tetrakis[3′,5′-bis(1″,1″-dimethyl-ethyl)phenyl] porphyrinato.

Figure 10 gives the ORTEP view of the molybdenum complex: the molybdenum atom is octahedrally coordinated by the four nitrogen atoms, the chlorine atom, and the $CB_1$ atom of the σ-bonded phenyl group. The average ⟨Mo–N⟩ distance of 2.070 ± 0.01 Å is equal to those found in other molybdenum(IV) porphyrins (2.074(5) Å in Mo(TPP)Cl₂)[120] and 2.070(5) Å in Mo(TPP)(NC₆H₅)₂[121]). The molybdenum carbon bond is 2.241(4) Å (2.05 Å in Rh(TPP)(C₆H₅)Cl[103]). The molybdenum atom lies at 0.089 Å out-of-plane of the four nitrogens towards the chlorine atom (the rhodium is in the plane in Rh(TPP)(C₆H₅)Cl) and 0.125 Å out-of-the average plane of the macrocycle. This thus results in a tight contact between the ortho hydrogen atoms of the σ phenyl ring ($HCB_2–N_2$ = 2.70 Å, $HCB_6–N_3$ = 2.75 Å) and the nitrogen atoms and imposes an almost staggered conformation. The dihedral angle between the $N_1$, $N_3$, $CB_1$ plane and that of the phenyl ring is 41°. Furthermore, the macrocycle possesses an $S_4$ ruffling with angles from 8.5 to 14.2° between adjacent pyrrole rings (Fig. 10). Finally the C–Mo distance, 2.241(4) Å (2.205 Å in the rhodium complex) appears typical of this type of bond[105].

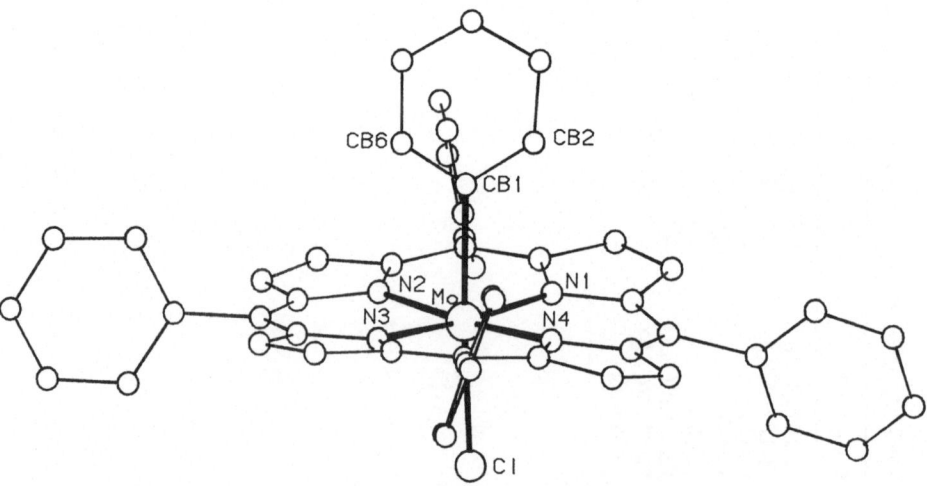

**Fig. 10.** ORTEP view of Mo(TPP)(C₆H₅)Cl (coordinates from Ref. 105)

The crystal structure of Ge(MGP)(CH$_3$)$_2$[27c] is poorly defined. The germanium atom lies on an inversion centre; the methyl-germanium bond length is 1.99(3) Å and the germanium-nitrogen distances are 2.02(4) and 2.03(4) Å. This latter complex exhibits antineoplastic activity against several types of tumors in vivo. This is most likely due to the activation of the methyl groups because the corresponding dichlorinated complex is inactive.

## C) Insertion of Small Molecules into the Metal-Carbon Bond

### C I) Carbon Dioxide

The nature of the porphyrin ring may influence reactions which are photoactivated or sensitized by visible light. Although electron transfer from metalloporphyrins to other substrates has been widely studied, examples for activation of the ligand bound to the metal upon irradiation have been very limited. The reactivity of σ-bonded aluminum and indium porphyrins towards carbon dioxide has been studied.

### C I 1) Insertion Reactions

The complex Al(TPP)Et is unreactive to carbon dioxide in the dark. However, it is activated by visible light and carbon dioxide is inserted into the aluminum carbon bond in the presence of 1-methylimidazole[55]. Inoue[55] demonstrated the formation of Al(TPP)CO$_2$Et by in situ infrared measurements and by characterization of the reaction products which are obtained by treatment of the reaction mixture with hydrogen chloride gas followed by 1-butanol or diazomethane (Eq. 17).

$$\text{Al(TPP)Et} \xrightarrow{\text{CO}_2} \text{Al(TPP)CO}_2\text{Et} \xrightarrow[-\text{Al(TPP)Cl}]{\text{HCl}} \text{EtCO}_2\text{H} \xrightarrow{\text{BuOH, }-\text{H}_2\text{O}} \text{EtCO}_2\text{Bu} \qquad (17)$$

$$\text{EtCO}_2\text{H} \xrightarrow[-\text{N}_2]{\text{CH}_2\text{N}_2} \text{EtCO}_2\text{Me}$$

Similar photochemical insertion reactions are observed for indium alkylporphyrins[122]. In the presence of pyridine and under irradiation by visible light, carbon dioxide inserts into the carbon-indium σ bond leading to stable carboxylate indium porphyrins (Scheme 10).

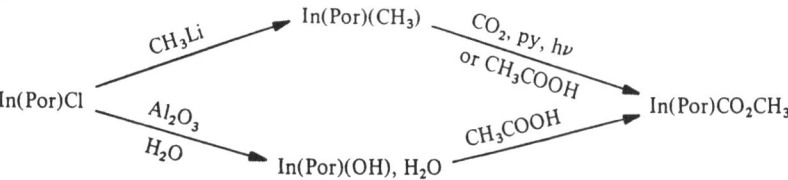

**Scheme 10**

Both irradiation by visible light and pyridine are required for the above insertion. The same carboxylate complexes can be prepared by reaction of acetic acid with either the alkyl(aryl)indium(III) porphyrins or the aquohydroxoindium(III) porphyrins.

## C12) Spectroscopic and Structural Data

As previously mentioned, the aluminum carboxylate derivative has been characterized in situ. The indium carboxylate complexes exhibit typical NMR spectra expected for hexa-coordinated systems. The X-ray crystal structure of $In(OEP)CO_2CH_3 \cdot 2CHCl_3$ has been solved[122]. Figure 11 presents the ORTEP view of this complex.

The out-of-plane position of the indium atom shown in Fig. 11 ($\Delta 4N = 0.61$ Å, $\Delta Por = 0.68$ Å) is less than that observed in the methyl complex ($\Delta 4N = 0.78$ Å, $\Delta Por = 0.92$ Å). The coordination of the indium atom can be described as either by a 4:1 or a 4:2 type. The indium is linked by the four nitrogen atoms of the macrocycle and by the two oxygen atoms of the acetate group giving rise to two different interactions. A strong bonding occurs between one oxygen atom and the metal atom ($In-O(2) = 2.14(1)$ Å) while a weaker interaction appears between the second oxygen atom and the metal ($In-O(1) = 2.60(1)$ Å). This results in a highly distorted coordination polyhedron with the $O(2)$ oxygen atom displaced from the ideal axial position of $16.3(6)°$ ($N(3)-In-O(2)$ $= 90.3(6)°$ instead of $106.6°$). Furthermore, the $O(1)$ oxygen atom interacts with one of the solvent molecules by means of a hydrogen bond. This coordination is intermediate between an end-on coordination like in $Fe(TpTP)CO_2CH_3$[123] and a bidentate coordination like in $Nb(TPP)(O)(CO_2CH_3)$[117, 118] or $Zr(OEP)(CO_2CH_3)_2$[112] (Fig. 12). All of these coordination schemes are in good agreement with the IR data.

In the IR spectrum of $In(OEP)CO_2CH_3$, $\nu_{as}(CO_2^-)$ and $\nu_s(CO_2^-)$ stretching frequencies are observed at 1570 and 1405 $cm^{-1}$. As pointed out earlier, bidentate type ligation produces a characteristic IR spectrum with $\Delta\nu(CO_2^-) = \nu_{as}(CO_2^-) - \nu_s(CO_2^-)$ close to or smaller than 150 $cm^{-1}$ [124] thus indicating a more ionic bonding.

When the acetate ligand is covalently bonded like in $Fe(TpTP)CO_2CH_3 \cdot 1/2$ $CH_3COOH$, the coordination of the acetate group is monodentate implying a $\Delta\nu$ of 409 $cm^{-1}$. For $In(OEP)CO_2CH_3$, the observed $\Delta\nu$ value is 165 $cm^{-1}$ which corresponds

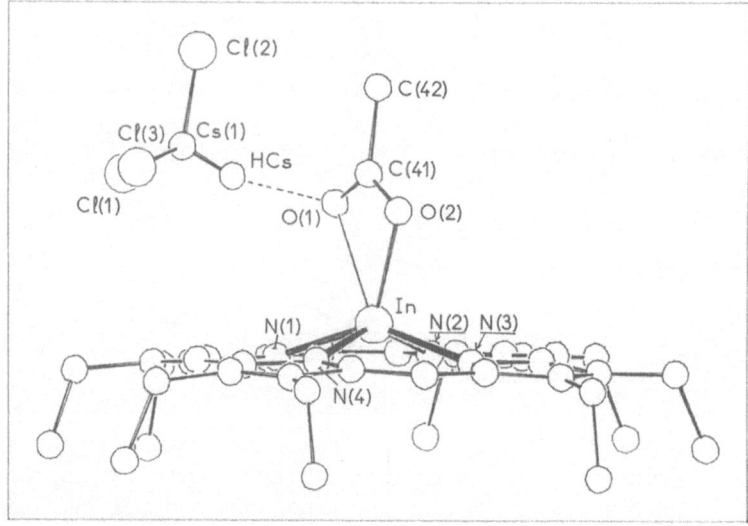

**Fig. 11.** ORTEP view of $In(OEP)CO_2CH_3 \cdot 2\,CHCl_3$[122]

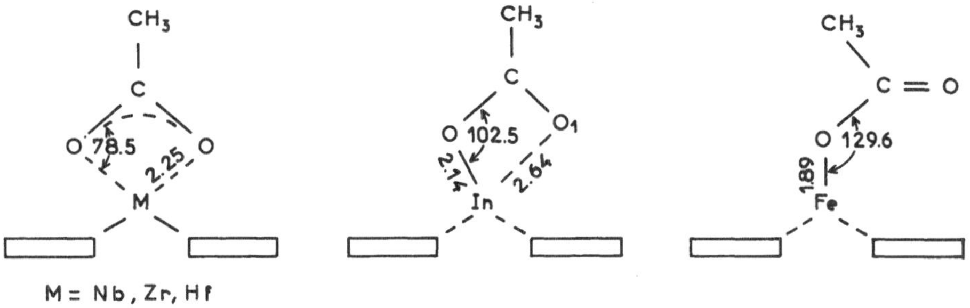

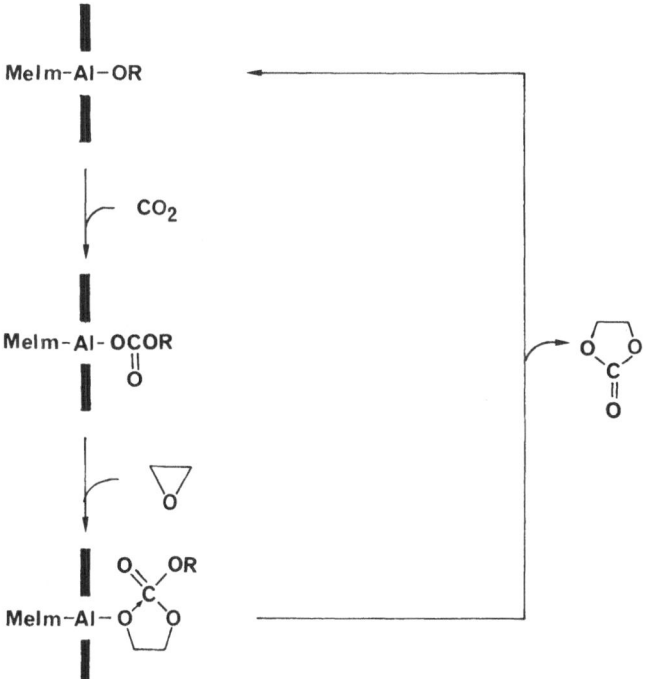

M = Nb, Zr, Hf

**Fig. 12.** Schematic view of the possible orientations of a bound acetate group in metalloporphyrin structural chemistry. The oxo ligand of the Nb complex and the second acetate groups of the Zr and Hf derivatives are omitted for clarity

to an intermediate ionic coordination. This may be due to an interaction with the chloroform molecule.

## C13) Mechanism of CO₂ Insertion

The presence of pyridine and irradiation by visible light are required for the insertion of carbon dioxide into the indium-carbon σ bond. However, the UV-visible spectrum of the starting complex is invariant with increasing amounts of pyridine, thus implying a five coordinate species in the media studied. These results contrast with those obtained for

**Scheme 11**

Al(TPP)(Et), where six-coordinate adducts were obtained upon addition of imidazole[55]. For the indium complexes, irradiation in benzene and pyridine leads to a reduced complex whose ESR trace suggests – at least partly – a metal centered reduction leading to radical species stabilized by pyridine. These radicals then add carbon dioxide to give the insertion products.

Another carbon dioxide insertion reaction is observed with Al (TPP)(OMe)[125]. This compound readily and reversibly traps carbon dioxide at room temperature in the presence of 1-methylimidazole. The trapped carbon dioxide is sufficiently activated to react with an epoxide at room temperature, thus producing the corresponding alkylcarbonate. As illustrated in Scheme 11, the cyclic carbonate is considered to be formed – at least partly – by nucleophilic attack on a linear intermediate. Thus the alkoxide aluminum porphyrin-methylimidazole system would be a good catalyst for synthesis of alkylene carbonates from carbon dioxide and epoxides under mild conditions.

## C II) Sulfur Dioxide

The insertion reaction of sulfur dioxide is of interest since it could lead to metal sulfinates which possess potential catalytic, bacterial, and plant growth regulator activities[126]. This reaction is known for iron, gallium, and indium porphyrins[127–130].

### C II 1) Synthesis

The insertion of sulfur dioxide between the metal atom and the alkyl or aryl group of indium[127, 128], gallium[129], and iron porphyrins[130] of the form M(Por)(R) gives rise to M(Por)SO$_2$R sulfinato derivatives (Eq. 18). The iron sulfinato compounds are air stable at room temperature, but can easily be oxidized by a stream of oxygen to give the corresponding sulfonato complexes.

$$M(Por)(R) \xrightarrow[CH_2Cl_2, -18\,°C]{SO_2} M(Por)SO_2R \xrightarrow[toluene, \Delta]{1/2\,O_2} M(Por)SO_3R \qquad (18)$$

M = In, Ga, Fe; R = alkyl or aryl group.

Sulfinato and sulfonato indium porphyrins are also isolated by oxydation of the corresponding thio In(Por)(SR) derivatives (Eq. 19).

$$In(Por)(SR) \xrightarrow[toluene]{O_2} \begin{cases} \xrightarrow{0°C} In(Por)SO_2R + In(Por)SO_3R \\ \xrightarrow{50-60°C} In(Por)SO_3R \end{cases} \qquad (19)$$

Acid hydrolysis of the μ-oxo-bridged dimers [Fe(Por)]$_2$O[130, 131] or the Ga–Cl bond of Ga(Por)Cl[129] by the appropriate sulfonic acid also leads to the corresponding sulfonato derivatives (Eq. 20).

Ga(Por)Cl

$\xrightarrow{\text{RSO}_3\text{H}}$ M(Por)SO$_3$R

[Fe(Por)]$_2$O

(20)

M = Ga, Fe

### C II 2) Spectral Characterization and Crystal Structures

Infrared spectroscopy is a powerful tool in the structural characterization of metallic sulfinates and sulfonates. The location[126, 132, 133] and wavenumber separation[134] of the symmetrical and asymmetrical vibrations of the SO$_2$ group depend on the nature of the metal-sulfinate linkage. For sulfinate complexes, $\nu_{sym}$(SO$_2$) and $\nu_{asym}$(SO$_2$) appear between 840 and 1100 cm$^{-1}$ with 240 < $\Delta\nu$ < 270 cm$^{-1}$ suggesting an unidentate O-sulfinato-arrangement[135].

The sulfonate structure is also confirmed by the position and intensity of the SO$_3$ bands[136, 137]. However, a choice between an O or an O,O' ligation is difficult since the bands due to the latter are broad and often overlap with those of the porphyrin skeleton.

Figure 13 shows an ORTEP view of Fe(TPP)SO$_2$C$_6$H$_5$[130]. The coordination polyhedron of the iron atom is of the 4 : 1 type in accordance with an O-sulfinato bonding. The iron atom lies 0.45(2) Å from both the four nitrogens and from the porphinato-plane. The mean iron-nitrogen distance is 2.05 ± 0.02 Å and the iron-oxygen bond length [1.92(1) Å] is statistically equal to that [1.898(3) Å] observed in Fe(TmTP)CO$_2$CH$_3$[123].

The $^1$H NMR data of Fe(Por)SO$_2$R and Fe(Por)SO$_3$R are characteristic of high spin species[130]. Furthermore, the diastereotopic methylene protons of the sulfonato complexes differ more in their chemical shift than those of the sulfinate complex suggesting a

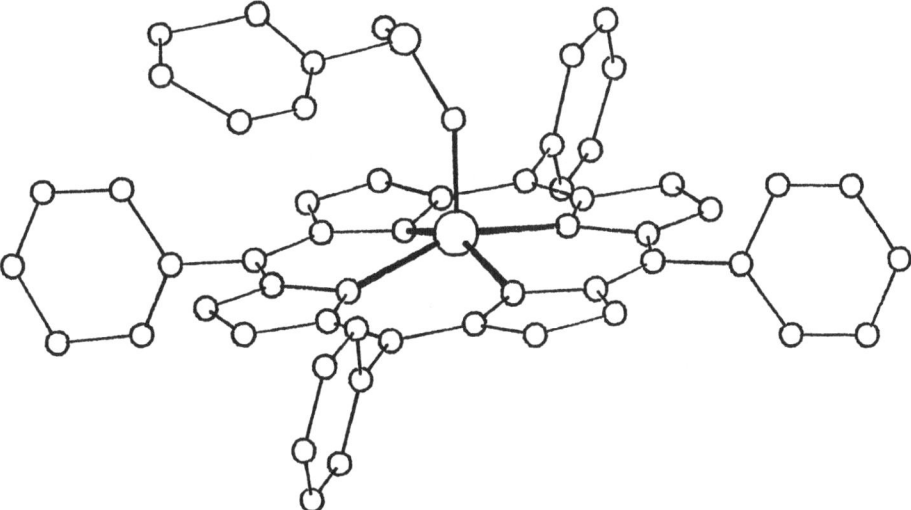

**Fig. 13.** ORTEP view of Fe(TPP)SO$_2$C$_6$H$_5$[130]

greater displacement of the metal center in the former complex. All of the NMR results agree with monodentate sulfinate bonding and bidentate coordination of the sulfonate group. ESR spectra with axial symmetry ($g_\perp \simeq 6$ and $g_\parallel \simeq 2$) are obtained for the two series of complexes, confirming their high-spin state ($S = 5/2$)[138]. A thermomagnetic study of $Fe(OEP)SO_2C_6H_5$, which rigorously follows a Curie Law between 4 and 300 K, gave $\mu_{eff} = 5.74$ B.M.[139].

Electronic spectra of the sulfinato and sulfonato complexes are similar to those of other high-spin iron(III) porphyrins and belong to the hyper class[140, 141]. The morphology of the NMR spectra and the chemical shifts of the indium and gallium sulfinato or sulfonato porphyrins are nearly identical to those for the halogenated derivatives[128, 129]. On the basis of IR data, an S-sulfinato structure can be eliminated but a choice between the arrangements O- and O,O'-sulfinate (sulfonate) is more difficult. The crystal structure of $Ga(OEP)SO_3CH_3$ has been established[129].

Figure 14 shows the O-sulfinate coordination of the gallium atom ($Ga–O_1 = 1.908(6)$ Å, $\langle Ga–N \rangle = 2.013(5)$ Å). The displacement of the oxygen atom linked to the gallium atom from the ideal axial position is important as shown by the values of N–Ga–O angles (N(1)–Ga–O(1) = 104.3(2)°, N(2)–Ga–O(1) = 102.1(2)°, N(3)–Ga–O(1) = 92.9(2)°, N(4)–Ga–O(1) = 97.2(2)°). This displacement from the ideal axial position also occurs for $SO_2C_6H_5$ in $Fe(TPP)SO_2C_6H_5$ and could be a characteristic of binding for the sulfinato or sulfonato group. Furthermore, the gallium atom lies 0.320(1) Å from the plane of the four nitrogens.

Figure 15 is a stereoscopic view of the $In(TPP)SO_3CH_3 \cdot 2 \, C_2H_4Cl_2$ complex and Fig. 16 represents the fundamental unit of this system and numbering scheme used[128]. In the solid state, $In(TPP)SO_3CH_3$ is an O,O'-sulfonato-intermolecular complex. It crys-

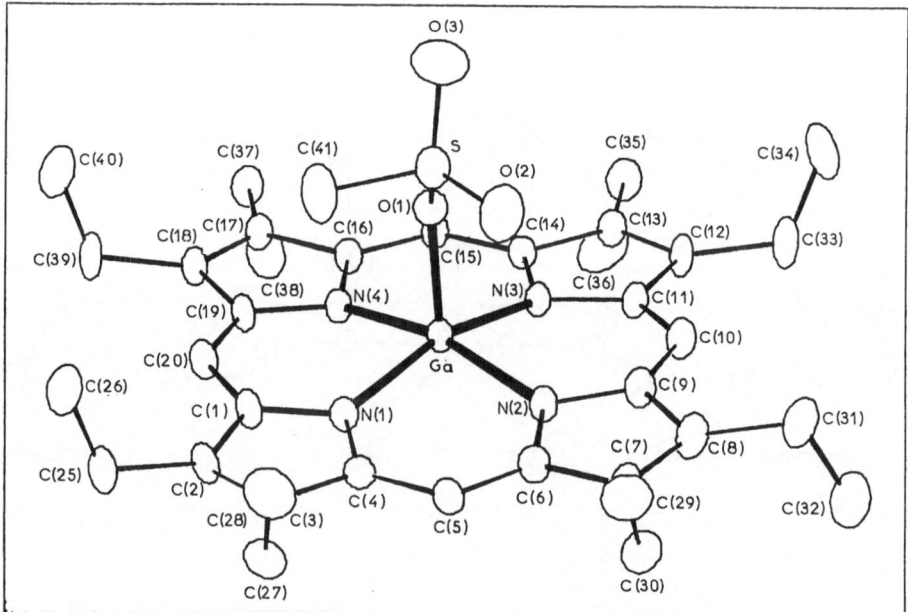

**Fig. 14.** ORTEP view of $Ga(OEP)SO_3CH_3$[129]

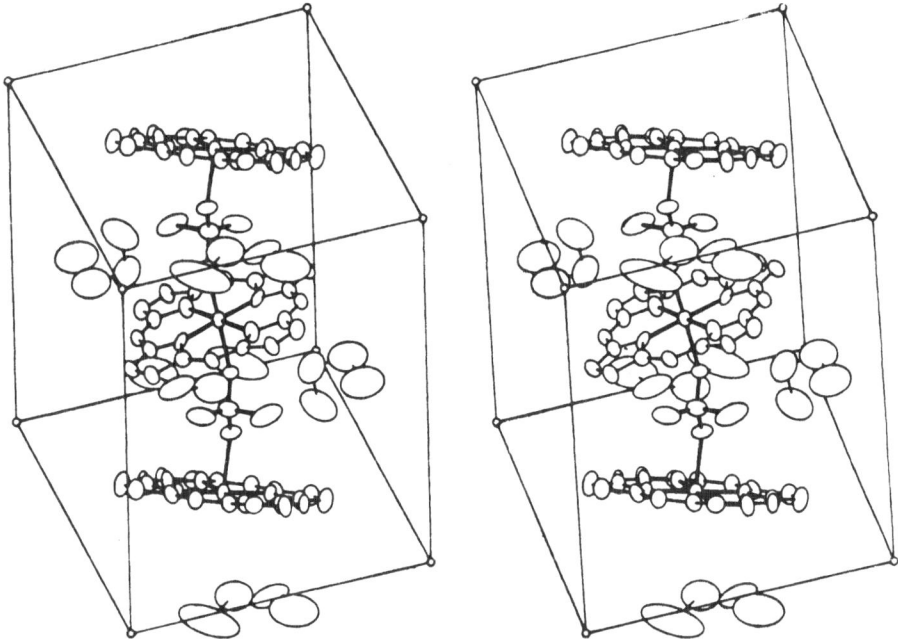

**Fig. 15.** Stereoscopic view of the unit cell for the In(TPP)SO$_3$CH$_3$ · 2 C$_2$H$_4$Cl$_2$ complex[128]

tallizes as a one-dimensional polymer, in a direction parallel to the č axis. The fundamental unit of the polymer (Fig. 16) consists of two In(TPP) entities bridged by a sulfonato group.

Both non-crystallographically equivalent macrocycles have an inversion center and are not parallel. The angle between the mean planes of these macrocycles is 28.6(1.2)°. The distance between the porphyrin cores, measured by the length In(1)–In(2), is c/2 [7.015(2) Å]. The sulfur atom is at the same distance from both metal ions (In(1)–S = 3.522(5), In(2)–S = 3.581(5) Å). This crystal structure is the first structural example of a polymer in metalloporphyrin chemistry.

Both octahedrally coordinated indium atoms lie on inversion centers (1/2, 1/2, 1/2, and 1/2, 1/2, 0) and are rigorously in the perfect plane of the nitrogen atoms. The thermal coefficients of the metal atoms are small (2.6(3) and 2.8(3) Å$^2$) so that these atoms are not statistically distributed above and below the porphyrin plane as is observed for Mn$^{II}$(TPP)[142]. Thus, to accomodate a large ion such as In$^{III}$, the porphyrin core must expand and the radius of the central hole (2.11(1) Å) is almost the same as observed in Sn(TPP)Cl$_2$. This was the first demonstration for the lack of influence of the ionic radii on the out-of-plane distance of the metal. Indium-nitrogen bond lengths in In(TPP)SO$_3$CH$_3$ are much smaller than those found in In(TPP)(CH$_3$)[104] (2.205(10) Å) where the indium atom is 0.79(1) Å above the porphyrin plane. Oxygen atoms O(1), O(2) are on the pseudo-quaternary axis of each macrocycle (87.5(4)° < N–In–O < 92.5(4)°). The sulfonate group is equidistant from both macrocycles and the oxygen-indium distances are statistically equal (2.352(12) and 2.357(11) Å). These latter distances are much larger than those observed in the unidentate sulfinato or sulfonato

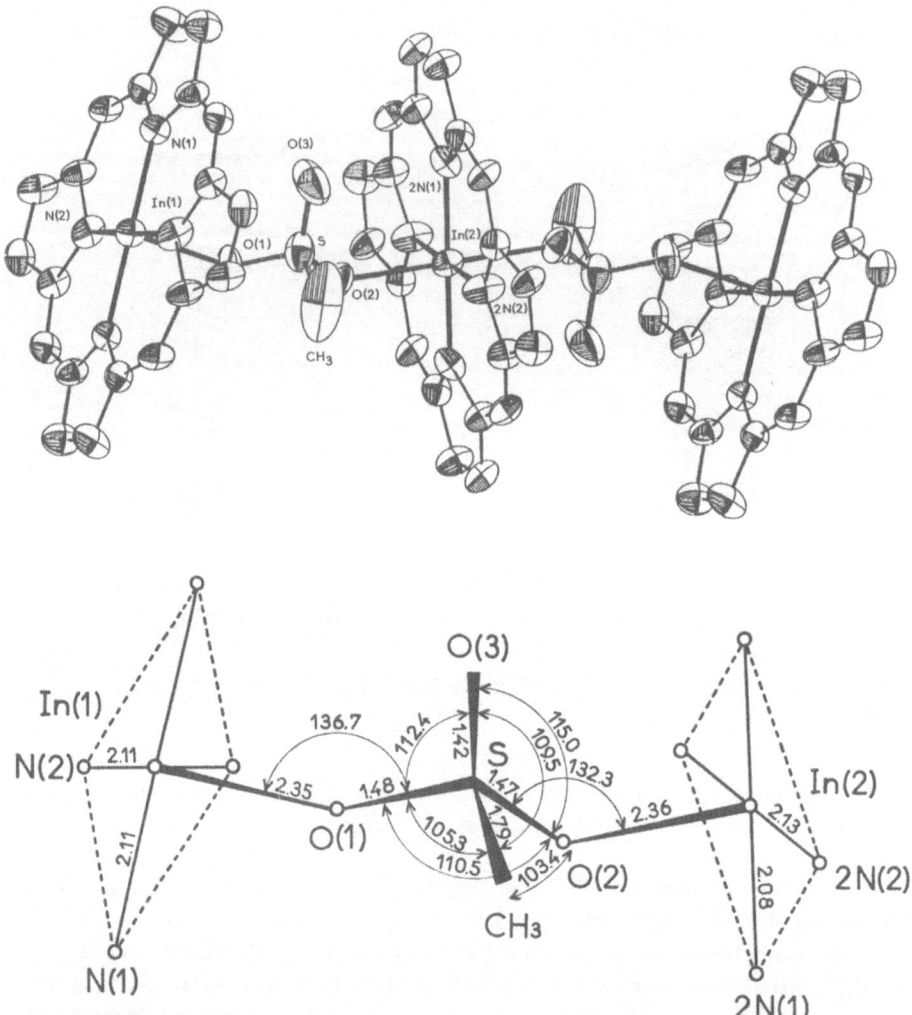

**Fig. 16.** ORTEP view and bonding characteristics of the [In(TPP)SO$_3$CH$_3$]$_n$ polymer

metalloporphyrins described above. The symmetrical position of both macrocycles, with respect to the SO$_3$CH$_3$ group, is not dictated by the space group since the sulfur atom is in a general position in the unit cell. This means that the crystal structure must be described by a close packing of [In(TPP)]$^+$ and [SO$_3$CH$_3$]$^-$ ions. This assumption is easily verified because the two macrocycles are not parallel and intersect along a straight line parallel to the O(3)–CH$_3$ ridge of the sulfonate tetrahedron.

The porphyrin core is highly distorted. The nitrogen atoms are not in the mean plane of the macrocycle, their distance from this plane being ± 0.10(2) Å. Thus the β-carbon atoms are out of the four nitrogen plane (0.39(2) for C(7), 0.38(2) for 2 C(3), 0.36(2) for C(8), and 0.30(2) for 2 C(2)). Finally, the angles between the plane of the four nitrogens and those of the pyrrole groups lie in the range of 4.7 to 10°.

## C II 3) *Monomeric-Polymeric Equilibrium of the Indium Derivatives*

The polymeric association of $In(TPP)SO_3CH_3$ is an ionic association between the $[In(TPP)]^+$ and $[SO_3CH_3]^-$ ions. Such a structure has been postulated for tin(IV) porphyrins bridged by dicarboxylic acids[143].

This structural arrangement observed in the solid state raises the question of whether the complex in solution is polymeric or monomeric. In this regard the NMR characteristics of different concentrations of $In(OEP)SO_3CH_3$ in $CDCl_3$ have been studied (Fig. 17). At low and medium concentrations, an $ABR_3$ multiplet is observed for the ethyl groups of the OEP ligands. This can be explained only by the presence of a non-octahedral monomeric structure in solution. Dissolving the complex destroys the polymeric chains and the structure of the obtained complex is similar to that of $In(TPP)(CH_3)$[104] for which the metal is 0.79 Å out of the porphyrin plane. In a saturated solution (0.1 mmol), the $^1H$ NMR spectrum of the methylene groups is a broad quadruplet which collapses to a singlet by decoupling the methyl groups. The spectrum is approximately an $A_2R_3$ type so that the magnetic environments of both faces of the macrocycle are equivalent. It thus follows that aggregates are formed similar to those described in the solid state. This assumption is supported by the fact that the signal width of the polymeric form is 12.5 Hz compared to 1.6 Hz in the monomeric form. These observations may certainly be explained by a more rigid structure of the $SO_3CH_3$ group induced by the polymeric form of the complex.

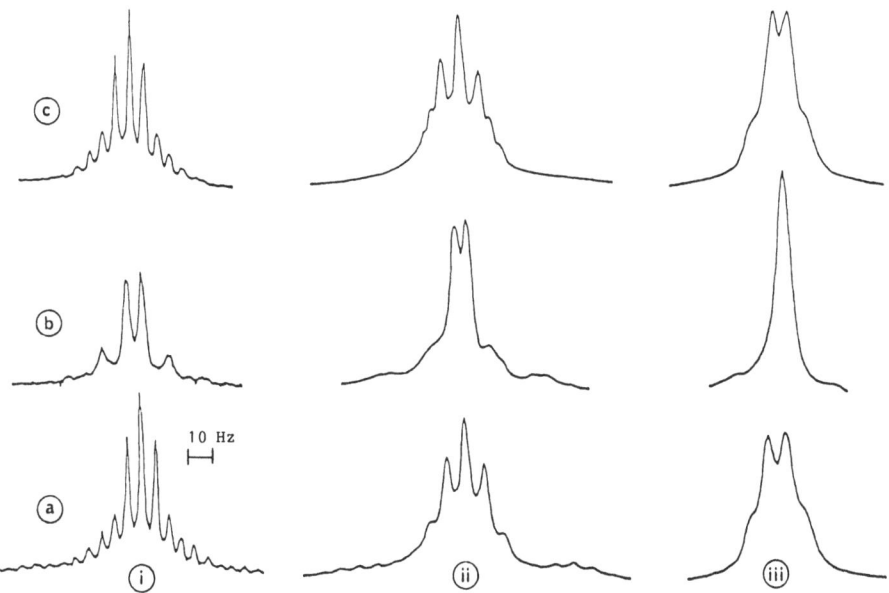

**Fig. 17.** $^1H$ NMR spectra of A and B methylenic protons of $In(OEP)SO_3CH_3$ in $CDCl_3$: a) experimental, b) decoupled, c) simulated ($\delta_{CH_2}$ = 4.18 ppm, $|^2J|$ = 14.5 Hz, $|^3J|$ = 7.6 Hz) at the following concentrations: (i) 0.0056 mol · $l^{-1}$ ($\delta_A$ = 4.166, $\delta_B$ = 4.234 ppm); (ii) 0.070 mol · $l^{-1}$ ($\delta_A$ = 4.176, $\delta_B$ = 4.240 ppm); (iii) 0.13 mol · $l^{-1}$ (mean $\delta$ = 4.100 ppm, $\delta_A$ − $\delta_B$ < 0.02 Hz)

*C II 4)  Electrochemistry of In(Por)SO₃R and In(Por)SO₃R(N-MeIm) in CH₂Cl₂
and CH₂Cl₂/py Mixtures*

The addition of N-methylimidazole to In(Por)SO₃R is given by Eqs. 21 and 22 where Por
= OEP or TPP, R = CH₃ or C₆H₅, and L = N-methylimidazole.

$$In(Por)SO_3R + L \rightleftarrows In(Por)SO_3R(L) \tag{21}$$

$$In(Por)SO_3R(L) + L \rightleftarrows In(Por)(L)_2^+ + SO_3R^- \tag{22}$$

The above two reactions were examined by $^1$H NMR, electronic absorption spectroscopy
and by conductivity monitored titrations of In(Por)SO₃R with N-MeIm[144]. The electrochemistry supports interpretations from the spectroscopic data and, in addition, gives
information about the singly and doubly reduced complexes[145].

A summary of the electrochemical data for In(Por)SO₃R reduction under several
solution conditions is presented in Table 16[145]. In the absence of N-MeIm,
In(TPP)SO₃R undergoes two reversible one electron reductions at $E_{1/2}$ = − 1.07 and
− 1.47 V. The addition of 0.8 equivalents of N-MeIm to In(TPP)SO₃R leads to the
appearance of two new reductions while the addition of 1.0 equivalent of N-MeIm to
In(TPP)SO₃R in CH₂Cl₂ results in an almost complete disappearance of the original
peaks as In(TPP)SO₃R(N-MeIm) is formed[145]. Finally, after the addition of four equivalents of N-MeIm, only two reversible peaks are present. Further addition of N-MeIm to
these solutions does not lead to the appearance of new waves, but rather to shifts in the
half-wave potentials for the two well defined peaks which were identified as due to the
stepwise reduction of [In(TPP)(N-MeIm)₂]⁺ at all N-MeIm/porphyrin ratios of more than
6 : 1[144].

The binding of pyridine to In(TPP)SO₃R was shown to be qualitatively similar to that
for the binding of N-MeIm by In(TPP)SO₃R[144, 145]. The stepwise formation of In(Por)
SO₃R(py) and [In(Por)(py)₂]⁺ occurs upon addition of pyridine to CH₂Cl₂ solutions of
In(Por)SO₃R. Stability constants for formation of In(Por)SO₃R(py) were estimated to be
greater than $10^6$ by $^1$H NMR studies while stability constants for formation of
[In(Por)(py)₂]⁺ from In(Por)SO₃R(py) ranged between log $K_2$ = 1.6 and 2.1[144].

The resulting electrochemistry of In(Por)SO₃R in CH₂Cl₂/py mixtures is similar to
that in CH₂Cl₂/N-methylimidazole mixtures but differs in how $E_{1/2}$ for the first reduction
of the complex shifts with increasing ligand concentration. The value of $E_{1/2}$ for this

**Table 16.** Half-wave (V vs. SCE) for the first and second reduction of In(Por)X in various solutions
containing 0.1 M TBAClO₄[a]

| Initial porphyrin | CH₂Cl₂ | | CH₂Cl₂ + 1 M N-methyl-imidazole[b] | | Pyridine[b] | |
|---|---|---|---|---|---|---|
| In(TPP)SO₃Ph | − 1.07 | − 1.47 | − 0.99 | − 1.44 | − 0.77 | − 1.30 |
| In(TPP)SO₃CH₃ | − 1.07 | − 1.47 | − 0.99 | − 1.44 | − 0.77 | − 1.30 |
| In(TPP)SO₃Ph | − 1.30 | − 1.81[c] | − 1.23 | − | − 1.04 | − 1.64 |
| In(TPP)SO₃CH₃ | − 1.30 | − 1.82[c] | − 1.23 | − | − 1.04 | − 1.64 |

[a]  Taken from Ref. 145;  [b]  Reactions correspond to reduction of [In(Por)(L)₂]⁺ and In(Por)(L)₂;
[c]  $E_p$ at scan rate = 0.10 V/s

reduction shifts positively from $-0.91$ V in $CH_2Cl_2$ containing 1 equivalent of pyridine to $-0.87$ V in 1 M pyridine. In contrast, a negative shift in $E_{1/2}$ is observed on going from $CH_2Cl_2$ containing 3 equivalents of N-MeIm ($-0.94$ V) to 1 M N-MeIm in $CH_2Cl_2$ ($-0.99$ V)[145].

The first reduction of In(Por)SO$_3$R shifted positively by only 40 mV with increase of the pyridine concentration in $CH_2Cl_2$/py mixtures, but $E_{1/2}$ for the second reduction shifted negatively by 65 mV for each 10-fold increase of the pyridine concentration. These shifts were accounted for by Scheme 12 where X = SO$_3$R$^-$ [145].

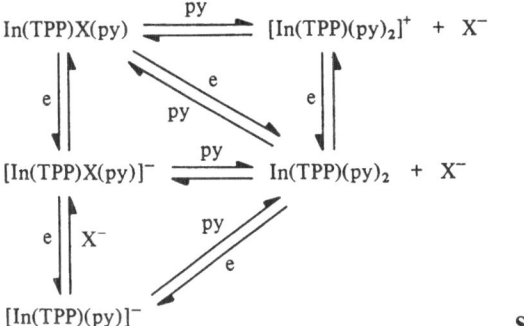

Scheme 12

Both In(TPP)SO$_3$(py) and [In(TPP)(py)$_2$]$^+$ are present in equilibrium when the concentration of pyridine is less than about 0.07 M in $CH_2Cl_2$. However, as the concentration of py in $CH_2Cl_2$ is increased above 0.1 M, the equilibrium is shifted and [In(TPP)(py)$_2$]$^+$ becomes the only In$^{III}$ species in solution[145].

Analysis of the electrochemical data led to the conclusion that the first reduced product in py/$CH_2Cl_2$ solutions was a mixture of [In(TPP)SO$_3$R(py)]$^-$ and In(TPP)(py)$_2$[145]. However, a bis-pyridine adduct was the ultimate product at longer time scales and this bis-pyridine complex was reduced in the second electron transfer step. During this reduction, In(TPP)(py)$_2$ loses a pyridine molecule to give [In(TPP)(py)]$^-$ and this loss of the second pyridine molecule resulted in a $-65$ mV shift of the second reduction peak per tenfold increase in py concentration[145].

The electrochemistry of In(Por)SO$_3$R complexes was also investigated in neat pyridine[145]. All four SO$_3$R derivatives (R = CH$_3$ and Ph) exhibited two reversible electron transfers and these reactions corresponded to the reductions of [In(Por)(py)$_2$]$^+$ and In(Por)(py)$_2$. Potentials for these reactions are given in Table 16.

# D) Polymetallic and Dimeric Derivatives

The use of metalloporphyrins which show metal-metal interactions can be an effective strategy for the synthesis of metal chain complexes. Some of these complexes exhibit high or unusual electric conductivities[146, 147]. Although much synthetic work has been devoted to metalloporphyrin chemistry[124, 140], only a few examples of porphyrin compounds containing metal-metal interactions have been published. Apart from the conducting polymeric porphyrins described by Ibers and Hoffman[146, 148], two classes of metal-metal bonded complexes are known. They are donor-acceptor complexes and the σ-bonded derivatives. The synthesis and characteristics of these systems are discussed below.

## D I) Donor-Acceptor Complexes

Bimetallic donor-acceptor complexes can be formed with tin[149–152] and germanium[150, 152] porphyrins. Two different methods of synthesis lead to these derivatives. The first is the reaction of $Sn(TPP)Cl_2$ with $Hg(Mn(CO)_5)_2$ or $M(Co(CO)_4)$ which gives $Sn(TPP)(Mn(CO)_4Hg(Mn(CO)_5)$ and $Sn(TPP)(Co(CO)_3MCo(CO)_4)$ where M = Zn, Cd, or Hg[149, 151] (Scheme 13).

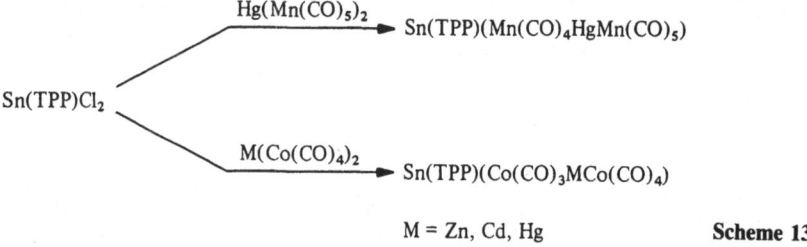

$$M = Zn, Cd, Hg \qquad \textbf{Scheme 13}$$

The second is the reaction of $Na_2Fe(CO)_4$ with dichloro tin(IV) and germanium(IV) porphyrins which produces the $M^{II}(Por)Fe(CO)_4$ complexes[150, 152] (Eq. 23).

$$M(Por)Cl_2 \xrightarrow[-2\,NaCl]{Na_2Fe(CO)_4} M(Por)Fe(CO)_4 \qquad\qquad (23)$$

M = Sn, Ge; Por = OEP, TmTP, TpTP.

The presence of the donor-acceptor bond has been established by $^{57}Fe$ Mössbauer spectroscopy. The relevant data are given in Table 17.

As shown in Table 17, the iron atom in $Sn(OEP)Fe(CO)_4$ and $Fe(CO)_5$ has a formal oxidation state of zero. $Sn(OEP)$ can be considered as a σ-donor ligand which enhances the 4s electron occupation of the iron atom. A quasi linear correlation has been found between the quadrupole splitting and the isomer shift for some $LFe(CO)_4$ complexes (L = triphenylphosphine, triethylphosphite, etc.) where the iron atom polyhedron corre-

**Table 17.** Mössbauer data of $Sn(OEP)Fe(CO)_4$ and $Fe(CO)_5$[150]

| Compound | T (K) | δ (mm · s⁻¹) | $\Delta E_Q$ (mm · s⁻¹) | $\Gamma_1$ (mm · s⁻¹) | $\Gamma_2$ (mm · s⁻¹) | $I_1$ | $I_2$ |
|---|---|---|---|---|---|---|---|
| $Sn^{II}(OEP)Fe(CO)_4$ | 297 | − 0.154 | 2.23 | 0.22 | 0.24 | 0.45 | 0.55 |
| $Sn^{II}(OEP)Fe(CO)_4$ | 95 | − 0.081 | 2.27 | 0.31 | 0.32 | 0.46 | 0.54 |
| $Fe(CO)_5$ | 77 | − 0.088 | 2.54 | | | | |

δ: relative to metal iron at 297 K; $I_1$: relative intensity of the lower energy line; $I_2$: relative intensity of the higher energy line

sponds to a $C_{3v}$ symmetry[153]. The $\Delta$ and $\delta$ parameters of $Sn(OEP)Fe(CO)_4$ agree with this linear relationship. Such an arrangement is not surprising since reduction of the $M^{IV}(R)_2Cl_2$ complexes (M = Ge, Sn, Pb) by Collman's complex produces dimeric complexes of the form $(M^{II}(R)_2Fe(CO)_4)_2$[154-156].

The IR data are in good agreement with a local $C_{3v}$ symmetry for the $M(CO)_4$ unit[157]. All of the $M(Por)Fe(CO)_4$ compounds exhibit two typical CO bands which are sharp (2019–1933 cm$^{-1}$) and one very broad and intense CO band which is lowest in energy (1933–1904 cm$^{-1}$).

Figure 18 shows the X-ray crystal structure of $Sn(OEP)Fe(CO)_4$[150]. The tin atom is pentacoordinated by the four nitrogen atoms ($\langle Sn-N \rangle$ = 2.187(3) Å) and the iron atom (Sn–Fe = 2.492(1) Å, Sn–Fe–C(52) = 179.1(1)°) is in the axial position. As expected, the tin-iron bond is short and represents the first example of a "carbenoïd" metal-metal bond in metalloporphyrin chemistry. Furthermore, the large out-of-plane distance of the porphyrin central metal ($\Delta 4N$ = 0.818(9) Å) is in good agreement with a formal $Sn^{II}$ oxidation state. The iron atom has a pseudo $C_{3v}$ symmetry and the average iron carbonyl distance is 1.754(5) Å.

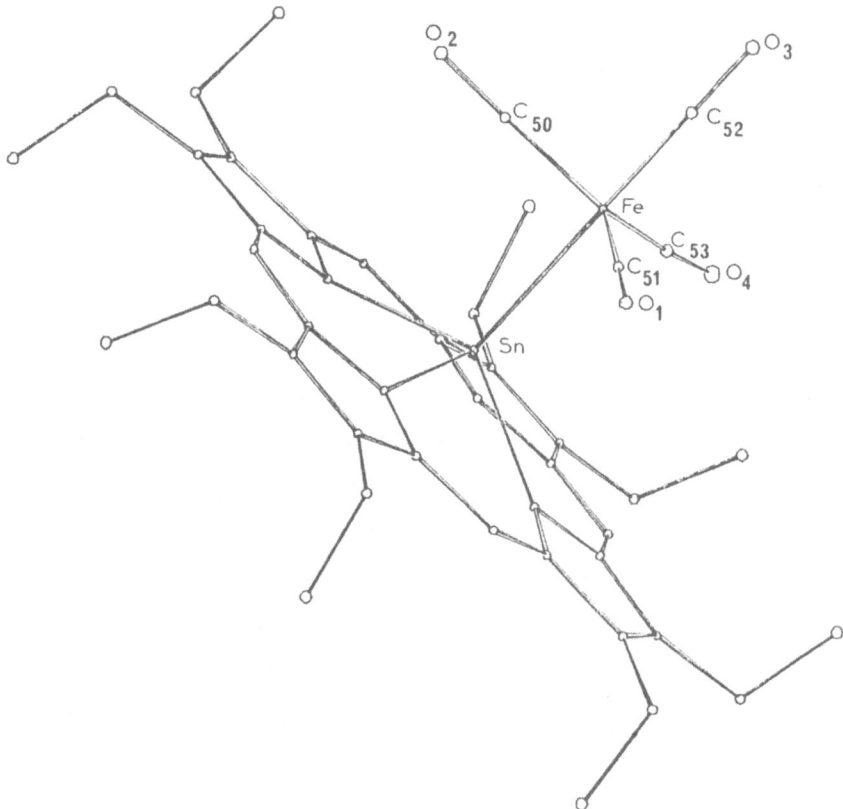

**Fig. 18.** Crystal structure of $Sn(OEP)Fe(CO)_4$[150]

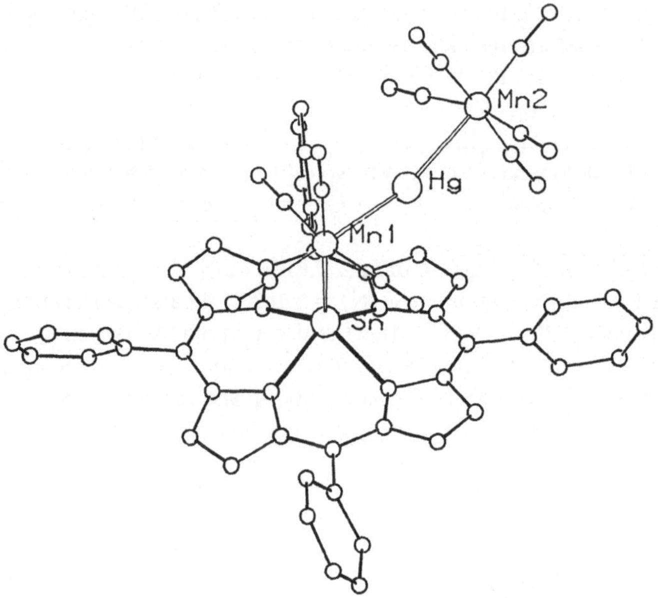

**Fig. 19.** ORTEP view of Sn(TPP)(Mn(CO)$_4$HgMn(CO)$_5$) · 0.5 CH$_2$Cl$_2$[151)]

Another structural example of a "carbenoïd" complex is given by Sn(TPP)(Mn(CO)$_4$HgMn(CO)$_5$) · 0.5 CH$_2$Cl$_2$ (Fig. 19). Similar to the structure of Sn(OEP)Fe(CO)$_4$, the tin-manganese distance is short (2.554(7) Å) and the tin atom lies 0.85 Å above the plane of the four nitrogens.

## D II) Single Bonded Complexes Containing One Porphyrin Unit

Various metal-metal single σ-bonded complexes have been obtained by the reaction of metal carbonyls with metal-carbon σ-bonded porphyrins or by the reaction of metal carbonyl anions and chlorometalloporphyrins[158, 159)] (Scheme 14). For example, the reaction of dimanganese carbonyl and methyl indium(III) porphyrin gives manganese pentacarbonyl indium porphyrin In(Por)Mn(CO)$_5$. The same compound is isolated when chloroindium porphyrin is allowed to react with the manganese pentacarbonyl monoanion. Various iron, cobalt, tungsten, and molybdenum complexes have been prepared by these two methods.

An intermediate In(Por)(Fe(CO)$_4$)(CH$_3$) species has been observed in the course of formation of the trinuclear complex (In(Por))$_2$Fe(CO)$_4$. In the first step, the reaction proceeds via a mechanism similar to that of an insertion reaction. Similar complexes with other metallate anions have been isolated following the same procedure[159)]. As shown in Table 18, these complexes generally exhibit three bands in the carbonyl stretching region (2100–1850 cm$^{-1}$). The UV-visible spectra of the octaethylporphyrin derivatives belong to the hyper class.

The crystal structure of In(OEP)Mn(CO)$_5$ has been solved[159)], and Fig. 20 reproduces the ORTEP view of the molecule. In contrast to Sn(OEP)Fe(CO)$_4$ (Fig. 18) the metal-

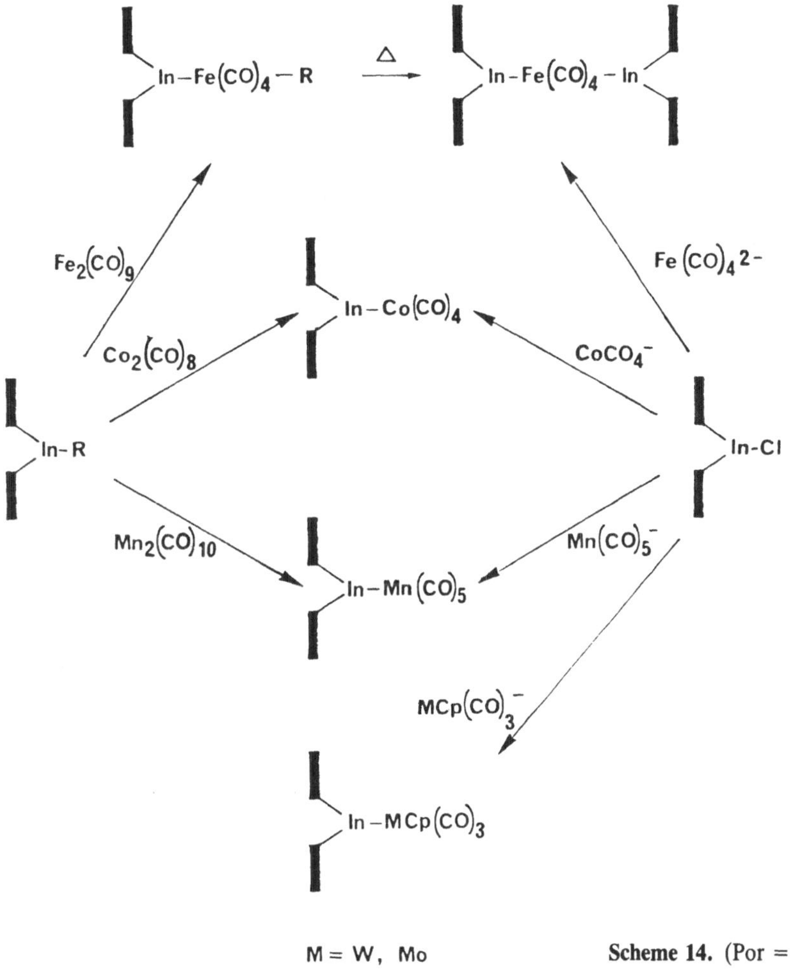

In–Fe(CO)$_4$–R $\xrightarrow{\Delta}$ In–Fe(CO)$_4$–In

In–Co(CO)$_4$

Fe$_2$(CO)$_9$

Co$_2$(CO)$_8$

Fe(CO)$_4{}^{2-}$

CoCO$_4{}^-$

In–R

In–Cl

Mn$_2$(CO)$_{10}$

Mn(CO)$_5{}^-$

In–Mn(CO)$_5$

MCp(CO)$_3{}^-$

In–MCp(CO)$_3$

M = W, Mo

**Scheme 14.** (Por = OEP, TPP)

**Table 18.** IR characteristics of In(Por)ML(CO)$_n$ complexes where L = CO or Cp[159]

| Por | ML(CO)$_n$ | $\nu(CO)cm^{-1}$ | | | | | | |
|-----|------------|------------------|---|---|---|---|---|---|
| | | Solution (THF) | | | Solid state (CsI) | | | |
| TPP | Mn(CO)$_5$ | 2878 | 1974 | | 2078 | 1996 | 1975 | 1965 |
| | Co(CO)$_4$ | 2070 | 1999 | 1969 | 2070 | 1999 | 1973 | 1961 |
| | CrCp(CO)$_3$ | 1977 | 1892 | 1872 | 1968 | 1898 | 1870 | |
| | MoCp(CO)$_3$ | 1980 | 1905 | 1882 | 1982 | 1905 | 1878 | |
| | WCp(CO)$_3$ | 1976 | 1898 | 1875 | 1979 | 1899 | 1871 | |
| OEP | Mn(CO)$_5$ | 2075 | 1972 | | 2071 | 1973 | 1965 | |
| | Co(CO)$_4$ | 2066 | 1999 | 1969 | 2066 | 2003 | 1977 | 1950 |
| | CrCp(CO)$_3$ | 1967 | 1897 | 1872 | 1964 | 1888 | 1866 | |
| | MoCp(CO)$_3$ | 1976 | 1899 | 1882 | 1976 | 1892 | 1874 | |
| | WCp(CO)$_3$ | 1973 | 1892 | 1877 | 1973 | 1885 | 1869 | |

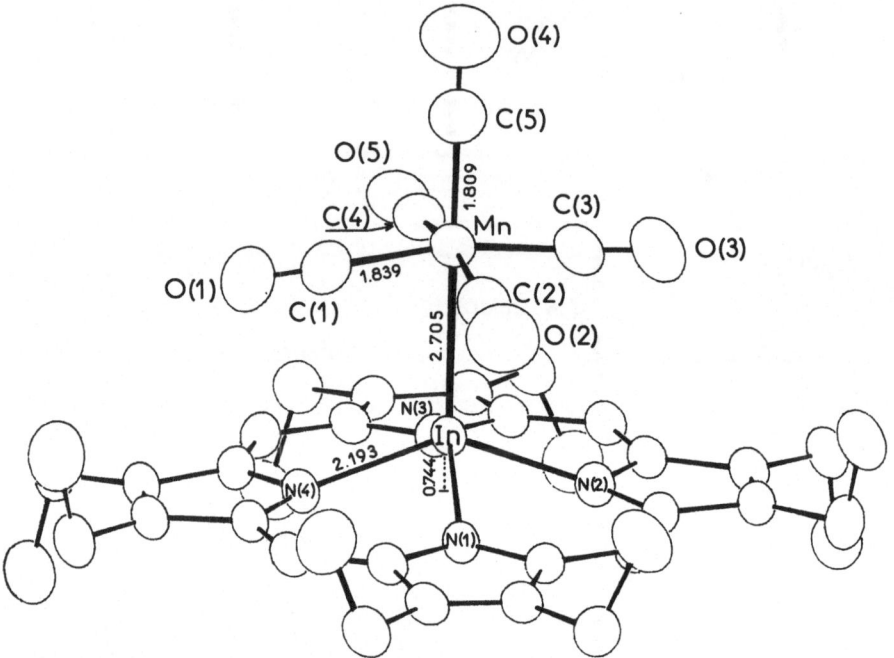

**Fig. 20.** ORTEP view of In(OEP)Mn(CO)$_5$[159)]

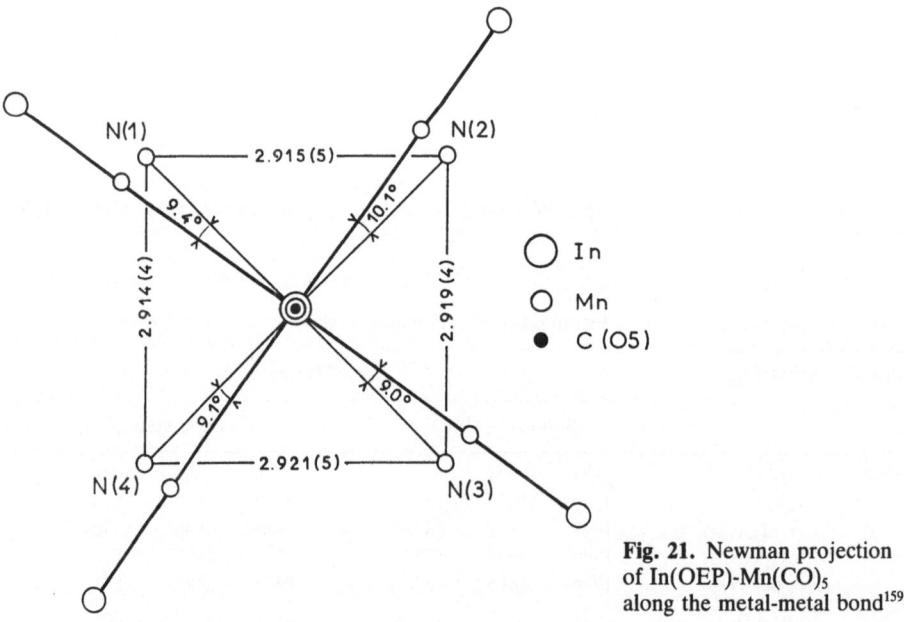

**Fig. 21.** Newman projection of In(OEP)-Mn(CO)$_5$ along the metal-metal bond[159)]

metal bond length is long (2.705(1) Å) and agrees well with a single covalent bond. Figure 21 is a Newman projection of the complex along the metal-metal bond which shows that the equatorial carbonyl groups almost eclipse the four nitrogen atoms. The torsional angles N(1)–In–Mn–C are between 9 and 10 degrees. The average Mn–C equatorial distance (1.839(4) Å) is longer than the axial distance (1.809(6) Å). This observation is in accordance with previous structural data related to metal carbonyls[160]. The metal atom is far out of the plane of the four nitrogens ($\Delta 4N = 0.744$ Å). This is generally observed for indium porphyrins. Furthermore, the manganese atom is also located out of the average equatorial carbonyl plane by 0.198 Å towards the axial carbonyl.

## D III) Single and Multiple Bonded Dimers

### D III 1) Single Bonded Homodimers

Di(porphyrinato)rhodium(II) [Rh(Por)]$_2$ described by Ogoshi[37, 161], Wayland[162], Collman[163], and Kadish[83] are the only known metalloporphyrin dimers containing one single metal-metal bond. Oxidative cleavage (Eq. 24) or thermal homolytic cleavage (Eq. 25) of the Rh–H bond of Rh(OEP)H can produce the [Rh(OEP)]$_2$ dimer[37, 161]. The Rh$^{II}$ dimer may also be formed by electrochemical reduction of some Rh$^{III}$ porphyrin complexes[86] (See Sect. D III 4).

$$2\ Rh(OEP)H \xrightarrow{\ 1/2\,O_2\ } [Rh(OEP)]_2 + H_2O \tag{24}$$

$$2\ Rh(OEP)H \xrightarrow{\ \Delta\ } [Rh(OEP)]_2 + H_2 \tag{25}$$

The same OEP dimer and its TpTP analog are obtained by photolysis of the hydride derivatives[162, 163]. The reaction of [Rh(OEP)]$_2$ with alkyl halides, unactivated olefins, and acetylenes gives the σ-bonded metal-carbon derivatives according to Scheme 15[161].

(a) R = Ph, CN, n-C$_3$H$_7$; (b) R = H, Ph

### D III 2) Double Bonded Homodimers

Double bonded ruthenium and osmium homodimers have been synthesized by Collman[163, 164]. These two metals are inserted into the porphyrin free base by using the corresponding metal chlorocarbonyl dimer [MCl$_2$(CO)$_3$]$_2$. Irradiation of the resulting carbonyl metal(II) porphyrin irradiated in pyridine solutions yields the bis(pyridine) metal(II) porphyrin and leads to the expected dimers by heating under vacuum. This is shown in Scheme 16.

$$[MCl_2(CO)_3]_2 \xrightarrow[\Delta]{H_2(Por)} M(Por)(CO)(S)$$

$$M(Por)(CO)(S) \xrightarrow{py, hv} M(Por)(py)_2$$

$$M(Por)(py)_2 \xrightarrow[vacuum]{\Delta} [M(Por)]_2$$

Por = OEP or TPP; M, S = Ru, MeOH or Os, py

**Scheme 16**

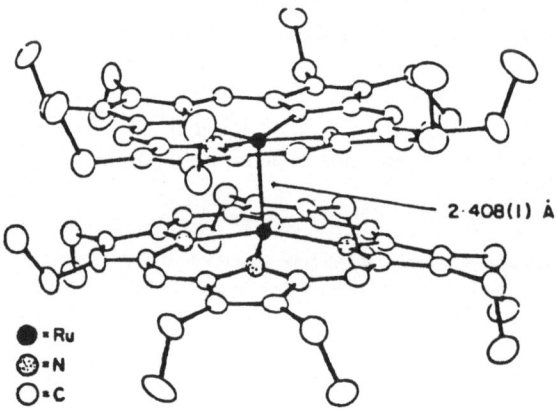

2·408(1) Å

● = Ru
◉ = N
○ = C

**Fig. 22.** ORTEP view of $(Rh(OEP))_2$[165]

NMR data of the above complexes are characteristic of paramagnetic species. For the octaethylporphyrin compounds the diastereotopic nature of the α-methylene protons indicates that the mirror symmetry of the porphyrin is lost and suggests a dimeric structure which is fully confirmed by X-ray diffraction. Figure 22 is the ORTEP drawing of $[Ru(OEP)]_2$. The crystal structure of $[Ru(OEP)]_2$ can be described as having two cofacial ruthenium octaethylporphyrin units linked together by a metal-metal bond. The short metal-metal distance (2.408(1) Å) agrees well with a Ru–Ru double bond. Each ruthenium atom is displaced by 0.30 Å from the plane of the four nitrogens towards the other metal atom. The two porphyrin cores are nearly parallel with an interplanar distance of 3.26 Å and are twisted by 23.8(1)° indicating possible orbital overlap between the aromatic systems.

## D III 3)  Quadruple Bonded Homodimers

Synthesis of the only prepared quadruple bonded homodimer[166] is schematically shown in Scheme 17.

$$H_2(OEP) \xrightarrow{Mo(CO)_6} Mo(OEP)(O)$$

$$Mo(OEP)(O) \xrightarrow{(CH_3)_3SiCl} Mo(OEP)Cl_2$$

$$Mo(OEP)Cl_2 \xrightarrow{PhC\equiv CPh/LiAlH_4} Mo(OEP)(PhC\equiv CPh)$$

$$2\ Mo(OEP)(PhC\equiv CPh) \xrightarrow[\text{vacuum}]{\Delta} [Mo(OEP)]_2$$

**Scheme 17**

The oxomolybdenum(IV) porphyrin treated with trimethylsilylchloride produces the trans dichloro derivative $Mo(OEP)Cl_2$ which gives $Mo(OEP)(PhC\equiv CPh)$ by reduction with $LiAlH_4$ in the presence of diphenylacetylene. The dimer is then obtained by vacuum pyrolysis of this five coordinate complex. Molybdenum(II) dimers are also obtained by heating a mixture of $MoCl_2(CO)_4$ and $H_2(Por)$ in oxygen-free toluene (Eq. 26) where Por is OEP, its 5-formyl-, 5-amino-, and 5-isocyanato derivative, or 5,15-dimethyletioporphyrin II.

$$MoCl_2(CO)_4 + H_2(Por) \rightarrow [Mo(Por)]_2 \tag{26}$$

The nature of the metal-metal bonding has been studied by variable temperature NMR spectroscopy and identical activation barriers were found for all complexes. This observation indicates that rotation about the metal-metal bond is the predominant process. The activation energy for this rotational process ($10.1 \pm 0.5$ kcal $\cdot$ mol$^{-1}$) is a measure of the $\delta$ bond strength and confirms the existence of the Mo–Mo quadruple bond.

## D III 4) Single and Multiple Bonded Heterodimers

Wayland[167] has synthesized Rh(OEP)In(OEP) by the reaction of (Na)[Rh(OEP)] with In(OEP)Cl[167]. This nucleophilic substitution of a metallate anion on a chloroindium(III) porphyrin is analogous to that reported for the synthesis of single bonded complexes containing one porphyrin unit[158, 159]. The similar morphology of the NMR spectra for Rh(OEP)In(OEP) and [Rh(OEP)]$_2$[161] suggests the existence of a metal-metal bond and this is confirmed by an X-ray study. Figure 23 shows the molecular structure of Rh(OEP)In(OEP) which consists of two face to face porphyrin units bound together via a Rh–In bond.

The interplanar separation of Rh(OEP)In(OEP) is 3.41 Å compared to 3.26 Å for [Ru(OEP)]$_2$[165]. As is also observed for the ruthenium dimer the In(OEP) core is twisted 21.8° relative to the Rh(OEP) group. Both the indium and rhodium atoms have an out-of-plane distance $\Delta 4N$ which is the same as that previously reported for the corresponding σ-bonded metal carbon complexes (given in Table 15 of part B V). A covalent radius of 1.36 Å may be estimated from the metal-carbon bond length in In(TPP)(CH$_3$). Similarly the covalent radius of the rhodium atom in Rh(OEP)(CH$_3$) is equal to 1.26 Å, giving a predicted covalent Rh–In distance of 2.62 Å. This latter value is in good agreement with the observed value (2.584(2) Å). In addition, the reactivity of the dimer

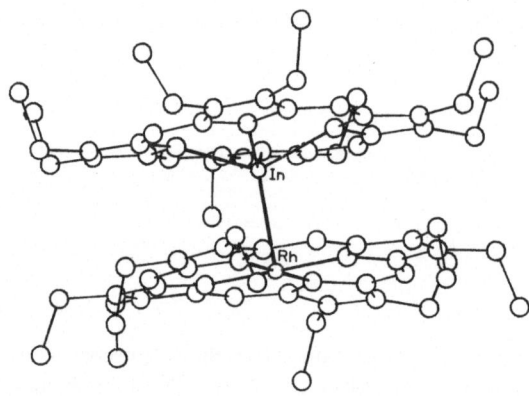

**Fig. 23.** ORTEP drawing of
Rh(OEP)In(OEP)[167]

towards alkyl halides agrees well with a polar covalent $Rh^I \rightarrow In^{III}$ bond. The $Rh^I$ may be held within the plane of its porphyrin ligand by $d_\pi \rightarrow p_\pi$ backbonding[109].

The double bonded Ru(OEP) = Os(OEP) heterodimer has been characterized by NMR spectroscopy but has not been isolated[163]. This complex is obtained by a coupling of the two corresponding metalloporphyrin monomers containing different metals in the same oxidation state.

## D IV) Electrochemistry of Bimetallic Metalloporphyrins

Electrochemistry has been reported for three types of bimetallic porphyrins. These are: (1) In(Por)M'L and Tl(Por)M'L complexes where Por = OEP or TPP and M'L = $Mn(CO)_5$, $Co(CO)_4$, $W(CO)_3Cp$, $Mo(CO)_3Cp$, or $Cr(CO)_3Cp^{159, 168-170}$; (2) M(Por)-$Fe(CO)_4$ complexes where M = Sn and Ge and Por = OEP, TmTP or $TpTP^{152-171}$; and (3) $[M(Por)]_2$ dimers where Por = OEP or TPP and Mn = $Rh^{83}$, $Ru^{61}$ or $Os^{61}$.

The existence of a metal-metal σ bond in the first group of In(Por)M'L and Tl(Por)M'L complexes has been demonstrated by IR, $^1H$ NMR spectroscopy and X-ray diffraction[158, 159]. The second group of bimetallic complexes may be considered as carbenoïd type complexes. On the basis of IR and Mössbauer spectroscopy[150, 171], and an X-ray structure of Sn(OEP) $Fe(CO)_4^{150}$, $Sn^{II}$ and $Fe^0$ oxidation states have been assigned[150]. By analogy, the $Ge(Por)Fe(CO)_4$ complexes are also postulated to have the main group metal in a low oxidation state and, as such, the metalloporphyrin would contain a $Ge^{II}$–$Fe^0$ metal-metal bond.

$Sn^{II}$ and $Ge^{II}$ oxidation states are extremely uncommon in metalloporphyrins. In fact, no $Ge^{II}$ porphyrins have ever been reported. Several spectral characterizations of $Sn^{II}(OEP)^{172}$ and $Sn^{II}(TPP)^{173}$ have appeared in the literature but it is only recently that a genuine air stable $Sn^{II}(TPP)$ complex has been isolated[174]. Thus, the carbenoïd metal-metal bonded metalloporphyrin complexes are unique in that the main group metal is air stable in the low oxidation state.

### D IV 1) Porphyrin Dimers of the Form [M(TPP)]₂

Three types of porphyrin dimers have been electrochemically investigated. These are $[Rh(TPP)]_2^{83}$, $[Ru(OEP)]_2^{61}$ and $[Os(OEP)]_2^{61}$. All three complexes contain + 2 metal ions. The [Rh(TPP)]₂ complex is formed in solution after the one electron reduction of a monomeric Rh$^{III}$ porphyrin[83]. The dimeric complex has not been isolated from the electrochemical reaction, but some of its electrochemical properties have been reported. [Rh(TPP)]₂ undergoes a reversible two electron reduction at $-1.95$ V vs. SCE in PhCN, 0.1 M TBAP. Similar reduction potentials are observed in THF ($-1.83$ V) and in py ($-1.90$ V). The peak to peak separation is $35 \pm 5$ mV, indicating two electrons are transferred in the reduction. Current integration also suggests a two electron reduction of the dimer. Spectroelectrochemical studies have demonstrated that this reduction is at the porphyrin ligand.

[Rh(TPP)]₂ is irreversibly oxidized at $E_p = -0.24$ V vs. SCE in PhCN, 0.1 M TBAP at a scan rate of 100 mV/s. No return reduction peak is coupled to the oxidation, and the products have been shown to be monomeric [Rh$^{III}$(Por)L]$^+$ species (L = solvent molecule). The same type of electrochemical reactivity is observed in THF ($E_p = -0.22$ V, scan rate = 100 mV/s) and in py ($E_p = -0.40$ V, scan rate = 100 mV/s).

The electrochemistry of [Ru(OEP)]₂ and [Os(OEP)]₂ has been reported in dimethoxyethane and CH₂Cl₂[61]. Each compound may undergo two reductions and four oxidations. The second reduction is not observed in CH₂Cl₂ due to the proximity of this reaction to the potential for solvent discharge.

The [Os(OEP)]₂ and [Ru(OEP)]₂ complexes contain Os$^{II}$ and Ru$^{II}$ central metals. According to the authors, reduction of the dimer leads to the stepwise formation of an anion radical and a dianion which are localized on both of the porphyrin rings. In contrast, oxidation of the two complexes is postulated to occur at the two central metals leading to the successive formation of dimers containing M$^{II}$/M$^{III}$, M$^{III}$/M$^{III}$, M$^{III}$/M$^{IV}$ and M$^{IV}$/M$^{IV}$ units. This behavior is quite different from that of [Rh(TPP)]₂ which undergoes only a single reduction step and cleaves to generate [Rh(TPP)]$^+$ after electro-oxidation[83].

### D IV 2) Ge(Por)Fe(CO)₄ and Sn(Por)Fe(CO)₄ Carbenoid Complexes

The electrochemistry of six M(Por)Fe(CO)₄ complexes was investigated in CH₂Cl₂ and PhCN containing 0.1 M TBA(PF₆)[152, 171]. Each dimer undergoes two reductions and one oxidation. The reductions are reversible to quasireversible while the oxidations are chemically and electrochemically irreversible. Values of peak and half-wave potentials for these reactions are summarized in Table 19 for oxidation and reduction at a gold electrode in PhCN. Values of $E_{1/2}$ or $E_p$ were also measured at a Pt electrode in CH₂Cl₂[171]. No significant differences were observed between the two electrodes or between the two solutions.

The potential difference between the first and second reduction peaks of M(Por)Fe(CO)₄ is 0.39 V for Sn(TpTP)Fe(CO)₄ and 0.40 V for Ge(TpTP)Fe(CO)₄ while for Sn(OEP)Fe(CO)₄ and Ge(OEP)Fe(CO)₄ these separations are increased to 0.47 and 0.48 V. Similar potential separations between the two reductions are measured for Ge(TmTP)Fe(CO)₄ and Sn(TmTP)Fe(CO)₄ (Table 19) and are in agreement with an average separation of $0.44 \pm 0.05$ V for metalloporphyrins in which the two reductions have occured at the porphyrin π ring system to give anion radicals and dianions[99, 175, 176].

The potentials for porphyrin π anion radical formation of M$^{II}$(OEP) complexes are linearly related to the electronegativity of the central metal ion[175, 176] and the most

**Table 19.** Peak and half-wave potentials (V vs. SCE) for the electrode reactions of M(Por)Fe(CO)$_4$ at a gold working electrode in PhCN containing 0.1 M TBA(PF$_6$)[a]

| Compound | Oxidation | Reduction | |
|---|---|---|---|
| | E$_p$(III)[b] | E$_{1/2}$(I) | E$_{1/2}$(II) |
| Ge(TpTP)Fe(CO)$_4$ | 0.51 | − 0.97 | − 1.43 |
| Sn(TpTP)Fe(CO)$_4$ | 0.65 | − 0.94 | − 1.39 |
| Ge(TmTP)Fe(CO)$_4$ | 0.47 | − 0.96 | − 1.42 |
| Sn(TmTP)Fe(CO)$_4$ | 0.66 | − 0.91 | − 1.36 |
| Ge(OEP)Fe(CO)$_4$ | 0.43 | − 1.22 | − 1.71 |
| Sn(OEP)Fe(CO)$_4$ | 0.50 | − 1.17 | − 1.64 |

[a] Data taken from Ref. 171
[b] Scan rate = 0.20 V/s

positive reduction potentials are found for complexes having metal ions with the highest electronegativity. The absolute differences in reduction potentials between Sn(Por)-Fe(CO)$_4$ and Ge(Por)Fe(CO)$_4$ complexes with the same porphyrin ring range between 20 and 70 mV with the former complexes being the most easy to reduce (see Table 19). This is consistent with the slightly higher electronegativity of Ge$^{II}$ (2.01) with respect to Sn$^{II}$ (1.96) and suggests that the site of electroreduction occurs at the porphyrin π ring system. This conclusion is also suggested by results of thin-layer spectroelectrochemistry and controlled potential electrolysis coupled with ESR[171].

Similar electrochemical behavior was observed for reduction of the six Sn(Por)-Fe(CO)$_4$ and Ge(Por)Fe(CO)$_4$ complexes. In all cases, addition of the first electron was at the π ring system of the metalloporphyrin. This is in contrast to other Fe(L)(CO)$_4$ complexes[177] and to In(Por)M(CO)$_3$Cp[168] where reduction occurs at the Fe atom[177] and at the axial ligand[168], respectively. The former reductions occur in a range of potentials between − 1.6 and 2.1 V and depend upon the donor character of the bound L group[177]. While one might expect M$^{II}$(Por)Fe(CO)$_4$ to exhibit these same electrode reactions, the addition of two electrons to the porphyrin ring of M$^{II}$(Por)Fe(CO)$_4$ will change the donor character of the metalloporphyrin such that extremely negative potentials would be needed for reduction of the iron carbonyl unit.

The stability of reduced [M(Por)Fe(CO)$_4$]$^{-1}$ and [M(Por)Fe(CO)$_4$]$^{-2}$ was investigated and comparisons were made to the electrochemistry of [Sn(Por)]$^{+2}$ or [Ge(Por)]$^{2+}$ (added as ClO$_4^-$ salts) and Na$_2$Fe(CO)$_4$ which are the possible decomposition products of the reduced bimetallic complexes[171]. The voltammograms of both [Sn(TpTP)]$^{+2}$ and Na$_2$Fe(CO)$_4$ are complicated[171]. The former species is reduced in several steps and the formation of a chlorin or a phlorin has been reported to be a product of the doubly reduced species[178]. In addition, a reduction peak is observed on the first negative potential scan which is not present on the second scan or at more negative potentials.

No peaks due to Sn(TpTP)(ClO$_4$)$_2$ were observed as reduction products of Sn(TpTP)Fe(CO)$_4$[171]. Likewise, no spectral bands associated with reduced Sn(TpTP)(ClO$_4$)$_2$ were observed upon controlled potential reduction of Sn(TpTP)Fe(CO)$_4$. It was noted, however, that the addition of Na$_2$Fe(CO)$_4$ to solutions of PhCN led to NaHFe(CO)$_4$ but no peaks for this species were observed during reduction of Sn(TpTP)Fe(CO)$_4$ nor were additional peaks observed in the cyclic voltammograms for reduction of any other M(Por)Fe(CO)$_4$ complexes[171].

The oxidation of M(Por)Fe(CO)$_4$ is irreversible in the sense that there are no coupled peaks associated with reduction of [M(Por)Fe(CO)$_4$]$^+$. Similar voltammograms were obtained for the six M(Por)Fe(CO)$_4$ complexes and peak potentials for the single irreversible oxidation at an Au electrode are listed in Table 19. As expected, the OEP complexes are easier to oxidize than the TpTP and TmTP complexes. Also, with a given porphyrin ligand the oxidation is easier for the Ge$^{II}$ complex than for the Sn$^{II}$ complex. This is due to the higher electronegativity of the germanium ion.

The peak potential for oxidation of each M(Por)Fe(CO)$_4$ complex shifts in a positive direction with increase in scan rate. The slope of $\Delta E_p/\Delta \log(v)$ varied between 51 and 65 mV and is 53 mV for the oxidation of Sn(TpTP)Fe(CO)$_4$ in benzonitrile, 0.1 M TBA(PF$_6$) indicating an irreversible electron transfer reaction in all cases[171]. Rotating disk voltammetry showed that the oxidation of each M(Por)Fe(CO)$_4$ complex was a diffusion controlled process. However, at a rotation rate of 1600 rpm the ratio of maximum current for the oxidation over the maximum current for reduction was generally larger than 1.0 for all of the compounds and only approached 1.0 as the rotation rate increased. The ratio of oxidation over reduction current by cyclic voltammetry was also equal to or greater than 1.0 depending on the compound and the scan rate, but in all cases the ratio increased to values larger than 1.0 as the scan rate was decreased. These data suggest an ECE type mechanism where the second electron transfer is cut off at high rotation rates (rotating disk voltammetry) or high potential scan rates (cyclic voltammetry).

Based on these data, the primary oxidation products of M(Por)Fe(CO)$_4$ were postulated to be [M$^{IV}$(Por)]$^{+2}$ and Fe$_3$(CO)$_{12}$[171]. These fragments are formed after the global abstraction of two electrons from M(Por)Fe(CO)$_4$ and cleavage of the metal-metal bond in the bimetallic complex. Evidence for the above oxidation products came from coulometric values of n and from identification of [M$^{IV}$(Por)]$^{2+}$ and Fe$_3$(CO)$_{12}$ as major products of the oxidation[171].

## D IV 3) Reduction of In(TPP)M(CO)$_3$Cp

Potentials have been measured for a number of In(Por)M'L complexes[159] and Tl(Por)M'L[170] but the mechanism for electroreduction has only been reported in detail for In(TPP)Mo(CO)$_3$Cp and In(TPP)W(CO)$_3$Cp[168]. The electrochemical characterization of In(TPP)M(CO)$_3$Cp was initially reported for complexes where M = Mo and W and Cp = $\eta^2$-C$_5$H$_5$[168]. At low temperature, the metal-metal bonded derivatives exhibit two reductions on the cyclic voltammetry time scale. The monoanion is extremely stable for the complex containing W while for the complex with Mo, the stability extends only to the cyclic voltammetry time scale. ESR spectra were presented for electrogenerated [In(TPP)W(CO)$_3$Cp]$^{-\cdot}$ and it was suggested that reduction might occur at the W(CO)$_3$Cp ligand, producing W$^{I}$[168].

A definitive oxidation state assignment of the neutral or reduced complexes has not been made. The neutral complexes may be considered as consisting of formal In$^{III}$(Por) Mo$^0$(CO)$_3$Cp units where Mo(CO)$_3$Cp is viewed as an axial ligand bound to the In$^{III}$ porphyrin by means of an ionic bond. However, ESR data of the reduced species indicate that substantial charge is transferred from the Mo or the W atom to the In$^{III}$ center and an alternative formulation[168] is $In^I(Por)Mo^{II}(CO)_3Cp$. This formulation is in good agreement with theoretical calculations of the electronic absorption spectra for Group 14 metalloporphyrins[141] and is consistent with the ESR data which suggest that the initial reduction site occurs at the tungsten atom.

A detailed study of $[In(TPP)Mo(CO)_3Cp]^-$ decomposition products was carried out by Kadish[159] and led to the pathway shown in Scheme 18 for reduction and cleavage of the metal-metal bond of $In(TPP)Mo(CO)_3Cp$ in $CH_2Cl_2$, 0.1 M TBAP.

$$In(TPP)Mo(CO)_3Cp \quad \overset{+e}{\rightleftarrows} \quad [In(TPP)Mo(CO)_3Cp]^{\cdot}$$

$$k \downarrow \text{ slow}$$

$$In(TPP)^{\cdot} + [Mo(CO)_3Cp]^-$$

**Scheme 18**

Cleavage of In–Mo bond in $[In(TPP)Mo(CO)_3Cp]^-$ leads to $In(TPP)^{\cdot}$ and $[Mo(CO)_3Cp]^-$. This was ascertained by monitoring the ESR signals during decomposition of $[In(TPP)Mo(CO)_3Cp]^{-}$ [159] as well as by recording cyclic voltammograms where the potential was scanned in a positive direction after formation of the singly reduced bimetallic complex. At low scan rates oxidation waves for both $In(TPP)^{\cdot}$ and $[Mo(CO)_3Cp]^-$ (to give $In(TPP)ClO_4$ and $[Mo(CO)_3Cp]_2$) were detected thus confirming the mechanism shown in Scheme 18.

## D IV 4) Oxidation of In(TPP)Mo(CO)₃Cp

The initial oxidation of $In(TPP)Mo(CO)_3Cp$ involves a one electron transfer and leads to a cleavage of the metal-metal bond[168]. Current voltage curves were analyzed during electro-oxidation of $In(TPP)Mo(CO)_3Cp$ at different temperatures and this data was combined with the electrochemistry of $In(TPP)ClO_4$, $[Mo(CO)_3Cp]^+$, and $[Mo(CO)_3Cp]_2$[159]. The products of each electro-oxidation were also monitored by ESR spectroscopy and on the basis of the data the mechanism given in Scheme 19 was proposed for oxidation of $In(TPP)Mo(CO)_3Cp$ in $CH_2Cl_2$, 0.1 M TBAP.

$$In(TPP)Mo(CO)_3Cp \quad \underset{0.85\,V}{\overset{-e}{\rightleftarrows}} \quad [In(TPP)Mo(CO)_3Cp]^{+}$$

$$k \downarrow \text{ fast; } + ClO_4^-$$

$$In(TPP)ClO_4 + Mo(CO)_3Cp^{\cdot}$$

**Scheme 19**

The key point in Scheme 19 is that cleavage of oxidized $In(TPP)Mo(CO)_3Cp$ leads to $In(TPP)ClO_4$ and $Mo(CO)_3Cp$. The $Mo(CO)_3Cp$ radical can either dimerize to give $[Mo(CO)_3Cp]_2$ or it may be oxidized directly at 0.85 V to give $[Mo(CO)_3Cp]^+$. The exact sequence of steps depends on the temperature and scan rate[170]. At − 70 °C the dimerization rate decreases and there is almost quantitative conversion of $Mo(CO)_3Cp$ to $[Mo(CO)_3Cp]^+$. Under these conditions the overall oxidation involves a two electron transfer at + 0.85 V. However, at room temperature large quantities of $Mo(CO)_3Cp$ are converted to $[Mo(CO)_3Cp]_2$ and an additional oxidation peak occurs at 1.0 V.

# E) Future Directions

Studies of organoiron porphyrin complexes as models for biological systems are of current interest. These synthesized molecules play a fundamental role in helping to elucidate the biochemical activity of hemoproteins. Physicochemical data of these σ-bonded complexes show that the alkyl or aryl axial ligands possess particular structural and electronic properties. These derivatives exhibit especially unusual behavior regarding spin state and magnetic susceptibility and an understanding of these data is essential to elucidate the nature of the biological processes and the structural characterization of intermediates exhibiting similar properties. For example, a comparison of various spectroscopic properties of chloroperoxidase with those of cytochrome P 450 has already led to the proposal that the heme environment in these two enzymes is similar[179–182].

Other transition and non-transition metal σ-bonded metalloporphyrin complexes have been prepared in order to investigate their chemical reactivity and to demonstrate the influence of the macrocyclic ligand on this reactivity when compared to that of other classes of organometallic derivatives. Some of the reactions observed for the σ-bonded metalloporphyrins (such as insertion reactions of sulfur dioxide or carbon dioxide) are usually found in organometallic chemistry. In contrast, the migration of a σ-bonded alkyl or aryl groups from the central iron or cobalt of the metalloporphyrin to one of the four porphyrin nitrogens is clearly due to the presence of the porphyrin unit. Such a migration might initially take place during the insertion reaction of certain small molecules. Furthermore, the apparent insertion of CO into the rhodium-hydride bond of Rh(OEP)H may also occur for other metalloporphyrin complexes via free radical pathways.

Only a few examples of σ-bonded metal-carbon porphyrins are now known, but it is theoretically possible to prepare and characterize representative complexes with almost all elements in the periodic table. These types of syntheses could lead to the related synthesis of numerous novel bimetallic or dimeric derivatives.

It can be predicted that future synthetic developments in porphyrin chemistry will lead to new complexes possessing metal-metal interactions. These types of complexes are of interest in biological processes[183], in synthetic organometallic chemistry, and in the preparation of metal-polymers. The known properties of metalloporphyrin dimers characterized to date also opens up the possibility for investigating a wide variety of organometallic type reactions with metalloporphyrins. It should also be possible to form metalloporphyrin complexes with metal-metal quadruple bonds as has been described for other non-porphyrin organometallic complexes. The combinations of metals needed to synthesize homo and heterometal dimers has been predicted in the pioneering work of Collman[163]. On the basis of a simple molecular orbital treatment, the bond order, the spin state, and the geometry of a given complex can be postulated.

Finally, it is hoped that this review will stimulate the synthesis of a wide variety of new dimers in order to determine exact relationships between bimetallic metalloporphyrin complex bond order and bond lengths between the two metals. Such studies should allow one to precisely determine the selective reactivity of each metalloporphyrin complex and at the same time to predict the potential catalytic activities of these type species.

*Acknowledgments.* Support from the Centre National de la Recherche Scientifique, the National Science Foundation, and the National Institute of Health are gratefully acknowledged by the authors. The authors would like to thank Drs. J. E. Anderson, J. M. Barbe, P. Cocolios, and A. Tabard for reading early drafts of this manuscript and Mr. D. Bayeul for technical help in the preparation of all the ORTEP views.

# F) Abbreviations and Nomenclature

## F I) Abbreviations

| | | | |
|---|---|---|---|
| X | halide | TBA(PF$_6$) | tetrabutylammonium hexa-fluorophosphate |
| Me | methyl | | |
| Et | ethyl | N–MeIm | N-methylimidazole |
| CH(CH$_3$)$_2$ | isopropyl | THF | tetrahydrofuran |
| (CH$_2$)$_3$CH$_3$ | butyl | py | pyridine |
| C(CH$_3$)$_3$ | isobutyl | 3-ClPy | 3-chloropyridine |
| CH=CHC$_6$H$_5$ | styryl | PhCN | benzonitrile |
| C≡C–C$_6$H$_5$ | phenylacetylenyl | DMF | dimethylformamide |
| C$_6$H$_5$ | phenyl | P$^{Me}$ | CH$_2$CH$_2$COOCH$_3$ |
| N–R | N-alkyl | P$^H$ | CH$_2$CH$_2$COOH |
| N–C$_6$H$_5$ | N-phenyl | V | vinyl |
| C$_6$F$_4$H | 4H-tetrafluorophenyl | DPEP | deoxophylloerythroetiopor-phyrinate |
| C$_6$F$_5$ | pentafluorophenyl | | |
| m-Me | m-methyl | Por | unspecified porphinate |
| p-Me | p-methyl | Δ 4 N | distance (Å) from the plane of the four nitrogen atoms |
| p-Et$_2$N | p-diethylamino | | |
| AcO | acetyl | Δ Por | distance (Å) from the mean plane of the C$_{20}$N$_4$ porphyrin core atoms |
| Cp | cyclopentadienyl | | |
| n-Bu$_4$ | n-tetrabutylammonium | | |
| TBA(ClO$_4$) | tetrabutylammonium per-chlorate | | |

## F II) Coordination Schemes

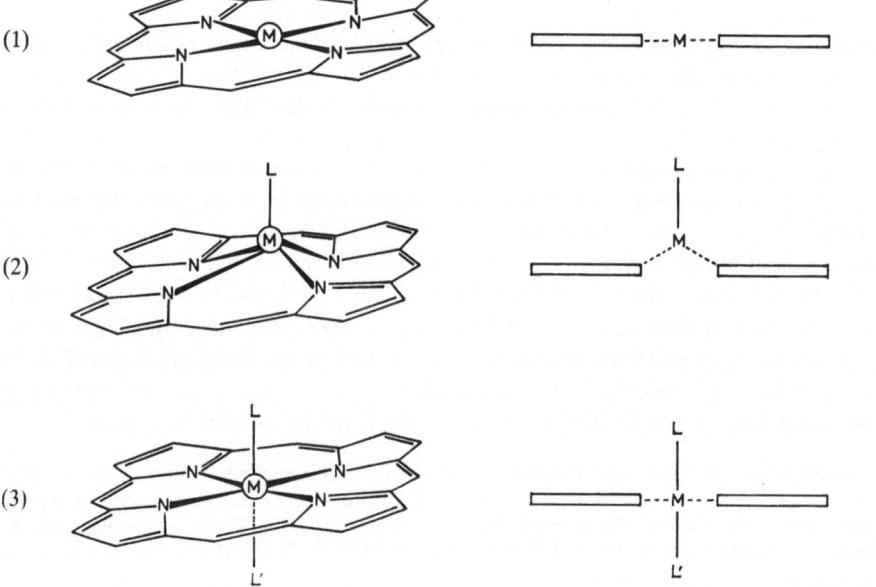

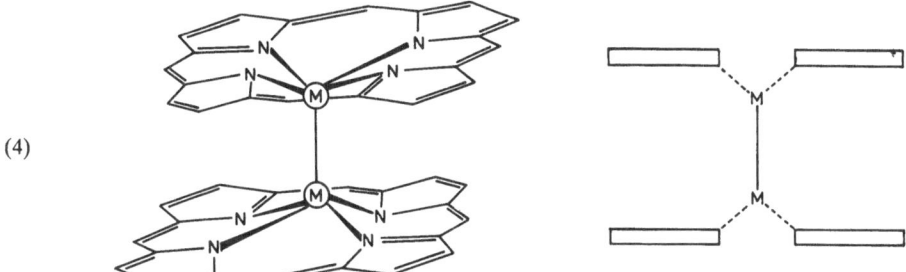

**Fig. 24.** Porphyrin coordination schemes: (1) square planar, (2) square pyramidal, (3) pseudo-octahedral, and (4) dimeric

## F III) Nomenclature

| Abbreviation | Name | Substituents | | | | | | | | |
|---|---|---|---|---|---|---|---|---|---|---|
| | | 2 | 3 | 7 | 8 | 12 | 13 | 17 | 18 | α, β, γ, δ |
| P | Porphine | H | H | H | H | H | H | H | H | H |
| OMP | Octamethylporphyrin | Me | Me | Me | Me | Me | Me | Me | Me | H |
| OEP | Octaethylporphyrin | Et | Et | Et | Et | Et | Et | Et | Et | H |
| TPP | Tetraphenylporphyrin | H | H | H | H | H | H | H | H | Ph |
| TmTP | tetra-*m*-tolylporphyrin | H | H | H | H | H | H | H | H | mT |
| TpTP | Tetra-*p*-tolylporphyrin | H | H | H | H | H | H | H | H | pT |
| (CN)$_4$TPP | Tetracyano-2,7,12,17-tetraphenylporphyrin | CN | H | CN | H | CN | H | CN | H | Ph |
| TpEt$_2$NPP | tetraparadiethylaminophenylporphyrin | H | H | H | H | H | H | H | H | pEt$_2$NPh |
| Etio-I | Etioporphyrin-I | Me | Et | Me | Et | Me | Et | Me | Et | H |
| Etio-II | Etioporphyrin-II | Me | Et | Et | Me | Me | Et | Et | Me | H |
| MPIX | Mesoporphyrin-IX | Me | Et | Me | Et | Me | $P^H$ | $P^H$ | Me | H |
| MPIXDME | Mesoporphyrin-IX-dimethylester | Me | Et | Me | Et | Me | Et | Me | Et | H |
| DP | Deuteroporphyrin-IX | Me | H | Me | H | Me | $P^H$ | $P^H$ | Me | H |
| DPDME | Deuteroporphyrin-IX-dimethylester | Me | H | Me | H | Me | $P^{Me}$ | $P^{Me}$ | Me | H |
| PPIX | Protoporphyrin-IX | Me | V | Me | V | Me | $P^H$ | $P^H$ | Me | H |
| PPIXDME | Protoporphyrin-IX-dimethylester | Me | V | Me | V | Me | $P^{Me}$ | $P^{Me}$ | Me | H |

# G) References

1. Mansuy, D., Nastainczyk, W., Ullrich, V.: Arch. Pharmakol. *285*, 315 (1974)
2. Ullrich, V.: Top. Curr. Chem. *83*, 68 (1979)
3. Ortiz de Montellano, P. R. (ed.): Cytochrome P-450: Structure, Mechanism, and Biochemistry, Plenum Press, New York 1986
4. Mansuy, D., Lange, M., Chottard, J. C., Bartoli J. F., Chevrier, B., Weiss, R.: Ang. Chem. Int. Ed. Engl. *17*, 781 (1978)
5. Augusto, O., Kunze, K. L., Ortiz de Montellano, P. R.: J. Biol. Chem. *257*, 6231 (1982)
6. Saito, S., Hano, H. A.: Proc. Natl. Acad. Sci. USA *78*, 5508 (1981)
7. Ortiz de Montellano, P. R., Kunze, K. L.: J. Am. Chem. Soc. *103*, 6334 (1981)
8. Miller, J. S. (ed.): Extended Linear Chain Compounds, New York, Plenum Press 1982
9. Clarke, D. A., Grigg, R., Johnson, A. W.: J. Chem. Soc., Chem. Commun. 208 (1966)
10. Clarke, D. A., Dolphin, D., Grigg, R., Johnson, A. W., Pinnock, H. A.: J. Chem. Soc. C, 881 (1968)
11. Reed, C. A., Mashiko, T., Bentley, S. P., Kastner, M. E., Scheidt, W. R., Spartalian, K., Lang, G.: J. Am. Chem. Soc. *101*, 2948 (1979)
12. Ortiz de Montellano, P. R., Kunze, K. L., Augusto, A.: ibid. *104*, 3545 (1982)
13. Cocolios, P., Laviron, E., Guilard, R.: J. Organomet. Chem. *228*, C39 (1982)
14. Ogoshi, H., Sugimoto, H., Yoshida, Z., Kobayashi, H., Sakai, H., Maeda, Y.: ibid. *234*, 185 (1982)
15. Mansuy, D., Battioni, J. P.: J. Chem. Soc., Chem. Commun., 638 (1982)
16. Cocolios, P., Lagrange, G., Guilard, R.: J. Organomet. Chem. *253*, 65 (1983)
17. Battioni, J. P., Lexa, D., Mansuy, D., Savéant, J. M.: J. Am. Chem. Soc. *105*, 207 (1983)
18. Ogoshi, H., Setsune, J., Omura, T., Yoshida, Z.: ibid. *97*, 6461 (1975)
19. Latour, J. M., Boreham, C. J., Marchon, J. C.: J. Organomet. Chem. *190*, C61 (1980)
20. Ogoshi, H., Setsune, J., Yoshida, Z.: ibid. *159*, 317 (1978)
21. Coutsolelos, A., Guilard, R.: ibid. *253*, 273 (1983)
22. Guilard, R., Cocolios, P., Fournari, P.: ibid. *129*, C11 (1977)
23. Cocolios, P., Guilard, R., Fournari, P.: ibid. *179*, 311 (1979)
24. Tabard, A., Zrineh, A., Guilard, R., Kadish, K. M.: Inorg. Chem. in press (1987)
25. Kadish, K. M., Xu, Q. Y., Barbe, J. M., Guilard, R.: submitted for publication
26. Maskasky, J. E., Kenney, M. E.: J. Am. Chem. Soc. *93*, 2060 (1971)
27. a) Maskasky, J. E., Kenney, M. E.: ibid. *93*, 1443 (1973)
    b) Miyamoto, T. K., Hasegawa, T., Takagi, S., Sasaki, Y.: Chem. Lett. 1181 (1983)
    c) Miyamoto, T. K., Sugita, N., Matsumoto, Y., Sasaki, Y., Konno, M.: ibid. 1695 (1983)
28. Cloutour, C., Lafargue, D., Richards, J. A., Pommier, J. C.: J. Organomet. Chem. *137*, 157 (1977)
29. Cloutour, C., Lafargue, D., Pommier, J. C.: ibid. *161*, 327 (1978)
30. Barbe, J. M., Guilard, R.: unpublished results
31. Clarke, D. A., Grigg, R., Johnson, A. W., Pinnock, H. A.: J. Chem. Soc., Chem. Commun. 309 (1967)
32. Momenteau, M., Fournier, M., Rougee, M.: J. Chim. Phys. *67*, 926 (1970)
33. a) Perree-Fauvet, M., Gaudemer, A., Boucly, P., Devynck, J.: J. Organomet. Chem. *120*, 439 (1976)
    b) Kadish, K. M., Tabard, A.: unpublished results
34. Ogoshi, H., Watanabe, E., Koketsu, N., Yoshida, Z.: Bull. Chem. Soc. Jpn. *49*, 2529 (1976)
35. Dolphin, D., Halko, D. J., Johnson, E.: Inorg. Chem. *20*, 4348 (1981)
36. Ogoshi, H., Omura, T., Yoshida, Z.: J. Am. Chem. Soc. *95*, 1666 (1973)
37. Ogoshi, H., Setsune, J., Yoshida, Z.: ibid. *99*, 3869 (1977)
38. Ogoshi, H., Setsune, J., Nanbo, Y., Yoshida, Z.: J. Organomet. Chem. *159*, 329 (1978)
39. Lexa, D., Mispelter, J., Savéant, J. M.: J. Am. Chem. Soc. *103*, 6806 (1981)
40. Lexa, D., Savéant, J. M.: ibid. *104*, 3503 (1982)
41. Grigg, R., Trocha-Grimshaw, J., Viswanatha, V.: Tetrahedron Lett. *4*, 289 (1976)
42. Abeysekera, A. M., Grigg, R., Trocha-Grimshaw, J., Viswanatha, V.: ibid. *36*, 3189 (1976)
43. Abeysekera, A. M., Grigg, R., Trocha-Grimshaw, J., Viswanatha, V.: J. Chem. Soc., Chem. Commun. 227 (1976)

44. Abeysekera, A. M., Grigg, R., Trocha-Grimshaw, J., Viswanatha, V.: J. Chem. Soc., Perkin I, 1395 (1977)
45. Parshall, G. W., Mrowea, J. J.: Adv. in Organomet. Chem. 7, 157 (1968)
46. Brault, D., Bizet, C., Morlière, P., Rougee, M., Land, E. J., Santus, R., Shallow, A. J.: J. Am. Chem. Soc. 102, 1015 (1980)
47. Brault, D., Neta, P.: ibid. 103, 2705 (1981)
48. Callot, H. J., Schaeffer, E.: J. Organomet. Chem. 145, 91 (1978)
49. Callot, H. J., Schaeffer, E.: J. Chem. Soc., Chem. Commun., 937 (1978)
50. Callot, H. J., Schaeffer, E.: Nouv. J. Chim. 4, 311 (1980)
51. Callot, H. J., Schaeffer, E.: J. Organomet. Chem. 193, 111 (1980)
52. Cohen, J. A., Chow, B. C.: Inorg. Chem. 13, 488 (1974)
53. Wayland, B. B., Woods, B. A.: J. Chem. Soc., Chem. Commun., 700 (1981)
54. Paonessa, R. S., Thomas, N. C., Halpern, J.: J. Am. Chem. Soc. 107, 4333 (1985)
55. Inoue, S., Takeda, N.: Bull. Chem. Soc. Jpn. 50, 984 (1977)
56. Takeda, N., Inoue, S.: ibid. 51, 3564 (1978)
57. Henrick, K., Matthews, R. W., Tasker, P. A.: Inorg. Chem. 16, 3293 (1977)
58. Brady, F., Henrick, K., Matthews, R. W.: J. Organomet. Chem. 210, 281 (1981)
59. Inoue, S., Murayama, H., Takeda, N., Ohkatsu, Y.: Chem. Lett., 317 (1982)
60. Collman, J. P., McElwee-White, L., Brothers, P. J., Rose, E.: J. Am. Chem. Soc. 107, 6110 (1985)
61. Collman, J. P., Prodolliet, J. W., Leidner, C. R.: ibid. 108, 2916 (1986)
62. Ratti, C., Guilard, R.: unpublished results
63. Kadish, K. M., Boisselier-Cocolios, B., Cocolios, P., Guilard, R.: Inorg. Chem. 24, 2139 (1985)
64. Kadish, K. M., Boisselier-Cocolios, B., Coutsolelos, A., Mitaine, P., Guilard, R.: ibid. 24, 4521 (1985)
65. Tabard, A., Cocolios, P., Lagrange, G., Gerardin, R., Hubsch, J., Lecomte, C., Zarembowitch, J., Guilard, R.: submitted for publication
66. Scheidt, W. R., Reed, C. A.: Chem. Rev. 81, 543 (1981)
67. Mitra, S.: Magnetic Susceptibility of Iron Porphyrins, Part. II, in: Iron Porphyrins (eds. Lever, A. B. P., Gray, H. B.), p. 1, Reading MA, Addison-Wesley 1983
68. Guilard, R., Boisselier-Cocolios, B., Tabard, A., Cocolios, P., Simonet, B., Kadish, K. M.: Inorg. Chem. 24, 2509 (1985)
69. La Mar, G. N., Walker, F. A.: Nuclear Magnetic Resonance of Paramagnetic Metalloporphyrins in: The Porphyrins, Vol. IV (ed. Dolphin, D.), p. 61, New York, Academic Press 1979
70. Tabard, A., Lagrange, G., Guilard, R., Jouan, M., Dao, N. Q.: J. Organomet. Chem. 308, 335 (1986)
71. Slichter, L. P., Drickamer, H. G.: J. Chem. Phys. 56, 2142 (1972)
72. La Mar, G. N., Eaton, G. R., Holm, R. H., Walker, F. A.: J. Am. Chem. Soc. 95, 63 (1973)
73. Cervone, E., Camassel, F. D., Luciani, M. L., Furlani, C.: J. Inorg. Nucl. Chem. 31, 1101 (1969)
74. Lang, G.: Applied Physics 38, 915 (1967)
75. Geiger, D. K., Scheidt, W. R.: Inorg. Chem. 23, 1970 (1984)
76. Reed, C. A.: Iron(I) and Iron(IV) Porphyrins, in: Electrochemical and Spectroelectrochemical Studies of Biological Redox Components (ed. Kadish, K. M.), p. 333, Washington, American Chemical Soc. 1982
77. Lançon, D., Cocolios, P., Guilard, R., Kadish, K. M.: J. Am. Chem. Soc. 106, 4472 (1984)
78. Lançon, D., Cocolios, P., Guilard, R., Kadish, K. M.: Organometallics 3, 1164 (1984)
79. Kadish, K. M., Lançon, D., Cocolios, P., Guilard, R.: Inorg. Chem. 23, 2372 (1984)
80. Guilard, R., Lagrange, G., Tabard, A., Lançon, D., Kadish, K. M.: ibid. 24, 3649 (1985)
81. Callot, H. J., Cromer, R., Louati, A., Gross, M.: Nouv. J. Chim. 8, 765 (1985)
82. Callot, H. J., Metz, F., Cromer, R.: ibid. 8, 759 (1984)
83. Kadish, K. M., Yao, C. L., Anderson, J. E., Cocolios, P.: Inorg. Chem. 24, 4515 (1985)
84. Kadish, K. M., Anderson, J. E., Yao, C. L., Guilard, R.: ibid. 25, 1277 (1986)
85. Anderson, J. E., Yao, C. L., Kadish, K. M.: ibid. 25, 718 (1986)
86. (a) Anderson, J. E., Yao, C. L., Kadish, K. M. : J. Am. Chem. Soc. 109, 1106 (1987)
    (b) Anderson, J. E., Yao, C. L., Kadish, K. M.: Organometallics 6, 706 (1987)
    (c) Yao, C. L.: PhD Thesis, University of Houston (1987)

87. Sugimoto, H., Ueda, N., Mori, M.: J. Chem. Soc., Dalton Trans., 1611 (1982)
88. Cornillon, J. L., Anderson, J. E., Swistak, C., Kadish, K. M.: J. Am. Chem. Soc. *108*, 7633 (1986)
89. Tabard, A., Guilard, R., Kadish, K. M.: Inorg. Chem. *25*, 4277 (1986)
90. Kadish, K. M., Xu, Q. Y., Barbe, J. M., Anderson, J. E., Wang, E., Guilard, R., submitted for publication
91. Kadish, K. M., Cornillon, J. L., Cocolios, P., Tabard, A., Guilard, R.: Inorg. Chem. *24*, 3645 (1985)
92. Momenteau, M., Loock, B.: J. Mol. Catal. *7*, 315 (1980)
93. Momenteau, M., Loock, B., Larollette, D., Tetreau, C., Mispelter, J.: J. Chem. Soc., Chem. Commun., 962 (1983)
94. Kadish, K. M.: J. Electroanal. Chem. *168*, 261 (1984)
95. Doppelt, P.: Inorg. Chem. *23*, 4009 (1984)
96. Furhrhop, J. H.: Redox reactions of metalloporphyrins, in: Structure and Bonding, Vol. 18, Chapter 1, pp. 1–68, New York, Springer 1974
97. Loew, G. H.: Theoretical investigations of iron porphyrins, in: Iron Porphyrins (eds. Lever, A. B. P., Gray, H. P.), Part I, Chapter 1, pp. 1–87, Reading, Mass., Addison-Wesley 1983
98. Lagrange, G., Cocolios, P., Guilard, R.: J. Organomet. Chem. *260*, C16 (1984)
99. Felton, R. H., Linschitz, H.: J. Am. Chem. Soc. *88*, 1113 (1966)
100. Walker, F. A., Beroiz, D., Kadish, K. M.: ibid. *98*, 3484 (1976)
101. Cloutour, C., Lafargue, D., Pommier, J. C.: J. Organomet. Chem. *190*, 35 (1980)
102. Takeda, A., Syal, S. K., Sasada, Y., Omura, T., Ogoshi, H., Yoshida, Z. Y.: Acta Cryst. *B32*, 62 (1976)
103. Fleischer, E. B., Lavallee, D.: J. Am. Chem. Soc. *89*, 7132 (1967)
104. Lecomte, C., Protas, J., Cocolios, P., Guilard, R.: Acta Cryst. *B36*, 2769 (1980)
105. Colin, J., Chevrier, B.: Organometallics *4*, 1090 (1985)
106. Troughton, P. G. H., Skapski, A. C.: J. Chem. Soc., Chem. Commun., 575 (1968)
107. Churchill, M. R.: Inorg. Chem. *4*, 1734 (1965)
108. Churchill, M. R.: Perfect. Struct. Chem. *3*, 91 (1970)
109. Buchler, J. W., Kokisch, W., Smith, P. D.: Struct. Bonding *34*, 79 (1978)
110. Scheidt, W. R., Lee, Y. L., Hatano, K.: Inorg. Chim. Acta, Bioinorg. Chem. *79*, 192 (1983)
111. Scheidt, W. R., Frisse, M. E.: J. Am. Chem. Soc. *97*, 17 (1975)
112. Hoard, J. L.: Stereochemistry of Porphyrins and Metalloporphyrins, in: Porphyrins and Metalloporphyrins (ed. Smith, K. M.), p. 317, Elsevier, Amsterdam 1975
113. Chevrier, B., Weiss, R., Lange, M., Chottard, J. C., Mansuy, D.: J. Am. Chem. Soc. *103*, 2899 (1981)
114. Ball, R. G., Lee, K. M., Marshall, A. G., Trotter, J.: Inorg. Chem. *19*, 1463 (1980)
115. Cullen, D. L., Meyer, E. F. Jr., Smith, K. M.: ibid. *16*, 1179 (1977)
116. Hoard, J. L., Cohen, G. H., Glick, M. D.: J. Am. Chem. Soc. *89*, 1992 (1967)
117. Lecomte, C., Protas, J., Guilard, R., Fournari, P.: J. Chem. Soc., Chem. Commun., 434 (1976)
118. Lecomte, C., Protas, J., Guilard, R., Fliniaux, B., Fournari, P.: J. Chem. Soc., Dalton Trans., 1306 (1979)
119. Lecomte, C., Protas, J., Richard, P., Barbe, J. M., Guilard, R.: ibid., 247 (1982)
120. Diebold, T., Chevrier, B., Weiss, R.: Inorg. Chem. *18*, 1193 (1979)
121. Colin, J., Butler, G., Weiss, R.: ibid. *19*, 3829 (1980)
122. Cocolios, P., Guilard, R., Bayeul, D., Lecomte, C.: ibid. *24*, 2058 (1985)
123. Oumous, H., Lecomte, C., Protas, J., Cocolios, P., Guilard, R.: Polyhedron *3*, 651 (1984)
124. Buchler, J. W.: Static Coordination Chemistry of Metalloporphyrins, in: Porphyrins and Metalloporphyrins (ed. Smith, K. M.), p. 157, Amsterdam, Elsevier 1975
125. Aida, T., Inoue, S.: J. Am. Chem. Soc. *105*, 1304 (1983)
126. Vitzthum, G., Lindner, E.: Angew. Chem., Int. Ed. Engl. *10*, 315 (1971)
127. Guilard, R., Cocolios, P., Fournari, P., Lecomte, C., Protas, J.: J. Organomet. Chem. *168*, C49 (1979)
128. Cocolios, P., Fournari, P., Guilard, R., Lecomte, C., Protas, J., Boubel, J. C.: J. Chem. Soc., Dalton Trans., 2081 (1980)
129. Boukhris, A., Lecomte, C., Coutsolelos, A., Guilard, R.: J. Organomet. Chem. *303*, 151 (1986)

130. Cocolios, P., Lagrange, G., Guilard, R., Oumous, H., Lecomte, C.: J. Chem. Soc., Dalton Trans., 567 (1984)
131. Philippi, M. A., Baenzinger, N., Goff, H. M.: Inorg. Chem. *20*, 3904 (1981)
132. Kitching, W., Fong, C. W.: Organomet. Chem. Rev. *A5*, 281 (1970)
133. Wojcicki, A.: Adv. Organomet. Chem. *12*, 31 (1974)
134. Deakon, G. B., Felder, P. W.: Aust. J. Chem. *22*, 549 (1969)
135. Cocolios, P.: Thèse de Doctorat d'Etat, p. 154, Dijon, France 1982
136. Mays, M. J., Bailey, J.: J. Chem. Soc., Dalton Trans., 578 (1977)
137. Olapinski, H., Weidlein, J., Hausen, H. D.: J. Organomet. Chem. *64*, 193 (1974)
138. Palmer, G.: Electron Paramagnetic Resonance of Hemoproteins, in: The Porphyrins, Vol. IV (ed. Dolphin, D.), p. 313, New York, Academic Press 1979
139. Oumous, H.: Thèse de Troisième Cycle, p. 58, Nancy, France 1983
140. Buchler, J. W.: Synthesis and Properties of Metalloporphyrins, in: The Porphyrins, Vol. I (ed. Dolphin, D.), p. 389, New York, Academic Press 1978
141. Gouterman, M.: Electronic Spectra, in: The Porphyrins, Vol. III (ed. Dolphin, D.), p. 1, New York, Academic Press 1978
142. Kirner, J. F., Reed, C. A., Scheidt, W. R.: J. Am. Chem. Soc. *99*, 1093 (1977)
143. Buchler, J. W., Habets, H., Van Kaam, J., Rohbock, K.: unpublished work
144. Cornillon, J. L., Anderson, J. E., Kadish, K. M.: Inorg. Chem. *25*, 991 (1986)
145. Cornillon, J. L., Anderson, J. E., Kadish, K. M.: ibid. *25*, 2611 (1986)
146. Ibers, J. A., Pace, L. J., Martinsen, J., Hoffman, B. M.: Struct. Bonding *50*, 1 (1982)
147. Hanack, M., Pawlowski, G.: Naturwiss. *69*, 266 (1982)
148. Hoffman, B. M., Ibers, J. A.: Acc. Chem. Res. *16*, 15 (1983)
149. Onaka, S., Kondo, Y., Toriumi, K., Ito, T.: Chemistry Letters, 1605 (1980)
150. Barbe, J. M., Guilard, R., Lecomte, C., Gerardin, R.: Polyhedron *3*, 889 (1984)
151. Onaka, S., Kondo, Y., Yamashita, M., Tatematsu, Y., Kato, Y., Goto, M., Ito, T.: Inorg. Chem. *24*, 1070 (1985)
152. Kadish, K. M., Boisselier-Cocolios, B., Swistak, C., Barbe, J. M., Guilard, R.: ibid. *25*, 121 (1986)
153. Collins, R. L., Petit, R.: J. Chem. Phys. *39*, 3433 (1963)
154. Marks, T. J., Newman, A. R.: J. Am. Chem. Soc. *95*, 769 (1973)
155. Cornwell, A. R., Harrison, P. G., Richards, J. A.: J. Organomet. Chem. *108*, 47 (1976)
156. Burnham, R. A., Lyle, M. A., Stobart, S. R.: ibid. *125*, 179 (1977)
157. Barbe, J. M.: Thèse de Troisième Cycle, p. 44, Dijon, France 1985
158. Cocolios, P., Moïse, C., Guilard, R.: J. Organomet. Chem. *228*, C43 (1982)
159. Guilard, R., Mitaine, P., Moïse, C., Lecomte, C., Boukhris, A., Swistak, C., Tabard, A., Lacombe, D., Cornillon, J.-L., Kadish, K. M.: Inorg. Chem. in press (1987)
160. (a) Martin, M., Rees, B., Mitschler, A.: Acta Cryst. *B38*, 6 (1982)
    (b) Zhu, N. J., Lecomte, C., Coppens, P., Keister, J. P.: ibid. *B38*, 1286 (1982)
161. Setsune, J., Yoshida, Z., Ogoshi, H.: J. Chem. Soc., Perkin I, 983 (1982)
162. Wayland, B. B., Newman, A.: Inorg. Chem. *20*, 3093 (1981)
163. Collman, J. P., Barnes, C. E., Woo, L. K.: Proc. Natl. Acad. Sci. USA *80*, 7684 (1983)
164. Collman, J. P., Barnes, C. E., Collins, T. J., Brothers, P. J.: J. Am. Chem. Soc. *103*, 7030 (1981)
165. Collman, J. P., Barnes, C. E., Swepston, P. N., Ibers, J. A.: ibid. *106*, 3500 (1984)
166. Collman, J. P., Woo, L. K.: Proc. Natl. Acad. Sci. USA *81*, 2592 (1984)
167. Jones, N. L., Carroll, P. J., Wayland, B. B.: Organometallics *5*, 33 (1986)
168. Cocolios, P., Chang, D., Vittori, O., Guilard, R., Moïse, C., Kadish, K. M.: J. Am. Chem. Soc. *106*, 572 (1984)
169. Guilard, R., Mitaine, P., Kadish, K. M.: manuscript in preparation
170. Guilard, R., Zrineh, A., Fechat, M., Tabard, A., Mitaine, P., Swistak, C., Kadish, K. M., submitted for publication
171. Kadish, K. M., Swistak, C., Boisselier-Cocolios, B., Barbe, J. M., Guilard, R.: Inorg. Chem., in press (1986)
172. Whitten, D. G., Yau, J. C., Carroll, F. A.: J. Am. Chem. Soc. *93*, 2291 (1971)
173. Edwards, L., Dolphin, D., Gouterman, M., Adler, A. D.: J. Mol. Spect. *38*, 16 (1971)
174. Landrum, J. T., Amini, M., Zuckerman, J. J.: Inorg. Chem. Acta *90*, L73 (1984)
175. Fuhrhop, J. H., Kadish, K. M., Davis, D. G.: J. Am. Chem. Soc. *95*, 5140 (1973)

176. Kadish, K. M.: Redox Reactions in Metalloporphyrins, in: Progress in Inorganic Chemistry (ed. Lippard, S.), Vol. 34, p. 435, New York, Academic Press 1986
177. Connelly, N. G., Geiger, W. E.: The Electron-Transfer Reactions of Mononuclear Organo-transition Metal Complexes, in: Advances in Organometallic Chemistry (eds. Stone, F. G. A., West, R.), Vol. 23, p. 24, New York, Academic Press 1984
178. Baral, S., Hambright, P., Neta, P.: J. Phys. Chem. *88*, 1595 (1984)
179. Cramer, S. P., Dawson, J. H., Hodgson, K. O., Hager, L. P.: J. Am. Chem. Soc. *100*, 7282 (1980)
180. Dolphin, D., James, B. R., Welborn, H. C.: Enzymatic and Electrochemical Reduction of Dioxygen, in: Electrochemical and Spectroelectrochemical Studies of Biological Redox Components (ed. Kadish, K. M.), p. 563, Washington, American Chemical Soc. 1982
181. Dawson, J. H., Trudell, J. R., Barth, G., Linder, R. E., Bunnenburg, E., Djerassi, C., Chiang, R., Hager, L. P.: J. Am. Chem. Soc. *98*, 3709 (1976)
182. English, D. R., Hendrickson, D. N., Suslick, K. S., Eigenbrot, C. W., Scheidt, W. R.: ibid. *106*, 7258 (1984)
183. Gunter, M. J., Mander, L. N., McLaughlin, G. M., Murray, K. S., Berry, K. J., Clark, P. E., Buckingham, D. A.: ibid. *102*, 1470 (1980)

# Author Index Volumes 1–64